T0297016

LONDON MATHEMATICAL SOCIETY LECTURE NOTE SERIES

Managing Editor: Professor M. Reid, Mathematics Institute,
University of Warwick, Coventry CV4 7AL, United Kingdom

The titles below are available from booksellers, or from Cambridge University Press at
http://www.cambridge.org/mathematics

330 Noncommutative localization in algebra and topology, A. RANICKI (ed)
331 Foundations of computational mathematics, Santander 2005, L.M. PARDO, A. PINKUS, E. SÜLI & M.J. TODD (eds)
332 Handbook of tilting theory, L. ANGELERI HÜGEL, D. HAPPEL & H. KRAUSE (eds)
333 Synthetic differential geometry (2nd Edition), A. KOCK
334 The Navier–Stokes equations, N. RILEY & P. DRAZIN
335 Lectures on the combinatorics of free probability, A. NICA & R. SPEICHER
336 Integral closure of ideals, rings, and modules, I. SWANSON & C. HUNEKE
337 Methods in Banach space theory, J.M.F. CASTILLO & W.B. JOHNSON (eds)
338 Surveys in geometry and number theory, N. YOUNG (ed)
339 Groups St Andrews 2005 I, C.M. CAMPBELL, M.R. QUICK, E.F. ROBERTSON & G.C. SMITH (eds)
340 Groups St Andrews 2005 II, C.M. CAMPBELL, M.R. QUICK, E.F. ROBERTSON & G.C. SMITH (eds)
341 Ranks of elliptic curves and random matrix theory, J.B. CONREY, D.W. FARMER, F. MEZZADRI & N.C. SNAITH (eds)
342 Elliptic cohomology, H.R. MILLER & D.C. RAVENEL (eds)
343 Algebraic cycles and motives I, J. NAGEL & C. PETERS (eds)
344 Algebraic cycles and motives II, J. NAGEL & C. PETERS (eds)
345 Algebraic and analytic geometry, A. NEEMAN
346 Surveys in combinatorics 2007, A. HILTON & J. TALBOT (eds)
347 Surveys in contemporary mathematics, N. YOUNG & Y. CHOI (eds)
348 Transcendental dynamics and complex analysis, P.J. RIPPON & G.M. STALLARD (eds)
349 Model theory with applications to algebra and analysis I, Z. CHATZIDAKIS, D. MACPHERSON, A. PILLAY & A. WILKIE (eds)
350 Model theory with applications to algebra and analysis II, Z. CHATZIDAKIS, D. MACPHERSON, A. PILLAY & A. WILKIE (eds)
351 Finite von Neumann algebras and masas, A.M. SINCLAIR & R.R. SMITH
352 Number theory and polynomials, J. MCKEE & C. SMYTH (eds)
353 Trends in stochastic analysis, J. BLATH, P. MÖRTERS & M. SCHEUTZOW (eds)
354 Groups and analysis, K. TENT (ed)
355 Non-equilibrium statistical mechanics and turbulence, J. CARDY, G. FALKOVICH & K. GAWEDZKI
356 Elliptic curves and big Galois representations, D. DELBOURGO
357 Algebraic theory of differential equations, M.A.H. MACCALLUM & A.V. MIKHAILOV (eds)
358 Geometric and cohomological methods in group theory, M.R. BRIDSON, P.H. KROPHOLLER & I.J. LEARY (eds)
359 Moduli spaces and vector bundles, L. BRAMBILA-PAZ, S.B. BRADLOW, O. GARCÍA-PRADA & S. RAMANAN (eds)
360 Zariski geometries, B. ZILBER
361 Words: Notes on verbal width in groups, D. SEGAL
362 Differential tensor algebras and their module categories, R. BAUTISTA, L. SALMERÓN & R. ZUAZUA
363 Foundations of computational mathematics, Hong Kong 2008, F. CUCKER, A. PINKUS & M.J. TODD (eds)
364 Partial differential equations and fluid mechanics, J.C. ROBINSON & J.L. RODRIGO (eds)
365 Surveys in combinatorics 2009, S. HUCZYNSKA, J.D. MITCHELL & C.M. RONEY-DOUGAL (eds)
366 Highly oscillatory problems, B. ENGQUIST, A. FOKAS, E. HAIRER & A. ISERLES (eds)
367 Random matrices: High dimensional phenomena, G. BLOWER
368 Geometry of Riemann surfaces, F.P. GARDINER, G. GONZÁLEZ-DIEZ & C. KOUROUNIOTIS (eds)
369 Epidemics and rumours in complex networks, M. DRAIEF & L. MASSOULIÉ
370 Theory of p-adic distributions, S. ALBEVERIO, A.YU. KHRENNIKOV & V.M. SHELKOVICH
371 Conformal fractals, F. PRZYTYCKI & M. URBAŃSKI
372 Moonshine: The first quarter century and beyond, J. LEPOWSKY, J. MCKAY & M.P. TUITE (eds)
373 Smoothness, regularity and complete intersection, J. MAJADAS & A. G. RODICIO
374 Geometric analysis of hyperbolic differential equations: An introduction, S. ALINHAC
375 Triangulated categories, T. HOLM, P. JØRGENSEN & R. ROUQUIER (eds)
376 Permutation patterns, S. LINTON, N. RUŠKUC & V. VATTER (eds)
377 An introduction to Galois cohomology and its applications, G. BERHUY
378 Probability and mathematical genetics, N. H. BINGHAM & C. M. GOLDIE (eds)
379 Finite and algorithmic model theory, J. ESPARZA, C. MICHAUX & C. STEINHORN (eds)
380 Real and complex singularities, M. MANOEL, M.C. ROMERO FUSTER & C.T.C WALL (eds)
381 Symmetries and integrability of difference equations, D. LEVI, P. OLVER, Z. THOMOVA & P. WINTERNITZ (eds)
382 Forcing with random variables and proof complexity, J. KRAJÍČEK
383 Motivic integration and its interactions with model theory and non-Archimedean geometry I, R. CLUCKERS, J. NICAISE & J. SEBAG (eds)
384 Motivic integration and its interactions with model theory and non-Archimedean geometry II, R. CLUCKERS, J. NICAISE & J. SEBAG (eds)
385 Entropy of hidden Markov processes and connections to dynamical systems, B. MARCUS, K. PETERSEN & T. WEISSMAN (eds)

386 Independence-friendly logic, A.L. MANN, G. SANDU & M. SEVENSTER
387 Groups St Andrews 2009 in Bath I, C.M. CAMPBELL *et al* (eds)
388 Groups St Andrews 2009 in Bath II, C.M. CAMPBELL *et al* (eds)
389 Random fields on the sphere, D. MARINUCCI & G. PECCATI
390 Localization in periodic potentials, D.E. PELINOVSKY
391 Fusion systems in algebra and topology, M. ASCHBACHER, R. KESSAR & B. OLIVER
392 Surveys in combinatorics 2011, R. CHAPMAN (ed)
393 Non-abelian fundamental groups and Iwasawa theory, J. COATES *et al* (eds)
394 Variational problems in differential geometry, R. BIELAWSKI, K. HOUSTON & M. SPEIGHT (eds)
395 How groups grow, A. MANN
396 Arithmetic differential operators over the p-adic integers, C.C. RALPH & S.R. SIMANCA
397 Hyperbolic geometry and applications in quantum chaos and cosmology, J. BOLTE & F. STEINER (eds)
398 Mathematical models in contact mechanics, M. SOFONEA & A. MATEI
399 Circuit double cover of graphs, C.-Q. ZHANG
400 Dense sphere packings: a blueprint for formal proofs, T. HALES
401 A double Hall algebra approach to affine quantum Schur–Weyl theory, B. DENG, J. DU & Q. FU
402 Mathematical aspects of fluid mechanics, J.C. ROBINSON, J.L. RODRIGO & W. SADOWSKI (eds)
403 Foundations of computational mathematics, Budapest 2011, F. CUCKER, T. KRICK, A. PINKUS & A. SZANTO (eds)
404 Operator methods for boundary value problems, S. HASSI, H.S.V. DE SNOO & F.H. SZAFRANIEC (eds)
405 Torsors, étale homotopy and applications to rational points, A.N. SKOROBOGATOV (ed)
406 Appalachian set theory, J. CUMMINGS & E. SCHIMMERLING (eds)
407 The maximal subgroups of the low-dimensional finite classical groups, J.N. BRAY, D.F. HOLT & C.M. RONEY-DOUGAL
408 Complexity science: the Warwick master's course, R. BALL, V. KOLOKOLTSOV & R.S. MACKAY (eds)
409 Surveys in combinatorics 2013, S.R. BLACKBURN, S. GERKE & M. WILDON (eds)
410 Representation theory and harmonic analysis of wreath products of finite groups, T. CECCHERINI-SILBERSTEIN, F. SCARABOTTI & F. TOLLI
411 Moduli spaces, L. BRAMBILA-PAZ, O. GARCÍA-PRADA, P. NEWSTEAD & R.P. THOMAS (eds)
412 Automorphisms and equivalence relations in topological dynamics, D.B. ELLIS & R. ELLIS
413 Optimal transportation, Y. OLLIVIER, H. PAJOT & C. VILLANI (eds)
414 Automorphic forms and Galois representations I, F. DIAMOND, P.L. KASSAEI & M. KIM (eds)
415 Automorphic forms and Galois representations II, F. DIAMOND, P.L. KASSAEI & M. KIM (eds)
416 Reversibility in dynamics and group theory, A.G. O'FARRELL & I. SHORT
417 Recent advances in algebraic geometry, C.D. HACON, M. MUSTAŢĂ & M. POPA (eds)
418 The Bloch–Kato conjecture for the Riemann zeta function, J. COATES, A. RAGHURAM, A. SAIKIA & R. SUJATHA (eds)
419 The Cauchy problem for non-Lipschitz semi-linear parabolic partial differential equations, J.C. MEYER & D.J. NEEDHAM
420 Arithmetic and geometry, L. DIEULEFAIT *et al* (eds)
421 O-minimality and Diophantine geometry, G.O. JONES & A.J. WILKIE (eds)
422 Groups St Andrews 2013, C.M. CAMPBELL *et al* (eds)
423 Inequalities for graph eigenvalues, Z. STANIĆ
424 Surveys in combinatorics 2015, A. CZUMAJ *et al* (eds)
425 Geometry, topology and dynamics in negative curvature, C.S. ARAVINDA, F.T. FARRELL & J.-F. LAFONT (eds)
426 Lectures on the theory of water waves, T. BRIDGES, M. GROVES & D. NICHOLLS (eds)
427 Recent advances in Hodge theory, M. KERR & G. PEARLSTEIN (eds)
428 Geometry in a Fréchet context, C. T. J. DODSON, G. GALANIS & E. VASSILIOU
429 Sheaves and functions modulo p, L. TAELMAN
430 Recent progress in the theory of the Euler and Navier-Stokes equations, J.C. ROBINSON, J.L. RODRIGO, W. SADOWSKI & A. VIDAL-LÓPEZ (eds)
431 Harmonic and subharmonic function theory on the real hyperbolic ball, M. STOLL
432 Topics in graph automorphisms and reconstruction (2nd Edition), J. LAURI & R. SCAPELLATO
433 Regular and irregular holonomic D-modules, M. KASHIWARA & P. SCHAPIRA
434 Analytic semigroups and semilinear initial boundary value problems (2nd Edition), K. TAIRA
435 Graded rings and graded Grothendieck groups, R. HAZRAT
436 Groups, graphs and random walks, T. CECCHERINI-SILBERSTEIN, M. SALVATORI & E. SAVA-HUSS (eds)
437 Dynamics and analytic number theory, D. BADZIAHIN, A. GORODNIK & N. PEYERIMHOFF (eds)
438 Random walks and heat kernels on graphs, M.T. BARLOW
439 Evolution equations, K. AMMARI & S. GERBI (eds)
440 Surveys in combinatorics 2017, A. CLAESSON *et al* (eds)
441 Polynomials and the mod 2 Steenrod algebra I, G. WALKER & R.M.W. WOOD
442 Polynomials and the mod 2 Steenrod algebra II, G. WALKER & R.M.W. WOOD
443 Asymptotic analysis in general relativity, T. DAUDÉ, D. HÄFNER & J.-P. NICOLAS (eds)
444 Geometric and cohomological group theory, P.H. KROPHOLLER, I.J. LEARY, C. MARTÍNEZ-PÉREZ & B.E.A. NUCINKIS (eds)

London Mathematical Society Lecture Note Series: 444

Geometric and Cohomological Group Theory

Edited by

PETER H. KROPHOLLER
University of Southampton

IAN J. LEARY
University of Southampton

CONCHITA MARTÍNEZ-PÉREZ
Universidad de Zaragoza

BRITA E.A. NUCINKIS
Royal Holloway, University of London

CAMBRIDGE
UNIVERSITY PRESS

University Printing House, Cambridge CB2 8BS, United Kingdom

One Liberty Plaza, 20th Floor, New York, NY 10006, USA

477 Williamstown Road, Port Melbourne, VIC 3207, Australia

314-321, 3rd Floor, Plot 3, Splendor Forum, Jasola District Centre, New Delhi - 110025, India

79 Anson Road, #06-04/06, Singapore 079906

Cambridge University Press is part of the University of Cambridge.

It furthers the University's mission by disseminating knowledge in the pursuit of education, learning and research at the highest international levels of excellence.

www.cambridge.org
Information on this title: www.cambridge.org/9781316623220
DOI: 10.1017/9781316771327

First published 2018

A catalogue record for this publication is available from the British Library

ISBN 978-1-316-62322-0 Paperback

Contents

List of Participants

Azer Akhmedov	NDSU
James Belk	Bard
Robert Bieri	Frankfurt
Collin Bleak	St Andrews
Brian Bowditch	Warwick
Martin Bridson	Oxford
Jeff Burdges	St Andrews
José Burillo	Barcelona
Kai-Uwe Bux	Bielefeld
Montserrat Casals-Ruiz	Oxford
Yu-Yen Chien	University of Southampton
Ian Chiswell	Queen Mary, London
Sean Cleary	CUNY
Ged Corob Cook	Royal Holloway
Michael Davis	Ohio State
Matthew Day	University of Arkansas
Dieter Degrijse	KU Leuven, Kortrijk
Dennis Dreesen	Southampton
Andrew Duncan	Newcastle
Martin Dunwoody	Southampton
Jan Dymara	Wroclaw
Ioannis Emmanouil	University of Athens
Daniel Farley	Miami University, Oxford, Ohio
Michal Ferov	Southampton
Ariadna Fossas Tenas	EPFL Lausanne

With affiliations at time of the Symposium.

Giovanni Gandini	Bonn
Łukasz Garncarek	Wroclaw
Alejandra Garrido	Oxford
Ross Geoghegan	Binghamton
Anne Giralt	Jussieu, Paris
Tadeusz Januszkiewicz	Inst. Math., Polish Academy of Sciences
Aditi Kar	Oxford
Martin Kassabov	Cornell
Ilya Kazachkov	Oxford
Dessislava Kochloukova	UNICAMP
Peter Kropholler	Southampton
Robert Kropholler	Oxford
Benno Kuckuck	Oxford
Ian Leary	Southampton
Yash Lodha	Cornell
Eric López Platón	UPC
John Mackay	Oxford
Conchita Martínez-Pérez	Zaragosa
Armando Martino	Southampton
Francesco Matucci	Paris-Sud 11
Jon McCammond	UC Santa Barbara
Sebastian Meinert	FU Berlin
Ashot Minasyan	Southampton
Justin Moore	Cornell
Volodymyr Nekrashevych	Texas A&M
Brita Nucinkis	Royal Holloway, London
Nansen Petrosyan	KU Leuven
Steve Pride	Glasgow
Piotr Przytycki	Warsaw
Sarah Rees	Newcastle
Holger Reich	FU Berlin
Claas Röver	NUI Galway
Colva Roney-Dougal	St Andrews
Mark Sapir	Vanderbilt
Dirk Schuetz	Durham
Marco Schwandt	Bielefeld
Wolfgang Steimle	Bonn
Melanie Stein	Trinity
Simon St.John-Green	Southampton
Zoran Šunić	Texas A&M
Jacek Świątkowski	Wroclaw
Olympia Talelli	Athens
Robert Tang	University of Warwick
Mark Ullmann	FU Berlin

Marco Varisco	University of Albany, SUNY
Alina Vdovina	Newcastle
Karen Vogtmann	Cornell
Kun Wang	Ohio State University
Christian Wegner	Bonn
John Wilson	Oxford
Henry Wilton	UC London
Stefan Witzel	Muenster
Daniel Woodhouse	McGill University
Xiaolei Wu	SUNY Binghamton
Pavel Zalesskii	Brasilia

Preface

This proceedings volume results from the fourth London Mathematical Society Durham Symposium of an influential series that belongs to the mathematical territory that we now see as part of Geometric Group Theory. Notably it was also the 100th in the entire series of LMS Durham Symposia. The first of these four meetings was held in 1976 organised by Scott and Wall, the second in 1994 organised by Kropholler and Stöhr (with Niblo as an additional editor for the proceedings volume), and the third in 2003 organised by Bridson, Kropholler and Leary. Proceedings volumes for these three meetings appeared in the *London Mathematical Society Lecture Notes* series as volumes 36, 252 and 358, and we are pleased to be able to continue this tradition.

This fourth meeting drew together some 80 mathematicians from around the world. It shared with the earlier meetings the high standards and significance of its main lecture series. These lecture series were delivered by Kai-Uwe Bux, Desi Kochloukova, Jon McCammond, Justin Moore, Piotr Przytycki, and Holger Reich. There was also room in the schedule for individual invited lectures from Azer Akhmedov, Collin Bleak, Brian Bowditch, Martin Bridson, Michael Davis, Ioannis Emmanouil, Dan Farley, Ross Geoghegan, Martin Kassabov, Conchita Martínez-Pérez, Volodymyr Nekrashevych, Nansen Petrosyan, Colva Roney-Dougal, Mark Sapir, Karen Vogtmann, Christian Wegner, John Wilson, Henry Wilton, and Stefan Witzel.

The titles of the four proceedings volumes that have flowed from these symposia have evolved in a way that mirrors the evolution of the subject: over a period of almost four decades, the role of geometry in group theory has grown hugely. This change was evident at the symposium and can also be seen in the present volume.

We thank all the authors who have contributed to this volume. We thank the London Mathematical Society and the Engineering and Physical Sciences Research Council for their support both in terms of advice and financially. The organisational burden that has in the past fallen on the scientific committee associated with an LMS Durham Symposium has now largely been replaced by the very supportive and tireless work of the administrative staff in the Mathematics Department of Durham University and we gratefully acknowledge this contribution, which has gone a long way to making these symposia run smoothly and ensuring that the focus is on the important science at the heart of our work.

The warm and friendly atmosphere of the meeting itself led to many useful interactions and a flow of ideas. We hope that the reader will find some of this excitement is reflected in the present volume.

Peter Kropholler, University of Southampton
Ian Leary, University of Southampton
Conchita Martínez-Pérez, University of Zaragoza
Brita Nucinkis, Royal Holloway, University of London

Obstructions for subgroups of Thompson's group V

José Burillo* Sean Cleary* Claas E. Röver*

Abstract

Thompson's group V has a rich variety of subgroups, containing all finite groups, all finitely generated free groups and all finitely generated abelian groups, the finitary permutation group of a countable set, as well as many wreath products and other families of groups. Here, we describe some obstructions for a given group to be a subgroup of V.

1 Introduction

Thompson constructed a finitely presented group now known as V as an early example of a finitely presented infinite simple group. The group V contains a remarkable variety of subgroups, such as the finitary infinite permutation group S_∞, and hence all (countable locally) finite groups, finitely generated free groups, finitely generated abelian groups, Houghton's groups, copies of Thompson's groups F, T and V, and many of their generalizations, such as the groups $G_{n,r}$ constructed by Higman [9]. Moreover, the class of subgroups of V is closed under direct products and restricted wreath products with finite or infinite cyclic top group.

In this short survey, we summarize the development of properties of V focusing on those which prohibit various groups from occurring as subgroups of V.

Thompson's group V has many descriptions. Here, we simply recall that V is the group of right-continuous bijections from the unit interval $[0, 1]$ to itself, which map dyadic rational numbers to dyadic rational numbers, which are differentiable except at finitely many dyadic rational numbers, and with slopes, when defined, integer powers of 2. The elements of this group can

*The authors are grateful for the hospitality of Durham University during the Symposium on Cohomological and Geometric Group Theory. The first author acknowledges support from MEC grant MTM2011–25955. The second author acknowledges support from the National Science Foundation and that this work was partially supported by a grant from the Simons Foundation (#234548 to Sean Cleary).

be described by reduced tree pair diagrams of the type (S, T, π) where π is a bijection between the leaves of the two finite rooted binary trees S and T.

Higman [9] gave a different description of V, which he denoted as $G_{2,1}$ in a family of groups generalizing V.

2 Obstructions

Higman [9] described several important properties of V which can serve as obstructions to subgroups occurring in V.

Theorem 2.1 ([9]) *An element of infinite order in V has only finitely many roots.*

This prevents all Baumslag-Solitar groups $B_{m,n} = \langle a, b \mid a^n b = ba^m \rangle$ from occurring as subgroups of V, if m properly divides n; see [13].

Theorem 2.2 ([9]) *Torsion free abelian subgroups of V are free abelian, and their centralizers have finite index in their normalizers in V.*

This prevents $GL_n(\mathbb{Z})$ from occurring as a subgroup of V for $n \geq 2$.

A group is *torsion locally finite* if every torsion subgroup is locally finite. That is, if every finitely generated torsion subgroup is finite. Röver [12] showed

Theorem 2.3 ([12]) *Thompson's group V is torsion locally finite.*

This rules out many branch groups from occurring as subgroups of V, including the Grigorchuk groups of intermediate growth [7] and the Gupta-Sidki groups [8]. It also rules out Burnside groups.

Holt and Röver [10] showed that V has indexed co-word problem.

Theorem 2.4 ([10]) *The set of words (over an arbitrary but fixed finite generating set) which do not represent the identity in V is an indexed language, and hence can be recognized by a nested-stack automaton.*

This property is not easy to verify, however. But it is inherited by finitely generated subgroups (see [10]), and hence groups which do not have an indexed co-word problem cannot occur as a subgroup of V.

Lehnert and Schweitzer [11] improved this result.

Theorem 2.5 ([11]) *The set of words (over an arbitrary but fixed finite generating set) which do not represent the identity in V is a context-free language, and hence can be recognized by a pushdown automaton.*

Again, this property is inherited by finitely generated subgroups, but the condition is still not easy to verify.

More recently, Bleak and Salaza-Díaz [4] and subsequently Corwin [6], using similar techniques showed

Theorem 2.6 ([4, 6]) *Neither the free product $\mathbb{Z} * \mathbb{Z}^2$ nor the standard restricted wreath product $\mathbb{Z} \wr \mathbb{Z}^2$ with $\mathbb{Z}^2$ as top group are subgroups of V.*

One theorem of Higman [9] together with a metric estimate of Birget [1] gives another obstruction.

Theorem 2.7 ([9]) *For any element v of infinite order in V, there is a power v^n such that for the reduced tree pair diagram (S, T, π) for v^n, there is a leaf i in the source tree S which is paired with a leaf j in the target tree T so that j is a child of of i.*

Theorem 2.8 ([1]) *For any finite generating set of V, There are constants C and C' such that word length $|v|$ of an element of V with respect to that generating set satisfies $Cn \leq |v| \leq C'n \log n$ where n is the size of the reduced tree pair diagram representing v.*

Since the powers of v^n will have length thus growing linearly, these two theorems give as a consequence the following.

Theorem 2.9 *Cyclic subgroups of V are undistorted.*

We note that this argument applies as well to generalizations of V where there is a linear lower bound on word length in terms of the number of carets, such as braided versions of V [5].

This last theorem has an obvious corollary.

Corollary 2.10 *If a group embeds in V, its cyclic subgroups must be undistorted.*

The reason for this is that in a chain of subgroups $G \supset H \supset K$ the distortion of K in H cannot be larger than the distortion of K in G.

We note that Bleak, Bowman, Gordon, Graham, Hughes, Matucci and J. Sapir [3] used Brin's methods of revealing pairs for elements of V to show that cyclic subgroups of V are undistorted.

This result excludes all Baumslag-Solitar groups with $|n| \neq |m|$, as these have distorted cyclic subgroups. It also rules out nilpotent groups which are not virtually abelian. An alternative argument excluding the Baumslag-Solitar groups is due to Bleak, Matucci and Neunhöffer [2].

References

[1] Jean-Camille Birget. The groups of Richard Thompson and complexity. *Internat. J. Algebra Comput.*, 14(5-6):569–626, 2004. International Conference on Semigroups and Groups in honor of the 65th birthday of Prof. John Rhodes.

[2] C. Bleak, F. Matucci, and M. Neunhöffer. Embeddings into Thompson's group V and $coC\mathcal{F}$ groups. *arXiv e-prints*, December 2013.

[3] Collin Bleak, Hannah Bowman, Alison Gordon Lynch, Garrett Graham, Jacob Hughes, Francesco Matucci, and Eugenia Sapir. Centralizers in the R. Thompson group V_n. *Groups Geom. Dyn.*, 7(4):821–865, 2013.

[4] Collin Bleak and Olga Salazar-Díaz. Free products in R. Thompson's group V. *Trans. Amer. Math. Soc.*, 365(11):5967–5997, 2013.

[5] José Burillo and Sean Cleary. Metric properties of braided Thompson's groups. *Indiana Univ. Math. J.*, 58(2):605–615, 2009.

[6] Nathan Corwin. *Embedding and non embedding results for R. Thompsons group V and related groups.* PhD thesis, University of Nebraska – Lincoln, 2013.

[7] R. I. Grigorčuk. On Burnside's problem on periodic groups. *Funktsional. Anal. i Prilozhen.*, 14(1):53–54, 1980.

[8] Narain Gupta and Saïd Sidki. On the Burnside problem for periodic groups. *Math. Z.*, 182(3):385–388, 1983.

[9] Graham Higman. *Finitely presented infinite simple groups.* Department of Pure Mathematics, Department of Mathematics, I.A.S. Australian National University, Canberra, 1974. Notes on Pure Mathematics, No. 8 (1974).

[10] Derek F. Holt and Claas E. Röver. Groups with indexed co-word problem. *Internat. J. Algebra Comput.*, 16(5):985–1014, 2006.

[11] J. Lehnert and P. Schweitzer. The co-word problem for the Higman-Thompson group is context-free. *Bull. Lond. Math. Soc.*, 39(2):235–241, 2007.

[12] Claas E. Röver. Constructing finitely presented simple groups that contain Grigorchuk groups. *J. Algebra*, 220(1):284–313, 1999.

[13] Claas E. Röver. *Subgroups of finitely presented simple groups.* PhD thesis, University of Oxford, 1999.

Groups of homological dimension one

Ioannis Emmanouil*

Abstract

We report on recent work concerning groups of homological dimension one and detail some methods that may be used in order to determine whether these groups are locally free.

0 Introduction

Stallings has established in [20] a characterization of finitely generated free groups, as those groups whose cohomological dimension is one. It is very easy to show that a free group has cohomological dimension one. Indeed, if G is a free group then the augmentation ideal I_G is a free $\mathbb{Z}G$-module; in fact, if G is freely generated by a subset S, then I_G is a free $\mathbb{Z}G$-module on the set $\{s-1 : s \in S\}$. The essence of Stallings' theorem is that the converse implication is also true, namely that any finitely generated group of cohomological dimension one is free. Bieri asked in [2] whether a (stronger) homological version of the latter result holds:

Is any finitely generated group of homological dimension one free?

Shortly after the publication of the proof of Stallings' theorem, Swan showed that the finite generation hypothesis is redundant therein, by proving that a (not necessarily finitely generated) group G is free if and only if $\operatorname{cd} G = 1$ (cf. [21]). In that direction, we note that Bieri's question may be equivalently formulated as follows:

Is any group of homological dimension one locally free?

Some interesting results concerning that problem have been obtained in [5] and [11], by embedding the integral group ring $\mathbb{Z}G$ of the group G into the associated von Neumann algebra $\mathcal{N}G$ and the algebra $\mathcal{U}G$ of unbounded operators which are affiliated to $\mathcal{N}G$.

We note that a group G is known to be finitely generated if and only if the augmentation ideal I_G is a finitely generated $\mathbb{Z}G$-module, whereas G has

*Research supported by a GSRT/Greece excellence grant, cofunded by the ESF/EU and National Resources.

homological dimension one if and only if I_G is a flat $\mathbb{Z}G$-module. In view of the above mentioned result of Stallings and Swan, the freeness of G is equivalent to the projectivity of I_G as a $\mathbb{Z}G$-module. Therefore, Bieri's question turns out to be equivalent to the following one:

If the augmentation ideal I_G is a finitely generated flat $\mathbb{Z}G$-module, then is it true that I_G is a projective $\mathbb{Z}G$-module?

If G is any countable group, then the augmentation ideal I_G is countably presented as a $\mathbb{Z}G$-module. It follows from a result of Lazard [12] that, in this case, the flatness of I_G implies that its projective dimension is ≤ 1 (and hence that $\operatorname{cd} G \leq 2$). The point is to show that if we strengthen the assumption on the group G and assume that it is finitely generated (and not just countable), then the flatness of I_G implies its projectivity. In this direction, we note that if G is the additive group of rational numbers, then the $\mathbb{Z}G$-module I_G is flat but not projective (i.e. $\operatorname{hd} G = 1$ but $\operatorname{cd} G = 2$); even though this countable group has homological dimension one, it isn't free.

In the present note, we shall follow [6] and elaborate on two methods that may be used in order to study this problem.

1 Projectivity of finitely generated flat modules

Let G be a group and consider the augmentation ideal I_G. As mentioned in the Introduction, we wish to prove that the $\mathbb{Z}G$-module I_G is projective, provided that we know it to be finitely generated and flat. We are only interested in the very particular $\mathbb{Z}G$-module I_G, but it may be the case that our group G is such that any finitely generated flat $\mathbb{Z}G$-module is projective. Even though it seems that we make the problem unnecessarily harder[1], we introduce the following class of groups:

Definition 1.1. Let $\mathfrak{S}$ be the class consisting of those groups G, which are such that any finitely generated flat $\mathbb{Z}G$-module is projective.

We shall begin by listing a couple of elementary properties of this class:

(i) $\mathfrak{S}$ contains all finite groups.

Proof. If G is a finite group, then the group ring $\mathbb{Z}G$ is (left) Noetherian and hence any finitely generated $\mathbb{Z}G$-module is finitely presented. In particular, any finitely generated flat $\mathbb{Z}G$-module is (finitely presented and flat and hence) projective. □

(ii) $\mathfrak{S}$ is subgroup-closed.

[1] See Remark 1.8(ii) below.

Proof. Let H be a subgroup of a group $G \in \mathfrak{S}$. In order to show that $H \in \mathfrak{S}$, let M be a finitely generated flat $\mathbb{Z}H$-module. Then, $\mathrm{ind}_H^G M$ is a finitely generated flat $\mathbb{Z}G$-module and hence, in view of the hypothesis made on G, it is $\mathbb{Z}G$-projective. Therefore, the $\mathbb{Z}H$-module $\mathrm{res}_H^G \mathrm{ind}_H^G M$ is projective. Since M is a direct summand of $\mathrm{res}_H^G \mathrm{ind}_H^G M$, it follows that M is a projective $\mathbb{Z}H$-module as well. □

(iii) Any $\mathfrak{S}$-group of homological dimension one is locally free.
Proof. Let H be a finitely generated subgroup of an $\mathfrak{S}$-group G with $\mathrm{hd}\, G = 1$. Then H has homological dimension one and hence the augmentation ideal I_H is a finitely generated flat $\mathbb{Z}H$-module. Since the class $\mathfrak{S}$ is subgroup-closed (cf. (ii) above), the group H is contained in $\mathfrak{S}$ and hence the $\mathbb{Z}H$-module I_H is projective. Invoking Stallings' theorem, we conclude that H is a free group, as needed. □

Remark 1.2. The class of those rings R over which any finitely generated flat left module is projective has been studied in [8] and [18]. It is shown therein that this property is equivalent to the ascending chain condition on certain sequences of principal left ideals in matrix rings over R. In particular, it turns out that a group G is an $\mathfrak{S}$-group if and only if for any $n \geq 1$ and any sequence of $n \times n$ matrices $(A_i)_i \in \mathbf{M}_n(\mathbb{Z}G)$ with $A_i A_{i+1} = A_i$ for all i, the ascending sequence of principal left ideals

$$\mathbf{M}_n(\mathbb{Z}G)A_0 \subseteq \mathbf{M}_n(\mathbb{Z}G)A_1 \subseteq \ldots \subseteq \mathbf{M}_n(\mathbb{Z}G)A_t \subseteq \cdots$$

of the matrix ring $\mathbf{M}_n(\mathbb{Z}G)$ is eventually constant.

The embedding of the integral group ring $\mathbb{Z}G$ of a group G into the associated von Neumann algebra $\mathcal{N}G$ enables one to define the von Neumann dimension $\dim_{\mathcal{N}G}(\mathcal{N}G \otimes_{\mathbb{Z}G} P) \in \mathbb{R}$ for any finitely generated projective $\mathbb{Z}G$-module P. The reader may consult Lück's book [14] for the details of the definition and several useful properties of this dimension theory.

For any group G we let Λ_G be the additive subgroup of $\mathbb{Q}$, which is generated by the inverses of the orders of the finite subgroups of G. The group G is said to satisfy Atiyah's conjecture if $\dim_{\mathcal{N}G}(\mathcal{N}G \otimes_{\mathbb{Z}G} P) \in \Lambda_G$ for any finitely generated projective $\mathbb{Z}G$-module P. Lück's book contains a thorough discussion of this conjecture and presents several classes of groups for which the conjecture is known to be true.

Definition 1.3. Let $\mathfrak{A}_{fin}$ be the class consisting of those groups that satisfy Atiyah's conjecture and have a finite upper bound on the orders of their finite subgroups.

The relevance of Atiyah's conjecture in the study of groups of homological dimension one was noticed by Kropholler, Linnell and Lück in [11]: In order to describe the link between the two themes, let us consider a finitely generated group G and fix a presentation of it as the quotient of a finitely generated free group Γ by a normal subgroup N. Then, as shown by Magnus in [16], the associated relation module $P = N/[N,N]$ fits into a short exact sequence of $\mathbb{Z}G$-modules

$$0 \longrightarrow P \longrightarrow F \longrightarrow I_G \longrightarrow 0, \tag{1}$$

where F is a finitely generated free $\mathbb{Z}G$-module (of rank equal to the rank of the free group Γ); see also [3, Chapter II, Proposition 5.4]. If the group G has homological dimension one, then the augmentation ideal I_G is flat as a $\mathbb{Z}G$-module. Invoking [12, Théorème I.3.2], we conclude that I_G has projective dimension ≤ 1 and hence P is a projective $\mathbb{Z}G$-module. Then, since the (finitely generated flat) $\mathbb{Z}G$-module I_G is projective if and only if it is finitely presented, in order to prove the conjectured freeness of the group G, we are reduced to showing that the relation module P is finitely generated.

It is well known that if we are dealing with vector spaces over a field, then any subspace of a finitely generated vector space V is also finitely generated, i.e. there is no strictly increasing sequence of subspaces of V. One may prove this property using the fact that the dimension of any such a subspace is bounded by the dimension of V and hence the dimensions of these subspaces may assume only finitely many values. In the same way, one may attempt to prove that the $\mathbb{Z}G$-module P in the exact sequence (1) above is finitely generated, using the formal properties of the von Neumann dimension (namely, its monotonicity and continuity). There is a problem though with this approach: The von Neumann dimension is a real number (not necessarily an integer) and a bounded set of real numbers need not be discrete. Atiyah's conjecture for groups with a finite upper bound on the orders of their finite subgroups guarantees that the set of the von Neumann dimensions that are involved is indeed discrete, making the classical linear algebra argument outlined above work.

In fact, we may state the following result, which is essentially due to Kropholler, Linnell and Lück (cf. [11]).

Theorem 1.4. *The class $\mathfrak{A}_{fin}$ is a subclass of $\mathfrak{S}$.*

Proof. Let G be a group contained in $\mathfrak{A}_{fin}$. In order to show that G is an $\mathfrak{S}$-group, it suffices, in view of [8, Proposition 3.5], to show that any finitely generated *and countably presented* flat $\mathbb{Z}G$-module is projective. To that end, let M be a finitely generated and countably presented flat $\mathbb{Z}G$-module. Being countably presented and flat, the module M has projective dimension ≤ 1 (cf. [12, Théorème I.3.2]) and hence the arguments used in the proof of [11, Lemma 4, Lemma 5 and Theorem 2] show that M is finitely presented and therefore projective. □

For any ideal I of a ring R we define the ideal $I^\omega = \bigcap_{n=1}^\infty I^n$. We also consider the ideals I^{ω^m}, $m \geq 1$, which are defined inductively by letting $I^{\omega^1} = I^\omega$ and $I^{\omega^m} = \left(I^{\omega^{m-1}}\right)^\omega$ for all $m > 1$. We say that the ideal I is residually nilpotent if $I^\omega = 0$.

Definition 1.5. For all $m \geq 1$ let $\mathfrak{X}_m$ be the class consisting of those groups G whose augmentation ideal I_G is such that $I_G^{\omega^m} = 0$. We also let $\mathfrak{X} = \bigcup_{m=1}^\infty \mathfrak{X}_m$.

The class $\mathfrak{X}_1$, which contains those groups whose augmentation ideal is residually nilpotent, has been studied extensively (cf. [17]). As shown by Lichtman in [13], $\mathfrak{X}_1$ contains as a subclass the class

$$\mathfrak{Y}_1 = \{\text{residually torsion-free nilpotent groups}\}.$$

In particular, $\mathfrak{X}_1$ contains all free groups (cf. [15, §I.10]) and all torsion-free abelian groups. As shown in [6, Proposition 5.2], the class $\mathfrak{X}$ is closed under subgroups and extensions.[2] It follows that $\mathfrak{X}$ contains as a subclass the class

$$\mathfrak{Y} = \{\text{iterated extensions of residually torsion-free nilpotent groups}\}.$$

The relevance of class $\mathfrak{X}$ in the study of $\mathfrak{S}$-groups (and therefore in the study of groups of homological dimension one) stems from the following lifting result, which is itself proved in [6, Proposition 5.3(ii)].

Theorem 1.6. *Let G be a group and $N \subseteq G$ a normal subgroup. If $N \in \mathfrak{X}$ and $G/N \in \mathfrak{S}$, then $G \in \mathfrak{S}$.* □

In particular, the trivial group being an $\mathfrak{S}$-group, it follows that $\mathfrak{X} \subseteq \mathfrak{S}$.

Corollary 1.7. *Let G be a group having a normal subgroup $N \subseteq G$, such that:*

1. *N is an iterated extension of residually torsion-free nilpotent groups and*
2. *G/N satisfies Atiyah's conjecture and has a finite upper bound on the orders of its finite subgroups.*

Then, $G \in \mathfrak{S}$. In particular, if G has homological dimension one, then G is locally free.

Proof. Since $N \in \mathfrak{Y} \subseteq \mathfrak{X}$ and $G/N \in \mathfrak{A}_{fin} \subseteq \mathfrak{S}$ (cf. Theorem 1.4), the result is an immediate consequence of Theorem 1.6. □

Remarks 1.8. (i) (cf. [6, Theorem 5.5]) If $\mathfrak{D}$ is any class of groups, let us denote by $\mathfrak{D}^\diamond$ the class consisting of the finitely generated $\mathfrak{D}$-groups of homological dimension one. For example, if we denote by $\mathfrak{F}$ the class of free

[2] In fact, the class $\mathfrak{X}_m$ is subgroup-closed for all m, whereas if N is a normal subgroup of a group G, such that $N \in \mathfrak{X}_m$ and $G/N \in \mathfrak{X}_{m'}$, then $G \in \mathfrak{X}_{m+m'}$.

groups, then $\mathfrak{F}^\diamond$ is the class consisting of the free groups of finite rank. With this notation, the chains of inclusions

$$\mathfrak{F} \subseteq \mathfrak{Y}_1 \subseteq \mathfrak{X}_1 \subseteq \mathfrak{X} \subseteq \mathfrak{S} \quad \text{and} \quad \mathfrak{F} \subseteq \mathfrak{Y}_1 \subseteq \mathfrak{Y} \subseteq \mathfrak{X} \subseteq \mathfrak{S}$$

induce equalities

$$\mathfrak{F}^\diamond = \mathfrak{Y}_1^\diamond = \mathfrak{X}_1^\diamond = \mathfrak{X}^\diamond = \mathfrak{S}^\diamond \quad \text{and} \quad \mathfrak{F}^\diamond = \mathfrak{Y}_1^\diamond = \mathfrak{Y}^\diamond = \mathfrak{X}^\diamond = \mathfrak{S}^\diamond.$$

Indeed, as we have noted above, a finitely generated $\mathfrak{S}$-group of homological dimension one is necessarily free.
(ii) Let G be a finitely generated group of homological dimension one, so that the augmentation ideal I_G is a finitely generated flat $\mathbb{Z}G$-module. If that particular module (namely I_G) is projective, then *all* finitely generated flat $\mathbb{Z}G$-modules are projective, i.e. G is an $\mathfrak{S}$-group. Indeed, the projectivity of I_G implies, in view of Stallings' theorem, that the group G is free and hence the inclusion $\mathfrak{F} \subseteq \mathfrak{S}$ (cf. (i) above) shows that $G \in \mathfrak{S}$.
(iii) It is known that $\mathfrak{Y}_1$-groups admit a left order. If G is a left orderable group of homological dimension one, then, as shown by Dicks and Linnell in [5, Corollary 6.12], any 2-generated subgroup of G is free. It follows from (i) above that the same result holds for any finitely generated subgroup of a $\mathfrak{Y}_1$-group of homological dimension one.

2 The cohomology group $H^2(G, \mathbb{Z}G)$

The relation between flatness and projectivity is a classical theme in homological algebra. There is a variety of conditions which, when imposed on a flat module, imply its projectivity. In this direction, it is worth mentioning a result due to Raynaud and Gruson. Even though they were mainly interested in modules over commutative rings (and, more generally, quasi-coherent sheaves on schemes), they established in [19] a criterion for a countably presented flat left module K over a not necessarily commutative ring R to be projective. As shown by Lazard [12, §*I*.3], such a module K can be expressed as the direct limit of a direct system $(K_n)_n$ of finitely generated free left R-modules. With this notation, Raynaud and Gruson proved that K is projective if and only if the inverse system of abelian groups $(\mathrm{Hom}_R(K_n, R))_n$ satisfies the Mittag-Leffler condition. The corresponding notion of Mittag-Leffler modules has been studied thoroughly and systematically in [1]. On the other hand, Gray has obtained in [9] a characterization of the Mittag-Leffler condition for inverse systems of *countable* abelian groups in terms of the vanishing of $\underleftarrow{\lim}^1$, the first derived functor of the inverse limit functor. Using this circle of ideas and the classical expression of the Ext-groups of K in terms of the Ext-groups of the K_n's, Jensen has obtained in [10] a simple projectivity test for a flat module, that we shall describe below (see also [6, Theorem 1.3]).

To that end, assume that R is a countable ring and consider a countable left R-module K. Then, Jensen's criterion asserts that K is projective if and only if K is flat and $\mathrm{Ext}^1_R(K,R)=0$. A stronger version of that result was obtained in [7, Theorem 2.22], where it is shown that

$$K \text{ is projective} \Longleftrightarrow \mathrm{pd}_R K \leq 1 \text{ and } \mathrm{Ext}^1_R(K,R)=0.$$

Let us now consider a non-zero countable left R-module M of finite projective dimension. Then, the functors $\mathrm{Ext}^i_R(M,\_)$ vanish identically for $i \gg 0$. In particular, $\mathrm{Ext}^i_R(M,R)=0$ for $i \gg 0$. Applying the result above to a suitable kernel in a countably generated free resolution of M (namely, to the $(n-1)$th kernel, where $n=\mathrm{pd}_R M$), we obtain the equality

$$\mathrm{pd}_R M = \max\{i \geq 0 : \mathrm{Ext}^i_R(M,R) \neq 0\}. \tag{2}$$

The latter equality is of some interest, as it shows that any countable module of finite projective dimension over a countable ring demonstrates a cohomological behavior reminiscent of the FP-condition. Indeed, if R is any ring and M any non-zero left R-module with $\mathrm{pd}_R M = n < \infty$, then the functor $\mathrm{Ext}^n_R(M,\_)$ does not vanish identically, whereas $\mathrm{Ext}^i_R(M,\_)=0$ for all $i>n$. In fact, if $P_* \longrightarrow M \longrightarrow 0$ is a projective resolution of M of length n, then it is easily seen that $\mathrm{Ext}^n_R(M,P_n)\neq 0$. Since P_n is a direct summand of a free left R-module F, it follows that we also have $\mathrm{Ext}^n_R(M,F)\neq 0$. The free module F is, in general, of infinite rank. In order to obtain some information on the minimal size of such an F, we note that:

(i) If κ is an infinite cardinal, such that any left ideal of R is generated by a set of cardinality κ and M is κ-generated, then one may choose P_n to be κ-generated. Therefore, it follows that $\mathrm{Ext}^n_R(M,R^{(\kappa)})\neq 0$.

(ii) In particular, if the ring R is left $\aleph_0$-Noetherian (i.e. if any left ideal of R is countably generated) and M is countably generated, then $\mathrm{Ext}^n_R(M,R^{(\mathbb{N})})\neq 0$.

(iii) If R is any ring and M is an FP-module, then P_n (and hence the free module F) may be chosen to be finitely generated; therefore, in this case, we have $\mathrm{Ext}^n_R(M,R)\neq 0$.

The essence of equality (2) is that, if R is a countable ring and M a countable left R-module of projective dimension n, then we may choose the free module F for which $\mathrm{Ext}^n_R(M,F)\neq 0$ to be the left regular module R. In other words, the non-triviality of the group $\mathrm{Ext}^n_R(M,R^{(\mathbb{N})})$ (which itself results from the fact that the pair (R,M) satisfies the hypotheses in (ii) above) may be strengthened to the conclusion that $\mathrm{Ext}^n_R(M,R)\neq 0$ (which is generically a consequence of the finiteness condition in (iii) above).

We now consider the special case where $R=\mathbb{Z}G$ is the integral group ring of a countable group G and $M=\mathbb{Z}$ (with the trivial action of G). Assuming that G has finite cohomological dimension, we obtain from (2) the equality

$$\mathrm{cd}\, G = \max\{i \geq 0 : H^i(G,\mathbb{Z}G)\neq 0\}.$$

If the countable group G has homological dimension one, then we have $\mathrm{cd}\, G \leq 2$ and hence it follows that, in this case,

$$\mathrm{cd}\, G = 1 \Longleftrightarrow H^2(G, \mathbb{Z}G) = 0.$$

The latter equivalence may be strengthened, as stated in the next result, which is proved in [6, Theorem 3.1].

Theorem 2.1. *Let G be a countable group of homological dimension one. Then, the following conditions are equivalent:*
(i) $\mathrm{cd}G = 1$ (i.e. G is free),
(ii) $H^2(G, \mathbb{Z}G) = 0$,
(iii) $H^2(G, \mathbb{Z}G)$ is a countable group. □

Corollary 2.2. *(cf. [6, Proposition 3.2]) Let G be a finitely generated group of homological dimension one, which contains a non-trivial finitely generated free group as a normal subgroup. Then, G is a free group.*

Proof (sketch). If $1 \neq N \subseteq G$ is a normal subgroup, which is finitely generated and free, then one may use the associated Lyndon-Hochschild-Serre spectral sequence in order to obtain an isomorphism of abelian groups $H^2(G, \mathbb{Z}G) \simeq H^1(G/N, H^1(N, \mathbb{Z}G))$. Since both groups N and G/N are finitely generated and $\mathbb{Z}G$ is countable, the abelian group $H^1(G/N, H^1(N, \mathbb{Z}G))$ turns out to be countable as well. The freeness of G then follows invoking Theorem 2.1. □

As an application of the result above, we can show that any group of homological dimension one must necessarily share certain properties with locally free groups. In order to state explicitly these properties, let us consider a group G of homological dimension one. Then, as shown in [6, §3], the following hold:
(i) For any non-trivial finitely generated and free subgroup $H \subseteq G$ the normalizer $N_G(H)$ of H is locally free and the quotient group $N_G(H)/H$ is locally finite.
(ii) If $H \subseteq G$ is an infinite cyclic subgroup, then the normalizer $N_G(H)$ coincides with the centralizer $C_G(H)$ and is locally infinite cyclic.
(iii) If G is non-abelian, then the center of G is trivial.[3]

References

[1] Angeleri-Hügel, L., Herbera, D.: Mittag-Leffler conditions on modules. Ind. Univ. Math. J. **57**, 2459-2517 (2008)

[2] Bieri, R.: Homological dimension of discrete groups. Queen Mary College Mathematics Notes. Queen Mary College, London 1976

[3]This result may be alternatively proved using the structure of groups of cohomological dimension ≤ 2; for such a proof, the reader may consult Cornick's unpublished notes [4].

[3] Brown, K.S.: Cohomology of Groups. (Grad. Texts Math. **87**) Berlin Heidelberg New York: Springer 1982

[4] Cornick, J.: On groups of homological dimension one. (unpublished manuscript)

[5] Dicks, W., Linnell, P.: L^2-Betti numbers of one-relator groups. Math. Ann. **337**, 855-874 (2007)

[6] Emmanouil, I., Talelli, O.: On the equality between homological and cohomological dimension of groups. J. reine angew. Math. **664**, 55-70 (2012)

[7] Emmanouil, I., Talelli, O.: On the flat length of injective modules. J. London Math. Soc. **84**, 408-432 (2011)

[8] Facchini, A., Herbera, D., Sakhajev, I.: Finitely generated flat modules and a characterization of semiperfect rings. Comm. Algebra **31**, 4195-4214 (2003)

[9] Gray, B.: Spaces of the same n-type, for all n. Topology **5**, 241-243 (1966)

[10] Jensen, C.U.: Variations on Whitehead's problem and the structure of Ext. Models, Modules and Abelian Groups, 407-414. (Edited by Gobel, R. and Goldsmith, B.) Walter de Gruyter: Berlin, New York 2008

[11] Kropholler, P., Linnell, P., Lück, W.: Groups of small homological dimension and the Atiyah conjecture. Geometry and Cohomology in Group Theory, Durham, July 2003, editors: Bridson, M., Kropholler, P.H. and Leary, I.J., LMS Lecture Notes Series **358**, 271-277, Cambridge University Press 2009

[12] Lazard, D.: Autour de la platitude. Bull. Soc. Math. France **97**, 81-128 (1969)

[13] Lichtman, A.I.: The residual nilpotency of the augmentation ideal and the residual nilpotency of some classes of groups. Israel J. Math. **26**, 276-293 (1977)

[14] Lück, W.: L^2-invariants: theory and applications to geometry and K-theory. Springer 2002

[15] Lyndon, R.C., Schupp, P.E.: Combinatorial group theory. Ergebnisse der Mathematik und ihrer Grenzgebiete **89**, Springer: Berlin, New York 1977

[16] Magnus, W.: On a theorem of Marshall Hall. Ann. Math. **40**, 764-768 (1939)

[17] Passi, I.B.S.: Group rings and their augmentation ideals. Lecture Notes in Mathematics **715**, Springer: Berlin 1979

[18] Puninski, G., Rothmaler, P.: When every finitely generated flat module is projective. J. Algebra **277**, 542-558 (2004)

[19] Raynaud, M., Gruson, L.: Critères de platitude et de projectivité. Invent. Math. **13**, 1-89 (1971)

[20] Stallings, J.R.: On torsion free groups with infinitely many ends. Ann. Math. **88**, 312-334 (1968)

[21] Swan, R.G.: Groups of cohomological dimension one. J. Algebra **12**, 585–610 (1969)

Braided diagram groups and local similarity groups

Daniel S. Farley   Bruce Hughes*

Abstract

Hughes defined a class of groups that act as local similarities on compact ultrametric spaces. Guba and Sapir had previously defined braided diagram groups over semigroup presentations. The two classes of groups share some common characteristics: both act properly by isometries on CAT(0) cubical complexes, and certain groups in both classes have type F_∞, for instance.

Here we clarify the relationship between these families of groups: the braided diagram groups over tree-like semigroup presentations are precisely the groups that act on compact ultrametric spaces via small similarity structures. The proof can be considered a generalization of the proof that Thompson's group V is a braided diagram group over a tree-like semigroup presentation.

We also prove that certain additional groups, such as the Houghton groups H_n, and $QAut(T_{2,c})$, lie in both classes.

Contents

1 Introduction

In [7], Hughes described a class of groups that act as homeomorphisms on compact ultrametric spaces. Fix a compact ultrametric space X. The essence

*The second-named author was supported in part by NSF Grant DMS–0504176.

of the idea was to associate to X a *finite similarity structure*, which is a function that associates to each ordered pair of balls $B_1, B_2 \subseteq X$ a finite set $\mathrm{Sim}_X(B_1, B_2)$ of surjective similarities from B_1 to B_2. (A *similarity* is a map that stretches or contracts distances by a fixed constant.) The finite sets $\mathrm{Sim}_X(B_1, B_2)$ are assumed to have certain desirable closure properties (such as closure under composition). A homeomorphism $h : X \to X$ is said to be *locally determined by* Sim_X if each $x \in X$ has a ball neighborhood B with the property that $h(B)$ is a ball and the restriction of h to B agrees with one of the local similarities $\sigma \in \mathrm{Sim}_X(B, h(B))$. The collection of all homeomorphisms that are locally determined by Sim_X forms a group under composition. We will call such a group an *FSS group* (finite similarity structure group) for short. Hughes [7] proved that each FSS group has the Haagerup property, and even acts properly on a CAT(0) cubical complex. In [5], the authors described a class of FSS groups that have type F_∞. That class includes Thompson's group V, and the main theorem of [5] is best understood as a generalization of [2], where Brown originally showed that V has type F_∞.

In earlier work, Guba and Sapir [6] had sketched a theory of braided diagram groups over semigroup presentations, and proved that Thompson's group V is a braided diagram group over the semigroup presentation $\langle x \mid x = x^2 \rangle$. Farley [4] showed that braided diagram groups over semigroup presentations act properly on CAT(0) cubical complexes.

The class $\mathcal{F}$ of FSS groups and the class $\mathcal{B}$ of braided diagram groups therefore have a common origin, as generalizations of Thompson's group V. Both classes also share other features in common (as noted above). It is therefore natural to wonder to what extent the two classes are the same. The main goal of this note is to prove Theorem 4.12, which says that the FSS groups determined by small similarity structures (Definition 4.6) are precisely the same as the braided diagram groups determined by tree-like semigroup presentations (Definition 4.1). It is even possible that Theorem 4.12 describes the precise extent of the overlap between $\mathcal{F}$ and $\mathcal{B}$, but we do not know how to prove this.

We give all relevant definitions, and our treatment is fairly self-contained as a result. A precise definition of braided diagram groups is given in Section 2, the precise definition of FSS groups appears in Section 3, and the main theorem is proved in Section 4. Along the way, we give additional examples in the class $\mathcal{F} \cap \mathcal{B}$, including the Houghton groups H_n and a certain group $QAut(T_{2,c})$ of quasi-automorphisms of the infinite binary tree. (These are Examples 4.3 and 4.4, respectively.)

This note has been adapted from the longer preprint [5]. The first part of the latter preprint (including roughly the first six sections) will be published elsewhere. The first author would like to thank the organizers of the Durham Symposium (August 2013) for the opportunity to speak. Example 4.3 first appeared as part of the first author's lecture. The idea of Example 4.4 occurred to the first author after listening to Collin Bleak's lecture at the

Symposium.

2 Braided diagram groups

In this section, we will recall the definition of braided diagram groups over semigroup presentations. Note that the theory of braided diagram groups was first sketched by Guba and Sapir [6]. A more extended introduction to braided diagram groups appears in [4].

Definition 2.1. Let Σ be a set, called an *alphabet*. The *free semigroup on* Σ, denoted Σ^+, is the collection of all positive non-empty strings formed from Σ, i.e.,

$$\Sigma^+ = \{u_1 u_2 \dots u_n \mid n \in \mathbb{N}, u_i \in \Sigma \text{ for } i \in \{1, \dots, n\}\}.$$

The *free monoid on* Σ, denoted Σ^*, is the union $\Sigma^+ \cup \{1\}$, where 1 denotes the empty string. (Here we assume that $1 \notin \Sigma$ to avoid ambiguity.) The operations in Σ^+ and Σ^* are concatenation.

We write $w_1 \equiv w_2$ if w_1 and w_2 are equal as words in Σ^*.

Definition 2.2. A *semigroup presentation* $\mathcal{P} = \langle \Sigma \mid \mathcal{R} \rangle$ consists of an alphabet Σ and a set $\mathcal{R} \subseteq \Sigma^+ \times \Sigma^+$. The elements of $\mathcal{R}$ are called *relations*.

Remark 2.3. A relation $(w_1, w_2) \in \mathcal{R}$ can be viewed as an equality between the words w_1 and w_2. We use ordered pairs to describe these equalities because we will occasionally want to make a distinction between the left and right sides of a relation.

A semigroup presentation $\mathcal{P}$ determines a semigroup $S_{\mathcal{P}}$, just as a group presentation determines a group. We will, however, make essentially no use of this semigroup $S_{\mathcal{P}}$. Our interest is in braided diagrams over $\mathcal{P}$ (see below).

Definition 2.4. (Braided Semigroup Diagrams) A *frame* is a homeomorphic copy of $\partial([0,1]^2) = (\{0,1\} \times [0,1]) \cup ([0,1] \times \{0,1\})$. A frame has a *top* side, $(0,1) \times \{1\}$, a *bottom* side, $(0,1) \times \{0\}$, and *left* and *right* sides, $\{0\} \times [0,1]$ and $\{1\} \times [0,1]$, respectively. The top and bottom of a frame have obvious left to right orderings.

A *transistor* is a homeomorphic copy of $[0,1]^2$. A transistor has top, bottom, left, and right sides, just as a frame does. The top and bottom of a transistor also have obvious left to right orderings.

A *wire* is a homeomorphic copy of $[0,1]$. Each wire has a bottom 0 and a top 1.

Let $\mathcal{P} = \langle \Sigma \mid \mathcal{R} \rangle$ be a semigroup presentation. Let $\mathcal{T}(\Delta)$ be a finite (possibly empty) set of transistors. Let $\mathcal{W}(\Delta)$ be a finite, nonempty set of wires. We let $F(\Delta) = \partial([0,1]^2)$ be a frame. We let $\ell_\Delta : \mathcal{W}(\Delta) \to \Sigma$ be an arbitrary function, called the *labelling function*.

For each wire $W \in \mathcal{W}(\Delta)$, we choose a point $t(W)$ on the bottom of a transistor, or on the top of the frame, and a point $b(W)$ on the top of a transistor, or on the bottom of the frame. The points $t(W)$ and $b(W)$ are called the *top* and *bottom contacts* of W, respectively.

We attach the top of each wire W to $t(W)$ and the bottom of W to $b(W)$. The resulting topological space Δ is called a *braided diagram over* $\mathcal{P}$ if the following additional conditions are satisfied:

1. If $W_i, W_j \in \mathcal{W}(\Delta)$, $t(W_i) = t(W_j)$ only if $W_i = W_j$, and $b(W_i) = b(W_j)$ only if $W_i = W_j$. In other words, the disjoint union of all of the wires maps injectively into the quotient.

2. We consider the top of some transistor $T \in \mathcal{T}(\Delta)$. Reading from left to right, we find contacts

$$b(W_{i_1}), b(W_{i_2}), \ldots, b(W_{i_n}),$$

where $n \geq 0$. The word $\ell_t(T) = \ell(W_{i_1})\ell(W_{i_2})\ldots\ell(W_{i_n})$ is called the *top label of* T. Similarly, reading from left to right along the bottom of T, we find contacts

$$t(W_{j_1}), t(W_{j_2}), \ldots, t(W_{j_m}),$$

where $m \geq 0$. The word $\ell_b(T) = \ell(W_{j_1})\ell(W_{j_2})\ldots\ell(W_{j_m})$ is called the *bottom label of* T. We require that, for any $T \in \mathcal{T}(\Delta)$, either $(\ell_t(T), \ell_b(T)) \in \mathcal{R}$ or $(\ell_b(T), \ell_t(T)) \in \mathcal{R}$. (We emphasize that it is not sufficient for $\ell_t(T)$ to be equivalent to $\ell_b(T)$ modulo the relation $\sim$ determined by $\mathcal{R}$. Note also that this condition implies that $\ell_b(T)$ and $\ell_t(T)$ are both non-empty, since $\mathcal{P}$ is a semigroup presentation. In particular, each transistor has wires attached to its top and bottom faces.)

3. We define a relation $\preceq$ on $\mathcal{T}(\Delta)$ as follows. Write $T_1 \preceq T_2$ if there is some wire W such that $t(W) \in T_2$ and $b(W) \in T_1$. We require that the transitive closure $\dot{\preceq}$ of $\preceq$ be a strict partial order on $\mathcal{T}(\Delta)$.

Definition 2.5. Let Δ be a braided diagram over $\mathcal{P}$. Reading from left to right across the top of the frame $F(\Delta)$, we find contacts

$$t(W_{i_1}), t(W_{i_2}), \ldots, t(W_{i_n}),$$

for some $n \geq 1$. The word $\ell(W_{i_1})\ell(W_{i_2})\ldots\ell(W_{i_n}) = \ell_t(\Delta)$ is called the *top label of* Δ. We can similarly define the *bottom label of* Δ, $\ell_b(\Delta)$. We say that Δ is a *braided* $(\ell_t(\Delta), \ell_b(\Delta))$*-diagram over* $\mathcal{P}$.

Remark 2.6. One should note that braided diagrams, despite the name, are not truly braided. In fact, two braided diagrams are equivalent (see Definition 2.10) if there is a certain type of marked homeomorphism between them. Equivalence therefore does not depend on any embedding into a larger space. Braided diagram groups (as defined in Theorem 2.13) also seem to have little in common with Artin's braid groups.

Example 2.7. Let $\mathcal{P} = \langle a, b, c \mid ab = ba, ac = ca, bc = cb \rangle$. Figure 1 shows an example of a braided $(aabc, acba)$-diagram over the semigroup presentation $\mathcal{P}$. The frame is the box formed by the dashed line. The wires that appear to cross in the figure do not really touch, and it is unnecessary to specify which wire passes over the other one. See Remark 2.6.

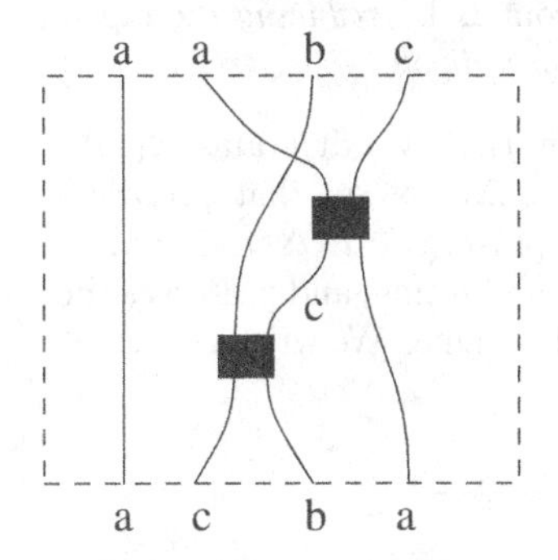

Figure 1: A braided $(aabc, acba)$-diagram over the semigroup presentation $\mathcal{P} = \langle a, b, c \mid ac = ca, ab = ba, bc = cb \rangle$.

Definition 2.8. (Concatenation of braided diagrams) Let Δ_1 and Δ_2 be braided diagrams over $\mathcal{P}$. We suppose that Δ_1 is a (w_1, w_2)-diagram and Δ_2 is a (w_2, w_3)-diagram. We can multiply Δ_1 and Δ_2 by stacking them. More explicitly, we remove the bottom of the frame of Δ_1 and the top of the frame of Δ_2, and then glue together the wires in order from left to right. This gluing is compatible with the labeling of the wires, since the bottom label of Δ_1 is the same as the top label of Δ_2. The result is a braided diagram $\Delta_1 \circ \Delta_2$, called the *concatenation* of Δ_1 and Δ_2.

Definition 2.9. (Dipoles) Let Δ be a braided semigroup diagram over $\mathcal{P}$. We say that the transistors $T_1, T_2 \in \mathcal{T}(\Delta)$, $T_1 \preceq T_2$, form a *dipole* if:

1. the bottom label of T_1 is the same as the top label of T_2, and
2. there are wires $W_{i_1}, W_{i_2}, \ldots, W_{i_n} (n \geq 1)$ such that the bottom contacts T_2, read from left to right, are precisely

$$t(W_{i_1}), t(W_{i_2}), \ldots, t(W_{i_n})$$

and the top contacts of T_1, read from left to right, are precisely

$$b(W_{i_1}), b(W_{i_2}), \ldots, b(W_{i_n}).$$

Define a new braided diagram as follows. Remove the transistors T_1 and T_2 and all of the wires $W_{i_1}, \ldots, W_{i_n}$ connecting the top of T_1 to the bottom of T_2. Let $W_{j_1}, \ldots, W_{j_m}$ be the wires attached (in that order) to the top of T_2, and let $W_{k_1}, \ldots, W_{k_m}$ be the wires attached to the bottom of T_1. We glue the bottom of W_{j_ℓ} to the top of W_{k_ℓ}. There is a natural well-defined labelling function on the resulting wires, since $\ell(W_{j_\ell}) = \ell(W_{k_\ell})$ by our assumptions. We say that the new diagram Δ' is obtained from Δ by *reducing the dipole* (T_1, T_2). The inverse operation is called *inserting a dipole*.

Definition 2.10. (Equivalent Diagrams) We say that two diagrams Δ_1, Δ_2 are *equivalent* if there is a homeomorphism $\phi : \Delta_1 \to \Delta_2$ that preserves the labels on the wires, restricts to a homeomorphism $\phi_| : F(\Delta_1) \to F(\Delta_2)$, preserves the tops and bottoms of the transistors and frame, and preserves the left to right orientations on the transistors and the frame. We write $\Delta_1 \equiv \Delta_2$.

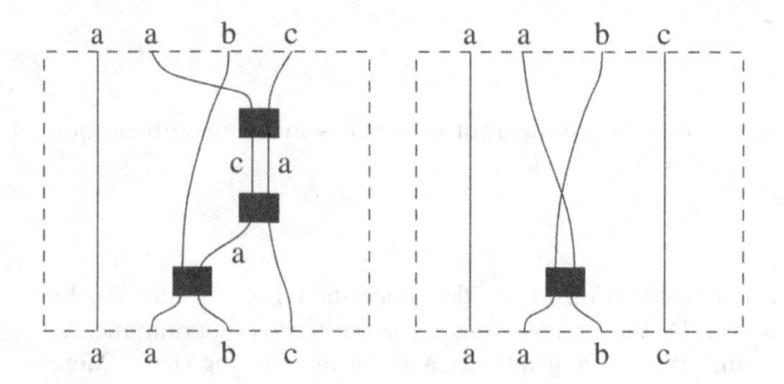

Figure 2: The diagram on the right is obtained from the one on the left by reduction of a dipole.

Definition 2.11. (Equivalent Modulo Dipoles; Reduced Diagram) We say that Δ and Δ' are *equivalent modulo dipoles* if there is a sequence $\Delta \equiv \Delta_1 \equiv \Delta_2 \equiv \ldots \equiv \Delta_n \equiv \Delta'$, where Δ_{i+1} is obtained from Δ_i by either inserting or removing a dipole, for $i \in \{1, \ldots, n-1\}$. We write $\Delta = \Delta'$. (The relation of equivalence modulo dipoles is indeed an equivalence relation – see [4].)

A braided diagram Δ over a semigroup presentation is called *reduced* if it contains no dipoles. Each equivalence class modulo dipoles contains a unique reduced diagram [4].

Example 2.12. In Figure 2, we have two braided diagrams over the semigroup presentation $\mathcal{P} = \langle a, b, c \mid ab = ba, ac = ca, bc = cb \rangle$. The two rightmost transistors in the diagram on the left form a dipole, and the diagram on the right is the result of reducing that dipole.

Theorem 2.13. *[4] Let $\mathcal{P} = \langle \Sigma \mid \mathcal{R} \rangle$ be a semigroup presentation, and let $w \in \Sigma^+$. We let $D_b(\mathcal{P}, w)$ denote the set of equivalence classes of braided (w, w)-diagrams modulo dipoles. The operation of concatenation induces a well-defined group operation on $D_b(\mathcal{P}, w)$. This group $D_b(\mathcal{P}, w)$ is called* the braided diagram group over $\mathcal{P}$ based at w.

3 Groups defined by finite similarity structures

3.1 Review of ultrametric spaces and finite similarity structures

We now give a quick review of finite similarity structures on compact ultrametric spaces, as defined in Hughes [7]. Most of this subsection is taken directly from [5].

Definition 3.1. An *ultrametric space* is a metric space (X, d) such that $d(x, y) \leq \max\{d(x, z), d(z, y)\}$ for all $x, y, z \in X$.

Lemma 3.2. *Let X be an ultrametric space.*

1. *Let $B_r(x)$ be an open metric ball in X. If $y \in B_r(x)$, then $B_r(x) = B_r(y)$.*
2. *If B_1 and B_2 are open metric balls in X, then either the balls are disjoint, or one is contained in the other.*
3. *Every open ball in X is a closed set, and every closed ball in X is an open set.*
4. *If X is compact, then each open ball B is contained in at most finitely many distinct open balls of X.*
5. *If X is compact, then each open ball in X is a closed ball (possibly of a different radius), and each closed ball is an open ball.*
6. *If X is compact and x is not an isolated point, then each open ball $B_r(x)$ is partitioned by its maximal proper open subballs, which are finite in number.*

Convention 3.3. We assume for the rest of the section that X is a compact ultrametric space. By Lemma 3.2(5), open balls are closed balls, and closed balls are open balls, so we can refer to both simply as balls, and we will follow this practice from now on.

Definition 3.4. Let $f : X \to Y$ be a function between metric spaces. We say that f is a *similarity* if there is a constant $C > 0$ such that $d_Y(f(x_1), f(x_2)) = C d_X(x_1, x_2)$, for all x_1 and x_2 in X.

Definition 3.5. A *finite similarity structure for X* is a function Sim_X that assigns to each ordered pair B_1, B_2 of balls in X a (possibly empty) set $\mathrm{Sim}_X(B_1, B_2)$ of surjective similarities $B_1 \to B_2$ so that whenever B_1, B_2, B_3 are balls in X, the following properties hold:

1. (Finiteness) $\mathrm{Sim}_X(B_1, B_2)$ is a finite set.
2. (Identities) $\mathrm{id}_{B_1} \in \mathrm{Sim}_X(B_1, B_1)$.
3. (Inverses) If $h \in \mathrm{Sim}_X(B_1, B_2)$, then $h^{-1} \in \mathrm{Sim}_X(B_2, B_1)$.
4. (Compositions) If $h_1 \in \mathrm{Sim}_X(B_1, B_2)$ and $h_2 \in \mathrm{Sim}_X(B_2, B_3)$, then $h_2 h_1 \in \mathrm{Sim}_X(B_1, B_3)$.
5. (Restrictions) If $h \in \mathrm{Sim}_X(B_1, B_2)$ and $B_3 \subseteq B_1$, then

$$h|B_3 \in \mathrm{Sim}_X(B_3, h(B_3)).$$

Definition 3.6. A homeomorphism $h\colon X \to X$ is *locally determined by* Sim_X provided that for every $x \in X$, there exists a ball B' in X such that $x \in B'$, $h(B')$ is a ball in X, and $h|B' \in \mathrm{Sim}(B', h(B'))$.

Definition 3.7. The *finite similarity structure* (*FSS*) *group* $\Gamma(\mathrm{Sim}_X)$ is the set of all homeomorphisms $h\colon X \to X$ such that h is locally determined by Sim_X.

Remark 3.8. The fact that $\Gamma(\mathrm{Sim}_X)$ is a group under composition is due to Hughes [7].

3.2 A description of the homeomorphisms determined by a similarity structure

In this subsection, we offer a somewhat simpler description of the elements in the groups $\Gamma(\mathrm{Sim}_X)$ (Proposition 3.11), which shows that elements $\gamma \in \Gamma(\mathrm{Sim}_X)$ can be described in a manner reminiscent of the tree pair representatives for elements in Thompson's group V (see [3]).

Definition 3.9. We define the *standard partitions* of X inductively as follows.

1. $\{X\}$ is a standard partition.
2. If $\mathcal{P} = \{\widehat{B}_1, \ldots, \widehat{B}_n\}$ is a standard partition, and $\{B_1, \ldots, B_m\}$ is the partition of $\widehat{B}_i$ into maximal proper subballs, then the following is also a standard partition: $(\mathcal{P} - \{\widehat{B}_i\}) \cup \{B_1, \ldots, B_m\}$.

Clearly, each standard partition is a partition of X into balls.

Lemma 3.10. *Every partition $\mathcal{P}$ of X into balls is standard.*

Proof. We prove this by induction on $|\mathcal{P}|$. It is clearly true if $|\mathcal{P}| = 1$. We note that compactness implies that each partition $\mathcal{P}$ of X into balls must be finite.

For an arbitrary ball $B \subseteq X$, we define the *depth* of B, denoted $d(B)$, to be the number of distinct balls of X that contain B. (This definition is similar to Definition 3.19 from [5].) We note that $d(B)$ is a positive integer by Lemma 3.2(4), and $d(X) = 1$.

Now we suppose that a partition $\mathcal{P}$ is given to us. We assume inductively that all partitions with smaller numbers of balls are standard. By finiteness of $\mathcal{P}$, there is some ball B having maximum depth m, where we can assume that $m \geq 2$. Let $\widehat{B}$ denote the ball containing B as a maximal proper subball. Clearly, $d(\widehat{B}) = m - 1$. We let $\{B_0, \ldots, B_k\}$ be the collection of maximal proper subballs of $\widehat{B}$, where $B = B_0$ and $k \geq 1$.

We claim that $\{B_0, B_1, B_2, \ldots, B_k\} \subseteq \mathcal{P}$. Choose $x \in B_i$. Our assumptions imply that x is in some ball B' of $\mathcal{P}$ such that $d(B') \leq m$. The only such balls are B_i, $\widehat{B}$, and any balls that contain $\widehat{B}$. (This uses an appeal to Lemma 3.2(2).) Since $\widehat{B} \cap B_0 \neq \emptyset$ and $B_0 = B \in \mathcal{P}$, the only possibility is that $B' = B_i$, since $\mathcal{P}$ is a partition. This proves that $\{B_0, \ldots, B_k\} \subseteq \mathcal{P}$.

Now we consider the partition $\mathcal{P}' = (\mathcal{P} - \{B_0, \ldots, B_k\}) \cup \{\widehat{B}\}$. This partition is standard by the induction hypothesis, and it follows directly that $\mathcal{P}$ itself is standard. □

Proposition 3.11. *Let* Sim_X *be a finite similarity structure on* X*, and let* $\gamma \in \Gamma(\mathrm{Sim}_X)$*. There exist standard partitions* $\mathcal{P}_1 = \{B_1, \ldots, B_n\}$ *and* $\mathcal{P}_2$ *of* X*, a bijection* $\phi : \mathcal{P}_1 \to \mathcal{P}_2$*, and elements* $\sigma_i \in \mathrm{Sim}(B_i, \phi(B_i))$ *such that* $\gamma_{|B_i} = \sigma_i$*, for* $i = 1, \ldots, n$*.*

Moreover, we can arrange that the balls B_i *are maximal in the sense that if* $B \subseteq X$ *and* $\gamma(B)$ *are balls such that* $\gamma_{|B} \in \mathrm{Sim}_X(B, \gamma(B))$*, then* $B \subseteq B_i$*, for some* $i \in \{1, \ldots, n\}$*.*

Proof. Since γ is locally determined by Sim_X, we can find an open cover of X by balls such that the restriction of γ to each ball is a local similarity in the Sim_X-structure. By compactness of X, we can pass to a finite subcover. An application of Lemma 3.2(2) allows us to pass to a subcover that is also a partition. We call this partition $\mathcal{P}_1$. We can then set $\mathcal{P}_2 = \gamma(\mathcal{P}_1)$. Both partitions are standard by Lemma 3.10.

The final statement is essentially Lemma 3.7 from [7]. □

4 Braided diagram groups and groups determined by finite similarity structures

4.1 Braided diagram groups over tree-like semigroup presentations

Definition 4.1. A semigroup presentation $\mathcal{P} = \langle \Sigma \mid \mathcal{R} \rangle$ is *tree-like* if,

1. every relation $(w_1, w_2) \in \mathcal{R}$ satisfies $|w_1| = 1$ and $|w_2| > 1$;
2. if $(a, w_1), (a, w_2) \in \mathcal{R}$, then $w_1 \equiv w_2$.

Example 4.2. The generalized Thompson's groups V_d are isomorphic to the braided diagram groups $D_b(\mathcal{P}, x)$, where $\mathcal{P} = \langle x \mid (x, x^d) \rangle$ is a tree-like semigroup presentation. This fact was already proved in [6] and [4], and it is also a consequence of Theorem 4.12.

Example 4.3. Consider the graph G_n made up from a disjoint union of n rays: $G_n = \{1, \ldots, n\} \times [0, \infty)$. We assume that each ray is given the standard CW-complex structure with a vertex at each integer. We define the *Houghton group* H_n to be the set of bijections h of the vertices G_n^0 such that

1. h preserves adjacency, with at most finitely many exceptions, and
2. h preserves ends: that is, for each $i \in \{1, \ldots, n\}$, there are $i_1, i_2 \in \mathbb{N}$ such that $h(\{i\} \times [i_1, \infty)) = \{i\} \times [i_2, \infty)$.

Ken Brown [2] showed that H_n is a group of type F_{n-1} but not of type F_n.

We will sketch a proof that each Houghton group H_n is a braided diagram group over a tree-like semigroup presentation. (It follows, in particular, that each of these groups is $\Gamma(\mathrm{Sim}_X)$, for an appropriate compact ultrametric space X and finite similarity structure Sim_X, by Theorem 4.12.) For $n \geq 2$, consider the semigroup presentation

$$\mathcal{P}_n = \langle a, r, x_1, \ldots, x_n \mid (r, x_1 x_2 x_3 \ldots x_n), (x_1, a x_1), (x_2, a x_2), \ldots, (x_n, a x_n) \rangle.$$

Similarly, we can define $\mathcal{P}_1 = \langle a, r \mid (r, ar) \rangle$. We claim that $D_b(\mathcal{P}_n, r)$ is isomorphic to H_n. We sketch the proof for $n \geq 2$; the proof in case $n = 1$ is very similar.

The elements of $D_b(\mathcal{P}_n, r)$ can be expressed in the form $\Delta_2 \circ \Delta_1^{-1}$, where each transistor in Δ_i is "positive"; i.e., the top label of the transistor is the left side of a relation in $\mathcal{P}_n$, and the bottom label is the right side. (This is proved as part of the proof of Theorem 4.12.) We can think of each diagram Δ_i as a recipe for separating G_n into connected components. The wires running between Δ_1 and Δ_2 in the concatenation $\Delta_2 \circ \Delta_1^{-1}$ describe how these connected components should be matched by the bijection $h \in H_n$. To put it more explicitly, the relations represent the following operations:

1. the relation $(r, x_1x_2 \ldots x_n)$ describes the initial configuration G_n of n disjoint rays. The letters x_i $(i \in \{1, \ldots, n\})$ represent the isomorphism types of the different rays. The different subscripts prevent different ends of G_n from being permuted nontrivially. (If we wish to remove the end-preserving condition above, we can simply replace the n distinct symbols x_1, x_2, …, x_n by the single symbol x.)

2. the relation (x_i, ax_i) (for $i \in \{1, \ldots, n\}$) represents the action of breaking the initial vertex away from the ray of isomorphism type x_i. The initial vertex of the ray gets the label a, and the new ray retains the label x_i (since it is of the same combinatorial type as the original ray, and the new ray is a permissible target for the original ray under the action of the Houghton group). The letter a thus represents a single floating vertex. Any two such vertices can be matched by an element of the Houghton group, which is why we use a single label for all of these vertices.

We illustrate how an (r, r)-diagram over $\mathcal{P}_2$ represents an element of H_2. Figure 3 depicts an (r, r)-diagram Δ_h over $\mathcal{P}_2$. This diagram Δ_h represents the element $h \in H_2$ that sends: $(2, n)$ to $(2, n-1)$, for each $n \geq 1$, $(1, n)$ to $(1, n+1)$ (for all n), and $(2, 0)$ to $(1, 0)$. Note that the bottom portion of the diagram represents a subdivision of the domain, and the top portion represents a subdivision of the range. It is straightforward to check that the indicated function does not change if we insert or remove dipoles.

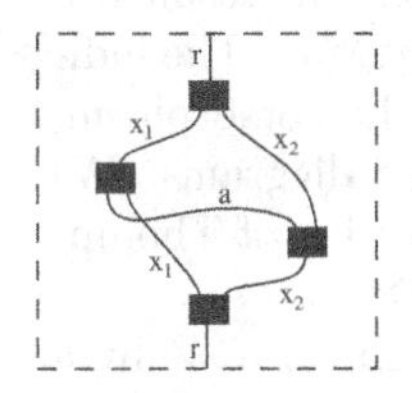

Figure 3: The given (r, r)-diagram over $\mathcal{P}_2$ represents an element of Houghton's group H_2.

Example 4.4. The same principle can be used to exhibit the group $\Gamma = \mathrm{QAut}(\mathcal{T}_{2,c})$ as a braided diagram group over a tree-like semigroup presentation. Here Γ (as defined in [1]) is the group of self-bijections h of the vertices of the infinite ordered rooted binary tree T such that

1. h preserves adjacency, with at most finitely many exceptions, and

2. h preserves the left-right ordering of the edges incident with and below a given vertex, again with at most finitely many exceptions.

Consider the semigroup presentation $\mathcal{P} = \langle a, x \mid (x, xax) \rangle$. We claim that $D_b(\mathcal{P}, x)$ is isomorphic to Γ. Much of the discussion from the previous example carries over identically. We will simply indicate how the single relation allows us to simulate breaking the binary tree into pieces. (Such a dissection of T would be represented by a positive diagram, as above. The wires connecting the bottoms of the positive diagrams Δ_1 and Δ_2 would again represent how the resulting pieces are matched by a bijection.)

The letter x represents the isomorphism type of the binary tree T. The relation represents breaking the tree at the root. The result of this operation yields a floating vertex (represented by the letter a), and two new rooted binary trees (both represented by x). The first x in xax represents the left branch, and the second x represents the right. The description of the isomorphism of $D_b(\mathcal{P}, x)$ with Γ now follows the general pattern of the previous example.

Remark 4.5. Example 4.3 shows that the F_∞ result of [5] cannot be extended to all groups determined by finite similarity structures (as defined in Section 3).

All of the groups in the above examples act properly on CAT(0) cubical complexes by a construction of [4].

We note also that the representation of the above groups as braided diagram groups suggests a method for producing embeddings into other groups, such as (perhaps most notably) Thompson's group V. For instance, the group from Example 4.4 can be embedded into V as follows. Given a braided diagram Δ over $\mathcal{P}$, systematically replace each a label with an x. The result is a braided diagram over the semigroup presentation $\mathcal{P}' = \langle x \mid (x, x^3) \rangle$. The indicated function $\phi : D_b(\mathcal{P}, x) \to D_b(\mathcal{P}', x)$ is easily seen to be a homomorphism, and ϕ is injective since it sends reduced diagrams to reduced diagrams. We can now appeal to the fact that $D_b(\mathcal{P}', x) \cong V_3$ (the 3-ary version of Thompson's group V), and the latter group embeds in V itself.

The group from Example 4.4 was previously known to embed in V by a result of [1].

4.2 Groups determined by small similarity structures

Definition 4.6. Let X be a compact ultrametric space. We say that the finite similarity structure Sim_X is *small* if, for every pair of balls B_1, B_2 in X, $|\mathrm{Sim}_X(B_1, B_2)| \leq 1$.

Definition 4.7. Let X be a compact ultrametric space endowed with a small similarity structure Sim_X. If $B \subseteq X$ is a ball in X that is not an isolated point, then a *local ball order* at B is an assignment of a linear order $<$ to the set $\{\widehat{B}_1, \ldots, \widehat{B}_n\}$ of maximal proper subballs of B. A *ball order* on X is an assignment of such a linear order to each ball $B \subseteq X$ that is not a singleton. The ball order is *compatible* with Sim_X if each $h \in \mathrm{Sim}_X(B_1, B_2)$ induces an order-preserving bijection of the maximal proper subballs of B_1 and B_2, for all choices of B_1 and B_2.

Lemma 4.8. *Let X be a compact ultrametric space endowed with a small similarity structure. There exists a ball order on X that is compatible with* Sim_X.

Proof. We recall a definition from [5]. Let $B \subseteq X$ be a metric ball. Let $[B] = \{B' \subseteq X \mid \mathrm{Sim}_X(B, B') \neq \emptyset\}$; $[B]$ is called the Sim_X*-class* of B.

From a given Sim_X-class of balls, choose a ball B. If B is not a singleton, then there exists a collection of maximal proper subballs B_1, ..., B_n of B. Choose a linear order on this collection of balls; without loss of generality, $B_1 < B_2 < \ldots < B_n$. If B' is another ball in $[B]$, then we can let h denote the unique element of $\mathrm{Sim}_X(B, B')$. This h carries the maximal proper subballs of B into maximal proper subballs of B', and thereby induces an order $h(B_1) < h(B_2) < \ldots < h(B_n)$ on the maximal proper subballs of B'. This procedure gives a local ball order to each ball $B' \in [B]$.

We repeat this procedure for each Sim_X-class of balls. The result is a ball order on X that is compatible with Sim_X. □

Remark 4.9. A ball order on X also determines a linear order on any given collection of pairwise disjoint balls in X. For let $\mathcal{C}$ be such a collection, and let $B_1, B_2 \in \mathcal{C}$. There is a unique smallest ball $B \subseteq X$ that contains both B_1 and B_2, by Lemma 3.2(4). Let $\{\widehat{B}_1, \ldots, \widehat{B}_n\}$ be the collection of maximal proper subballs of B. By minimality of B, we must have that B_1 and B_2 are contained in distinct maximal proper subballs of B; say $B_1 \subseteq \widehat{B}_1$ and $B_2 \subseteq \widehat{B}_2$. We write $B_1 < B_2$ if and only if $\widehat{B}_1 < \widehat{B}_2$. This defines a linear order on $\mathcal{C}$. The verification is straightforward.

Definition 4.10. Let X be a compact ultrametric space with a small similarity structure Sim_X and a compatible ball order. Define a semigroup presentation $\mathcal{P}_{\mathrm{Sim}_X} = \langle \Sigma \mid \mathcal{R} \rangle$ as follows. Let

$$\Sigma = \{[B] \mid B \text{ is a ball in } X\}.$$

If $B \subseteq X$ is a ball, let $B_1, \ldots, B_n$ be the maximal proper subballs of B, listed in order. If B is a point, then $n = 0$. We set

$$\mathcal{R} = \{([B], [B_1][B_2] \ldots [B_n]) \mid n \geq 1, B \text{ is a ball in } X\}.$$

Remark 4.11. We note that $\mathcal{P}_{\mathrm{Sim}_X}$ will always be a tree-like semigroup presentation, for any choice of compact ultrametric space X, small similarity structure Sim_X, and compatible ball order.

4.3 The main theorem

Theorem 4.12. *If X is a compact ultrametric space with a small similarity structure* Sim_X *and compatible ball order, then*

$$\Gamma(\mathrm{Sim}_X) \cong D_b(\mathcal{P}_{\mathrm{Sim}_X}, [X]).$$

Conversely, if $\mathcal{P} = \langle \Sigma \mid \mathcal{R} \rangle$ is a tree-like semigroup presentation, and $x \in \Sigma$, then there is a compact ultrametric space $X_{\mathcal{P}}$, a small finite similarity structure $\mathrm{Sim}_{X_{\mathcal{P}}}$*, and a compatible ball order such that*

$$D_b(\mathcal{P}, x) \cong \Gamma(\mathrm{Sim}_{X_{\mathcal{P}}}).$$

Proof. If $\gamma \in \Gamma(\mathrm{Sim}_X)$, then, by Proposition 3.11, there are standard partitions $\mathcal{P}_1$, $\mathcal{P}_2$ of X into balls, and a bijection $\phi : \mathcal{P}_1 \to \mathcal{P}_2$ such that, for any $B \in \mathcal{P}_1$, $\gamma(B) = \phi(B)$ and $\gamma|_B \in \mathrm{Sim}_X(B, \gamma(B))$. Since $|\mathrm{Sim}_X(B, \gamma(B))| \leq 1$, the triple $(\mathcal{P}_1, \mathcal{P}_2, \phi)$ determines γ without ambiguity. We call $(\mathcal{P}_1, \mathcal{P}_2, \phi)$ a *defining triple* for γ. Note that a given γ will usually have many defining triples. Let $\mathcal{D}$ be the set of all defining triples, for γ running over all of $\Gamma(\mathrm{Sim}_X)$.

We will now define a map $\psi : \mathcal{D} \to D_b(\mathcal{P}_{\mathrm{Sim}_X}, [X])$. To a partition $\mathcal{P}$ of X into balls, we first assign a braided diagram $\Delta_{\mathcal{P}}$ over $\mathcal{P}_{\mathrm{Sim}_X}$. There is a transistor $T_B \in \mathcal{T}(\Delta_{\mathcal{P}})$ for each ball B which properly contains some ball of $\mathcal{P}$. There is a wire $W_B \in \mathcal{W}(\Delta_{\mathcal{P}})$ for each ball B which contains a ball of $\mathcal{P}$. The wires are attached as follows:

1. If $B = X$, then we attach the top of W_B to the top of the frame. If $B \neq X$, then the top of the wire W_B is attached to the bottom of the transistor $T_{\widehat{B}}$, where $\widehat{B}$ is the (unique) ball that contains B as a maximal proper subball.

   Moreover, we attach the wires in an "order-respecting" fashion. Thus, if $\widehat{B}$ is a ball properly containing balls of $\mathcal{P}$, we let $B_1, B_2, \ldots, B_n$ be the collection of maximal proper subballs of $\widehat{B}$, listed in order. We attach the wires $W_{B_1}, W_{B_2}, \ldots, W_{B_n}$ so that $t(W_{B_i})$ is to the left of $t(W_{B_j})$ on the bottom of $T_{\widehat{B}}$ if $i < j$.

2. The bottom of the wire W_B is attached to the top of T_B if B properly contains a ball of $\mathcal{P}$. If not (i.e., if $B \in \mathcal{P}$), then we attach the bottom of W_B to the bottom of the frame. We can arrange, moreover, that the wires are attached in an order-respecting manner to the bottom of the frame. (Thus, if $B_1 < B_2$ $(B_1, B_2 \in \mathcal{P})$, we have that $b(W_{B_1})$ is to the left of $b(W_{B_2})$.)

The labelling function $\ell : \mathcal{W}(\Delta_{\mathcal{P}}) \to \Sigma$ sends W_B to $[B]$. It is straightforward to check that the resulting $\Delta_{\mathcal{P}}$ is a braided diagram over $\mathcal{P}_{\mathrm{Sim}_X}$. The top label of $\Delta_{\mathcal{P}}$ is $[X]$.

Given a bijection $\phi : \mathcal{P}_1 \to \mathcal{P}_2$, where $\mathcal{P}_1$ and $\mathcal{P}_2$ are partitions of X into balls and $[B] = [\phi(B)]$, we can define a braided diagram Δ_ϕ over $\mathcal{P}_{\mathrm{Sim}_X}$ as follows. We let $\mathcal{T}(\Delta_\phi) = \emptyset$, and $\mathcal{W}(\Delta_\phi) = \{W_B \mid B \in \mathcal{P}_1\}$. We attach the top of each wire to the frame in such a way that $t(W_{B_1})$ is to the left of $t(W_{B_2})$ if $B_1 < B_2$. (Here $<$ refers to the ordering from Remark 4.9.) We attach the bottom of each wire to the bottom of the frame in such a way that $b(W_{B_1})$ is to the left of $b(W_{B_2})$ if $\phi(B_1) < \phi(B_2)$.

Now, for a defining triple $(\mathcal{P}_1, \mathcal{P}_2, \phi) \in \mathcal{D}$, we set $\psi((\mathcal{P}_1, \mathcal{P}_2, \phi)) = \Delta_{\mathcal{P}_2} \circ \Delta_{\phi^{-1}} \circ \Delta_{\mathcal{P}_1}^{-1} \in \mathcal{D}_b(\mathcal{P}_{\mathrm{Sim}_X}, [X])$.

We claim that any two defining triples $(\mathcal{P}_1, \mathcal{P}_2, \phi)$, $(\mathcal{P}_1', \mathcal{P}_2', \phi')$ for a given $\gamma \in \Gamma(\mathrm{Sim}_X)$ have the same image in $D_b(\mathcal{P}_{\mathrm{Sim}_X}, [X])$, modulo dipoles. We begin by proving an intermediate statement. Let $(\mathcal{P}_1, \mathcal{P}_2, \phi)$ be a defining triple. Let $B \in \mathcal{P}_1$, and let $\widehat{B}_1, \ldots, \widehat{B}_n$ be the collection of maximal proper subballs of B, listed in order. We let $B' = \phi(B)$ and let $\widehat{B}_1', \ldots, \widehat{B}_n'$ be the collection of maximal proper subballs of B'. (Note that $[B'] = [B]$ by our assumptions, so both have the same number of maximal proper subballs, and in fact $[\widehat{B}_i] = [\widehat{B}_i']$ for $i = 1, \ldots, n$, since $\gamma|_B \in \mathrm{Sim}_X(B, B')$ and the elements of $\mathrm{Sim}_X(B, B')$ preserve order.) We set $\widehat{\mathcal{P}}_1 = (\mathcal{P}_1 - \{B\}) \cup \{\widehat{B}_1, \ldots, \widehat{B}_n\}$, $\widehat{\mathcal{P}}_2 = (\mathcal{P}_2 - \{B'\}) \cup \{\widehat{B}_1', \ldots, \widehat{B}_n'\}$, and $\widehat{\phi}|_{\mathcal{P}_1 - \{B\}} = \phi|_{\mathcal{P}_1 - \{B\}}$, $\widehat{\phi}(\widehat{B}_i) = \widehat{B}_i'$. We say that $(\widehat{\mathcal{P}}_1, \widehat{\mathcal{P}}_2, \widehat{\phi})$ is obtained from $(\mathcal{P}_1, \mathcal{P}_2, \phi)$ by *subdivision* at (B, B'). A straightforward argument shows that $\psi((\widehat{\mathcal{P}}_1, \widehat{\mathcal{P}}_2, \widehat{\phi}))$ is in fact obtained from $\psi((\mathcal{P}_1, \mathcal{P}_2, \phi))$ by inserting a dipole. We omit the details, which rely on the fact that each element of the Sim_X-structure preserves the local ball order.

Now suppose that $(\mathcal{P}_1, \mathcal{P}_2, \phi)$ and $(\mathcal{P}_1', \mathcal{P}_2', \phi')$ are defining triples for the same element $\gamma \in \Gamma(\mathrm{Sim}_X)$. We can find a common refinement $\mathcal{P}_1''$ of $\mathcal{P}_1$ and $\mathcal{P}_1'$. Using the fact that all partitions of X into balls are standard (Lemma 3.10), we can pass from $(\mathcal{P}_1, \mathcal{P}_2, \phi)$ to $(\mathcal{P}_1'', \widehat{\mathcal{P}}_2, \widehat{\phi})$ by repeated subdivision (for some partition $\widehat{\mathcal{P}}_2$ of X into balls and some bijection $\widehat{\phi} : \mathcal{P}_1'' \to \widehat{\mathcal{P}}_2$). Since subdivision does not change the values of ψ modulo dipoles, $\psi((\mathcal{P}_1, \mathcal{P}_2, \phi)) = \psi((\mathcal{P}_1'', \widehat{\mathcal{P}}_2, \widehat{\phi}))$ modulo dipoles. Similarly, we can subdivide $(\mathcal{P}_1', \mathcal{P}_2', \phi')$ repeatedly in order to obtain $(\mathcal{P}_1'', \widehat{\mathcal{P}}_2', \widehat{\phi}')$, where $\psi((\mathcal{P}_1', \mathcal{P}_2', \phi')) = \psi((\mathcal{P}_1'', \widehat{\mathcal{P}}_2', \widehat{\phi}'))$ modulo dipoles. Both $(\mathcal{P}_1'', \widehat{\mathcal{P}}_2', \widehat{\phi}')$ and $(\mathcal{P}_1'', \widehat{\mathcal{P}}_2, \widehat{\phi})$ are defining triples for γ, so we are forced to have $\widehat{\phi} = \widehat{\phi}'$ and $\widehat{\mathcal{P}}_2 = \widehat{\mathcal{P}}_2'$. It follows that $\psi((\mathcal{P}_1, \mathcal{P}_2, \phi)) = \psi((\mathcal{P}_1', \mathcal{P}_2', \phi'))$, so ψ induces a function from $\Gamma(\mathrm{Sim}_X)$ to $D_b(\mathcal{P}_{\mathrm{Sim}_X}, [X])$. We will call this function $\widehat{\psi}$.

Now we will show that $\widehat{\psi} : \Gamma(\mathrm{Sim}_X) \to D_b(\mathcal{P}_{\mathrm{Sim}_X}, [X])$ is a homomorphism. Let $\gamma, \gamma' \in \Gamma(\mathrm{Sim}_X)$. After subdividing as necessary, we can choose defining triples $(\mathcal{P}_1, \mathcal{P}_2, \phi)$ and $(\mathcal{P}_1', \mathcal{P}_2', \phi')$ for γ and γ' (respectively) in such a way that $\mathcal{P}_2 = \mathcal{P}_1'$. It follows easily that $(\mathcal{P}_1, \mathcal{P}_2', \phi'\phi)$ is a defining triple for

$\gamma'\gamma$. Therefore, $\widehat{\psi}(\gamma'\gamma) = \Delta_{\mathcal{P}_2'} \circ \Delta_{(\phi'\phi)^{-1}} \circ \Delta_{\mathcal{P}_1}^{-1}$. Now

$$\begin{aligned}\widehat{\psi}(\gamma') \circ \widehat{\psi}(\gamma) &= \Delta_{\mathcal{P}_2'} \circ \Delta_{(\phi')^{-1}} \circ \Delta_{\mathcal{P}_1'}^{-1} \circ \Delta_{\mathcal{P}_2} \circ \Delta_{\phi^{-1}} \circ \Delta_{\mathcal{P}_1}^{-1} \\ &= \Delta_{\mathcal{P}_2'} \circ \Delta_{(\phi')^{-1}} \circ \Delta_{\phi^{-1}} \circ \Delta_{\mathcal{P}_1}^{-1} \\ &= \Delta_{\mathcal{P}_2'} \circ \Delta_{(\phi'\phi)^{-1}} \circ \Delta_{\mathcal{P}_1}^{-1}\end{aligned}$$

Therefore, $\widehat{\psi}$ is a homomorphism.

We now show that $\widehat{\psi} : \Gamma(\mathrm{Sim}_X) \to D_b(\mathcal{P}_{\mathrm{Sim}_X}, [X])$ is injective. Suppose that $\widehat{\psi}(\gamma) = 1$. Using the final statement of Proposition 3.11, we choose a defining triple $(\mathcal{P}_1, \mathcal{P}_2, \phi)$ for γ with the property that, if $B \subseteq X$ is a ball, $\gamma(B)$ is a ball, and $\gamma|_B \in \mathrm{Sim}_X(B, \gamma(B))$, then B is contained in some ball of $\mathcal{P}_1$. We claim that $\psi((\mathcal{P}_1, \mathcal{P}_2, \phi))$ is a reduced diagram. If there were a dipole (T_1, T_2), then we would have $T_1 \in \mathcal{T}(\Delta_{\mathcal{P}_1}^{-1})$ and $T_2 \in \mathcal{T}(\Delta_{\mathcal{P}_2})$, since it is impossible for $\Delta_{\mathcal{P}}$ to contain any dipoles, for any partition $\mathcal{P}$ of X into balls. Thus $T_1 = T_{B_1}$ and $T_2 = T_{B_2}$, where $[B_1] = [B_2]$ and the wires from the bottom of T_{B_2} attach to the top of T_{B_1}, in order. This means that, if $\widehat{B}_1, \ldots, \widehat{B}_n$ are the maximal proper subballs of B_1, and $\widehat{B}_1', \ldots, \widehat{B}_n'$ are the maximal proper subballs of B_2, then $\gamma(\widehat{B}_i) = \widehat{B}_i'$, where the latter is a ball, and $\gamma|_{\widehat{B}_i} \in \mathrm{Sim}_X(\widehat{B}_i, \widehat{B}_i')$.

Now, since $[B_1] = [B_2]$, there is $h \in \mathrm{Sim}_X(B_1, B_2)$. Since Sim_X is closed under restrictions and h preserves order, we have $h_i \in \mathrm{Sim}_X(\widehat{B}_i, \widehat{B}_i')$ for $i = 1, \ldots, n$, where $h_i = h|_{\widehat{B}_i}$. It follows that $\gamma|_{\widehat{B}_i} = h_i$, so, in particular, $\gamma|_{B_1} = h$. Since B_1 properly contains some ball in $\mathcal{P}_1$, this is a contradiction. Thus, $\psi((\mathcal{P}_1, \mathcal{P}_2, \phi))$ is reduced.

We claim that $\psi((\mathcal{P}_1, \mathcal{P}_2, \phi))$ contains no transistors (due to the condition $\widehat{\psi}(\gamma) = 1$). We have shown that $\psi((\mathcal{P}_1, \mathcal{P}_2, \phi))$ is a reduced diagram in the same class as the identity $1 \in D_b(\mathcal{P}_{\mathrm{Sim}_X}, [X])$. The identity can be represented as the (unique) $([X], [X])$-diagram Δ_1 with only a single wire, W_X, and no transistors. We must have $\psi((\mathcal{P}_1, \mathcal{P}_2, \phi)) \equiv \Delta_1$. Thus, there is no ball that properly contains a ball of $\mathcal{P}_1$. It can only be that $\mathcal{P}_1 = \{X\}$, so we must have $\gamma \in \mathrm{Sim}_X(X, X)$. This forces $\gamma = 1$, so $\widehat{\psi}$ is injective.

Finally we must show that $\widehat{\psi} : \Gamma(\mathrm{Sim}_X) \to D_b(\mathcal{P}_{\mathrm{Sim}_X}, [X])$ is surjective. Let Δ be a reduced $([X], [X])$-diagram over $\mathcal{P}_{\mathrm{Sim}_X}$. A transistor $T \in \mathcal{T}(\Delta)$ is called *positive* if its top label is the left side of a relation in $\mathcal{P}_{\mathrm{Sim}_X}$, otherwise (i.e., if the top label is the right side of a relation in $\mathcal{P}_{\mathrm{Sim}_X}$) the transistor T is *negative*. It is easy to see that the sets of positive and negative transistors partition $\mathcal{T}(\Delta)$. We claim that, if Δ is reduced, then we cannot have $T_1 \dot{\preceq} T_2$ when T_1 is positive and T_2 is negative. If we had such $T_1 \dot{\preceq} T_2$, then we could find $T_1' \preceq T_2'$, where T_1' is positive and T_2' is negative. Since T_1' is positive, there is only one wire W attached to the top of T_1'. This wire must be attached to the bottom of T_2', since $T_1' \preceq T_2'$, and it must be the only wire attached to the bottom of T_2', since T_2' is negative and $\mathcal{P}_{\mathrm{Sim}_X}$ is a tree-like semigroup presentation by Remark 4.11. Suppose that $\ell(w) = [B]$. By the definition of $\mathcal{P}_{\mathrm{Sim}_X}$, $[B]$ is the left side of exactly one relation, namely $([B], [B_1][B_2] \ldots [B_n])$, where the B_i are maximal proper subballs of B, listed in order. It follows that the bottom label of T_1' is $[B_1][B_2] \ldots [B_n]$ and the top label of T_2' is $[B_1][B_2] \ldots [B_n]$. Therefore (T_1', T_2') is a dipole. This proves the claim.

A diagram over $\mathcal{P}_{\mathrm{Sim}_X}$ is *positive* if all of its transistors are positive, and *negative* if all of its transistors are negative. We note that Δ is positive if and only if Δ^{-1} is negative, by the description of inverses in the proof of Theorem 2.13. The above reasoning shows that any reduced $([X],[X])$-diagram over $\mathcal{P}_{\mathrm{Sim}_X}$ can be written $\Delta = \Delta_1^+ \circ \left(\Delta_2^+\right)^{-1}$, where Δ_i^+ is a positive diagram for $i = 1, 2$.

We claim that any positive diagram Δ over $\mathcal{P}_{\mathrm{Sim}_X}$ with top label $[X]$ is $\Delta_{\mathcal{P}}$ (up to a reordering of the bottom contacts), where $\mathcal{P}$ is some partition of X. There is a unique wire $W \in \mathcal{W}(\Delta)$ making a top contact with the frame. We call this wire W_X. Note that its label is $[X]$ by our assumptions. The bottom contact of W_X lies either on the bottom of the frame, or on top of some transistor. In the first case, we have $\Delta = \Delta_{\mathcal{P}}$ for $\mathcal{P} = \{X\}$ and we are done. In the second, the bottom contact of W_X lies on top of some transistor T, which we call T_X. Since the top label of T_X is $[X]$, the bottom label must be $[B_1] \ldots [B_k]$, where $B_1, \ldots, B_k$ are the maximal proper subballs of X. Thus there are wires $W_1, \ldots, W_k$ attached to the bottom of T_X, and we have $\ell(W_i) = [B_i]$, for $i = 1, \ldots, k$. We relabel each of the wires $W_{B_1}, \ldots, W_{B_k}$, respectively. Note that $\{B_1, \ldots, B_k\}$ is a partition of X into balls. We can continue in this way, inductively labelling each wire with a ball $B \subseteq X$. If we let $\overline{B}_1, \ldots, \overline{B}_m$ be the resulting labels of the wires which make bottom contacts with the frame, then $\{\overline{B}_1, \ldots, \overline{B}_m\} = \mathcal{P}$ is a partition of X into balls, and $\Delta = \Delta_{\mathcal{P}}$ by construction, up to a reordering of the bottom contacts.

We can now prove surjectivity of $\widehat{\psi}$. Let $\Delta \in D_b(\mathcal{P}_{\mathrm{Sim}_X}, [X])$ be reduced. We can write $\Delta = \Delta_2^+ \circ \left(\Delta_1^+\right)^{-1}$, where Δ_i^+ is positive, for $i = 1, 2$. It follows that $\Delta_i^+ = \Delta_{\mathcal{P}_i} \circ \sigma_i$, for $i = 1, 2$, where $\mathcal{P}_i$ is a partition of X into balls and σ_i is diagram containing no transistors. Thus, $\Delta = \Delta_{\mathcal{P}_2} \circ \sigma_2 \circ \sigma_1^{-1} \circ \Delta_{\mathcal{P}_1}^{-1} = \psi((\mathcal{P}_1, \mathcal{P}_2, \phi))$, where $\phi : \mathcal{P}_1 \to \mathcal{P}_2$ is a bijection determined by $\sigma_2 \circ \sigma_1^{-1}$. Therefore, $\widehat{\psi}$ is surjective.

Now we must show that if $\mathcal{P} = \langle \Sigma \mid \mathcal{R} \rangle$ is a tree-like semigroup presentation, $x \in \Sigma$, then there is a compact ultrametric space $X_{\mathcal{P}}$, a small similarity structure $\mathrm{Sim}_{X_{\mathcal{P}}}$, and a compatible ball order, such that $D_b(\mathcal{P}, x) \cong \Gamma(\mathrm{Sim}_{X_{\mathcal{P}}})$. Construct a labelled ordered simplicial tree $T_{(\mathcal{P},x)}$ as follows. Begin with a vertex $*$, the root, labelled by $x \in \Sigma$. By the definition of tree-like semigroup presentation (Definition 4.1), there is at most one relation in $\mathcal{R}$ having the word x as its left side. Let us suppose first that $(x, x_1 x_2 \ldots x_k) \in \mathcal{R}$, where $k \geq 2$. We introduce k children of the root, labelled $x_1, \ldots, x_k$ (respectively), each connected to the root by an edge. The children are ordered from left to right in such a way that we read the word $x_1x_2 \dots x_k$ as we read the labels of the children from left to right. If, on the other hand, x is not the left side of any relation in $\mathcal{R}$, then the tree terminates – there is only the root. We continue similarly: if x_i is the left side of some relation $(x_i, y_1y_2 \dots y_m) \in \mathcal{R}$ $(m \geq 2)$, then this relation is unique and we introduce a labelled ordered collection of children, as above. If x_i is not the left side of any relation in $\mathcal{R}$, then x_i has no children. This builds a labelled ordered tree $T_{(\mathcal{P},x)}$. We note that if a vertex $v \in T_{(\mathcal{P},x)}$ is labelled by $y \in \Sigma$, then the subcomplex $T_v \leq T_{(\mathcal{P},x)}$ spanned by v and all of its descendants is isomorphic to $T_{(\mathcal{P},y)}$, by a simplicial isomorphism which preserves the labelling and the order.

We let $\mathrm{Ends}(T_{(\mathcal{P},x)})$ denote the set of all edge-paths p in $T_{(\mathcal{P},x)}$ such that: i) p is without backtracking; ii) p begins at the root; iii) p is either infinite, or p terminates at a vertex without children. We define a metric on $\mathrm{Ends}(T_{(\mathcal{P},x)})$ as follows. If $p, p' \in \mathrm{Ends}(T_{(\mathcal{P},x)})$ and p, p' have exactly m edges in common, then we set $d(p,p') = e^{-m}$. This metric makes $\mathrm{Ends}(T_{(\mathcal{P},x)})$ a compact ultrametric space, and a ball order is given by the ordering of the tree. We can describe the balls in $\mathrm{Ends}(T_{(\mathcal{P},x)})$ explicitly. Let v be a vertex of $T_{(\mathcal{P},x)}$. We set $B_v = \{p \in \mathrm{Ends}(T_{(\mathcal{P},x)}) \mid v \text{ lies on } p\}$. Every such set is a ball, and every ball in $\mathrm{Ends}(T_{(\mathcal{P},x)})$ has this form. We can now describe a finite similarity structure $\mathrm{Sim}_{X_\mathcal{P}}$ on $\mathrm{Ends}(T_{(\mathcal{P},x)})$. Let B_v and $B_{v'}$ be the balls corresponding to the vertices $v, v' \in T_{(\mathcal{P},x)}$. If v and v' have different labels, then we set $\mathrm{Sim}_{X_\mathcal{P}}(B_v, B_{v'}) = \emptyset$. If v and v' have the same label, say $y \in \Sigma$, then there is label- and order-preserving simplicial isomorphism $\psi : T_v \to T_{v'}$. Suppose that p_v is the unique edge-path without backtracking connecting the root to v. Any point in B_v can be expressed in the form $p_v q$, where q is an edge-path without backtracking in T_v. We let $\widehat{\psi} : B_v \to B_{v'}$ be defined by the rule $\widehat{\psi}(p_v q) = p_{v'}\psi(q)$. The map $\widehat{\psi}$ is easily seen to be a surjective similarity. We set $\mathrm{Sim}_{X_\mathcal{P}}(B_v, B_{v'}) = \{\widehat{\psi}\}$. The resulting assignments give a small similarity structure $\mathrm{Sim}_{X_\mathcal{P}}$ on the compact ultrametric space $\mathrm{Ends}(T_{(\mathcal{P},x)})$ that is compatible with the ball order.

Now we can apply the first part of the theorem: for $X_{(\mathcal{P},x)} = \mathrm{Ends}(T_{(\mathcal{P},x)})$, we have $\Gamma(\mathrm{Sim}_{X_{(\mathcal{P},x)}}) \cong D_b(\mathcal{P}_{\mathrm{Sim}_{X_{(\mathcal{P},x)}}}, [X_{(\mathcal{P},x)}]) \cong D_b(\mathcal{P}, x)$, where the first isomorphism follows from the forward direction of the theorem, and the second isomorphism follows from the canonical identification of the semigroup presentation $\mathcal{P}_{\mathrm{Sim}_{X_{(\mathcal{P},x)}}}$ with $\mathcal{P}$. □

References

[1] Collin Bleak, Francesco Matucci, and Max Neunhöffer. Embeddings into Thompson's group V and *coCF* groups. *arXiv:1312.1855*.

[2] Kenneth S. Brown. Finiteness properties of groups. *J. Pure Appl. Algebra*, 44(1-3):45–75, 1987.

[3] J. W. Cannon, W. J. Floyd, and W. R. Parry. Introductory notes on Richard Thompson's groups. *Enseign. Math. (2)*, 42(3-4):215–256, 1996.

[4] Daniel S. Farley. Actions of picture groups on CAT(0) cubical complexes. *Geom. Dedicata*, 110:221–242, 2005.

[5] Daniel S. Farley and Bruce Hughes. Finiteness properties of some groups of local similarities. *arXiv:1206.2692*.

[6] Victor Guba and Mark Sapir. Diagram groups. *Mem. Amer. Math. Soc.*, 130(620):viii+117, 1997.

[7] Bruce Hughes. Local similarities and the Haagerup property, with an appendix by Daniel S. Farley. *Groups Geom. Dyn.*, 3:299–315, 2009.

On Thompson's group T and algebraic K-theory

Ross Geoghegan   Marco Varisco

Abstract

Using a theorem of Lück-Reich-Rognes-Varisco, we show that the Whitehead group of Thompson's group T is infinitely generated, even when tensored with the rationals. To this end we describe the structure of the centralizers and normalizers of the finite cyclic subgroups of T, via a direct geometric approach based on rotation numbers. This also leads to an explicit computation of the source of the Farrell-Jones assembly map for the rationalized higher algebraic K-theory of the integral group ring of T.

1 Introduction and statement of results

Thompson's groups F and T are well-known groups having both type F_∞ and infinite cohomological dimension. Recall that F and T can be defined as the groups of orientation-preserving dyadic piecewise-linear homeomorphisms of the closed unit interval $I = [0, 1]$ and of the circle $S^1 = \mathbb{R}/\mathbb{Z}$; see Section 2 (and [6] for a comprehensive introduction).

Essentially nothing has been known about the algebraic K-theory of these groups. Here we show that the Whitehead group of T is infinitely generated, even when tensored with the rationals. More precisely, our main theorem is the following.

Theorem 1.1. *The Farrell-Jones assembly map in algebraic K-theory induces an injective homomorphism*

$$\operatorname*{colim}_{k\in\mathbb{N}} Wh(C_k) \underset{\mathbb{Z}}{\otimes} \mathbb{Q} \longrightarrow Wh(T) \underset{\mathbb{Z}}{\otimes} \mathbb{Q} \tag{1.2}$$

and in particular $Wh(T) \otimes_{\mathbb{Z}} \mathbb{Q}$ is an infinite dimensional $\mathbb{Q}$-vector space.

On the left-hand side of (1.2) $C_k = \mathbb{Z}/k\mathbb{Z}$ denotes the finite cyclic group of order k. It is well known that $Wh(C_k) \otimes_{\mathbb{Z}} \mathbb{Q} \neq 0$ for all $k \notin \{1, 2, 3, 4, 6\}$; see for example [17, top of page 6]. The colimit in (1.2) is taken over the poset $\mathbb{N}$ with respect to the divisibility relation, and the homomorphisms $Wh(C_k) \to Wh(C_\ell)$ induced by $C_k = \mathbb{Z}/k\mathbb{Z} \to C_\ell = \mathbb{Z}/\ell\mathbb{Z}$, $1 \mapsto \frac{\ell}{k}$ whenever $k \mid \ell$.

The map in (1.2) is induced by identifying C_k with the cyclic subgroup $\langle\gamma_k\rangle$ of T generated by the pseudo-rotation of order k from Example 2.1; see the proof of Corollary 3.11.

Theorem 1.1 is a direct application to T of the paper [12]. That work and its applicability here are discussed in Section 3, where we also obtain results about the higher algebraic K-theory groups of the integral group ring of T. The ingredients about T needed for this application are summarized in the following theorem.

Theorem 1.3. *Every finite subgroup of T is cyclic, and for every integer $k \geq 0$ there is exactly one conjugacy class in T of cyclic subgroups of order k. Moreover, for every finite cyclic subgroup C of T, the centralizer $Z_T C$ and the normalizer $N_T C$ of C in T are equal, and there is a short exact sequence*

$$1 \longrightarrow C \longrightarrow Z_T C \longrightarrow T \longrightarrow 1\,. \tag{1.4}$$

We proved Theorem 1.3 in 2007 and only lately became aware that essentially the same result (but without the observation about normalizers, which is important for our application) appeared in Matucci's 2008 thesis [16, Theorem 7.1.5], and was subsequently generalized in [15, 14]. The full details of our proof of Theorem 1.3 are given in Section 2.

The group T is of type F_∞ by a theorem of Brown and Geoghegan; see for example [4, Remark 2 on page 56], where this is shown to follow immediately from the same result for the group F [5]. Thus the short exact sequence (1.4) of Theorem 1.3 has the following corollary (see for example [7, Section 7.2]).

Corollary 1.5. *For every finite cyclic subgroup C of T the centralizer $Z_T C$ of C in T is of type F_∞. Moreover, for every $s \in \mathbb{N}$, the abelian group $H_s(BZ_T C;\mathbb{Z})$ is finitely generated, and $H_s(BZ_T C;\mathbb{Q}) \cong H_s(BT;\mathbb{Q})$.*

The rational homology groups $H_s(BT;\mathbb{Q})$ of T (and hence also of $Z_T C$) are completely known thanks to a theorem of Ghys and Sergiescu. In fact, in [8, Corollaire C on pages 187–188] it is proved that $H^1(T;\mathbb{Z}) = 0$, $H^2(T;\mathbb{Z}) \cong \mathbb{Z} \oplus \mathbb{Z}$ with natural generators α and χ, and $H^*(T;\mathbb{Q}) \cong \mathbb{Q}[\alpha,\chi]$.

2 Thompson's group T and centralizers of finite subgroups

In this section we recall the definition of Thompson's groups F and T, and then prove Theorem 1.3; see Theorem 2.3 and Corollary 2.6 below.

We say that an interval $I \subset \mathbb{R}$ is *dyadic* if its endpoints are dyadic rationals. If I and J are closed dyadic intervals, a homeomorphism $f\colon I \to J$ is called *dyadic piecewise linear*, or *DPL* for short, if f is piecewise linear, the breakpoints occur at dyadic rational points, and the slopes are integer powers of 2. Notice that the inverse of a DPL homeomorphism is again DPL. *Thomspon's group* F is defined as the group of orientation-preserving DPL homeomorphisms of $[0,1]$.

We define an $\mathbb{R}$-*space* to be a pair (X,p) where X is a topological space and $p\colon \mathbb{R} \to X$ is a covering map. In other words, X is a connected 1-dimensional manifold together with a chosen universal covering map p. We consider every $\mathbb{R}$-space to be oriented via p. The primary example is of course $X = S^1 = \mathbb{R}/\mathbb{Z}$ together with the usual universal covering map $u\colon \mathbb{R} \to S^1$.

Let (X,p) and (Y,q) be $\mathbb{R}$-spaces, and let $f\colon X \to Y$ be a map. We say that f is *locally DPL* (short for local dyadic piecewise linear homeomorphism) if for every $x \in X$ there exist closed dyadic intervals I, J in $\mathbb{R}$ such that:

- $p_{|I}$ and $q_{|J}$ are embeddings;
- x belongs to the interior of $p(I)$ and $f(x)$ belongs to the interior of $q(J)$;
- f induces a homeomorphism $f_{|p(I)}\colon p(I) \to q(J)$;
- and the composition
$$I \xrightarrow{p_|} p(I) \xrightarrow{f_|} q(J) \xrightarrow{q_|^{-1}} J$$
is a DPL homeomorphism.

If (X,p) is an $\mathbb{R}$-space, then we define $H(X,p) = H(X)$ to be the group of all orientation-preserving homeomorphisms of X, and $T(X,p)$ to be the subgroup of $H(X,p)$ consisting of those orientation-preserving homeomorphisms that are locally DPL. *Thompson's group* T is defined as $T = T(S^1,u)$. Similarly we write $H = H(S^1,u)$. Thompson's group F can then be identified with the subgroup of T fixing a base point.

Example 2.1 (pseudo-rotations)**.** The following elements of T play an important role in our work. Given $q \geq 2$ we denote by $\gamma_q \in T$ be the pseudo-rotation of order q, i.e., the element of T (called C_{q-2} in [6, pages 236–237]) that cyclically permutes the images of the q intervals

$$\left[0,\tfrac{1}{2}\right], \left[\tfrac{1}{2},\tfrac{3}{4}\right], \ldots, \left[1-\tfrac{1}{2^j}, 1-\tfrac{1}{2^{j+1}}\right], \ldots, \left[1-\tfrac{1}{2^{q-1}}, 1\right] \tag{2.2}$$

and is affine on each of them.

The main result in this section is the following.

Theorem 2.3. *Let C be a finite subgroup of H. Then C is cyclic, the centralizer $Z_H C$ and the normalizer $N_H C$ of C in H are equal, and there is a short exact sequence*

$$1 \longrightarrow C \longrightarrow Z_H C \longrightarrow H \longrightarrow 1\,.$$

Moreover, if $C < T$, then $Z_T C = N_T C$, and there is a short exact sequence

$$1 \longrightarrow C \longrightarrow Z_T C \longrightarrow T \longrightarrow 1\,.$$

The proof of Theorem 2.3 uses Poincaré rotation numbers. We now recall their definition and basic properties, and we refer to [9, Chapter 11] for more details and proofs.

Given $h \in H$, choose a lift $\overline{h}\colon \mathbb{R} \to \mathbb{R}$ such that $u\overline{h} = hu$, and choose a point $x \in \mathbb{R}$. Define

$$\rho(h) = \mathbb{Z} + \lim_{n\to\infty} \frac{\overline{h}^n(x) - x}{n} \in \mathbb{R}/\mathbb{Z}.$$

Then $\rho(h) \in \mathbb{R}/\mathbb{Z}$ is independent of the choices of $\overline{h}$ and x (see [9, Proposition 11.1.1]), and it is called the *rotation number* of h.

Proposition 2.4. *Let $h, g \in H$ and let m be an integer.*

(i) If $h(x) = x + \vartheta \mod \mathbb{Z}$, i.e., if h is a rotation by $2\pi\vartheta$, then $\rho(h) = \vartheta$. In particular, $\rho(\mathrm{id}_{S^1}) = 0$.

(ii) $\rho(h^m) = m\rho(h)$.

(iii) $\rho(hgh^{-1}) = \rho(g)$.

(iv) If $\rho(h) = 0$, then h has a fixed point.

(v) If $h \neq \mathrm{id}_{S^1}$ has finite order, then $\rho(h) \in \mathbb{Q}/\mathbb{Z}$ and $\rho(h) \neq 0$. Let $\rho(h) = \frac{p}{q}$ with $(p, q) = 1$ and $0 < p < q$. Then the order of h is q; for every $x \in S^1$ the ordering of $\{x, h(x), h^2(x), \ldots, h^{q-1}(x)\}$ in S^1 is the same as that of $\{0, \frac{p}{q}, \frac{2p}{q}, \frac{(q-1)p}{q}\}$; and h is conjugate to the rotation by $2\pi\frac{p}{q}$.

Proof. Statements (i) and (ii) follow immediately from the definition, whereas (iii) and (iv) are proved in [9, Propositions 11.1.3 and 11.1.4].

(v) Let $h \neq \mathrm{id}_{S^1}$ have finite order. From [9, Proposition 11.1.1] we have that $\rho(h) \in \mathbb{Q}/\mathbb{Z}$. From (iv) and Lemma 2.5 below we conclude that $\rho(h) \neq 0$. So let $\rho(h) = \frac{p}{q}$ with $(p, q) = 1$ and $0 < p < q$. Suppose that the order of h is m. Then, using (i) and (ii), $0 = \rho(\mathrm{id}_{S^1}) = \rho(h^m) = m\rho(h) = m\frac{p}{q}$, and therefore $q|m$ since $(p, q) = 1$. On the other hand $\rho(h^q) = q\rho(h) = q\frac{p}{q} = 0 \in \mathbb{R}/\mathbb{Z}$, and therefore from (iv) and Lemma 2.5 we conclude that $h^m = \mathrm{id}_{S^1}$ and hence $m|q$. So the order of h is q. The last statements then follow from [9, Proposition 11.2.1]. □

Lemma 2.5. *If $h \in H$ has finite order and has a fixed point, then $h = \mathrm{id}_{S^1}$.*

Proof. If h has a fixed point, then h induces an orientation-preserving homeomorphism of a closed interval. Since the group of orientation-preserving homeomorphisms of a closed interval is torsion-free, if h also has finite order then $h = \mathrm{id}_{S^1}$. □

Corollary 2.6. *Any two cyclic subgroups of H (respectively, of T) with the same order are conjugate in H (respectively, in T).*

Proof. Let C be a cyclic subgroup of H with order q. Proposition 2.4(ii) implies that C has a unique generator g with rotation number $\frac{1}{q}$, and (v) implies that g is conjugate in H to the rotation by $\frac{2\pi}{q}$, and therefore the corollary is true for H. So assume that $g \in T$. By Proposition 2.4(v), the ordering of $\mathcal{O} = \{0, g(0), g^2(0), \dots, g^{q-1}(0)\}$ in S^1 is the same as that of $\{0, \frac{p}{q}, \frac{2p}{q}, \frac{(q-1)p}{q}\}$, and so each $g^k(0)$ is a dyadic rational. Think of $\mathcal{O}$ as a dyadic subdivision of S^1. Then there is a finer dyadic subdivision $\mathcal{O}'$ such that g is affine on each segment of $\mathcal{O}'$. Now let $\gamma_q \in T$ be the pseudo-rotation of order q from Example 2.1. Define $\mathcal{O}''$ to be the dyadic subdivision of (2.2) corresponding to $\mathcal{O}'$, so that γ_q is also affine on each segment of $\mathcal{O}''$, and define $h \in T$ to be the locally DPL homeomorphism that maps each segment of $\mathcal{O}'$ affinely onto the corresponding segment of $\mathcal{O}''$. Then $hgh^{-1} = \gamma_q$, and therefore the corollary is also true for T. □

We are now ready to prove Theorem 2.3.

Proof of Theorem 2.3. Let C be a finite subgroup of H. Assume that $C \neq 1$, otherwise there is nothing to prove. Define $S^1_0 = C\backslash S^1$ to be the quotient, and denote by $q\colon S^1 \to S^1_0$ the quotient map.

We first show that C is cyclic. By Lemma 2.5, if $g \in C$ has a fixed point, then $g = \mathrm{id}_{S^1}$. It follows that $q\colon S^1 \to S^1_0$ is a covering map and that S^1_0, being a closed 1-dimensional manifold, is homeomorphic to S^1, and therefore C is cyclic by covering space theory.

Notice that S^1_0 together with the composition qu is an $\mathbb{R}$-space. We abbreviate $H_0 = H(S^1_0, qu)$ and $T_0 = T(S^1_0, qu)$.

Fix a generator g of C. By Proposition 2.4(v), we know that $\rho(g) = \frac{p}{q} \neq 0$, with $(p,q) = 1$ and $0 < p < q$, and q is the order of C. Let s and t be such that $sp + tq = 1$. Then $g^s(1)$ is the element in the orbit of $1 \in S^1$ coming directly after 1 in the cyclic order. Let ℓ be the length in S^1 of $[1, g^s(1)]$. Then S^1_0 can be identified with $\mathbb{R}/\ell\mathbb{Z}$, and multiplication by $\frac{1}{\ell}$ induces a homeomorphism $f\colon S^1_0 \to S^1$. It follows that conjugation by f yields an isomorphism $H_0 \cong H$. Moreover, if $g \in T$, then ℓ is a dyadic rational and therefore the homeomorphism f is locally DPL, so conjugation by f restricts to an isomorphism $T_0 \cong T$.

To prove that $N_H C = Z_H C$, let $h \in N_H C$ be given. Then $hgh^{-1} = g^m$ for some integer m. By Proposition 2.4(ii)-(iii) we see that $\rho(g) = m\rho(g)$, and by Proposition 2.4(iv) and Lemma 2.5 we see that $\rho(g) \neq 0$. Therefore $m = 1$ and so $h \in Z_H C$. Since obviously $Z_H C \leq N_H C$, we conclude that $N_H C = Z_H C$.

Since $Z_H C$ acts on the quotient $S^1_0 = C\backslash S^1$, we get a group homomorphism $\pi\colon Z_H C \to H_0$. We are going to show next that there is a short exact sequence

$$1 \longrightarrow C \longrightarrow Z_H C \xrightarrow{\ \pi\ } H_0 \longrightarrow 1\,, \tag{2.7}$$

i.e., that $\ker \pi = C$ and that π is surjective. Since $H_0 \cong H$, as observed above, this will prove the first part of the theorem.

To show that $\ker \pi = C$, let $h \in \ker \pi$ be given. Then for any $x \in S^1$, $h(x) = g^{m(x)}(x)$ for some integer $m(x)$. By continuity and since S^1 is connected, it follows that $m(x)$ is constant, i.e., that $h \in C$. Since obviously $C \leq \ker \pi$, we conclude that $\ker \pi = C$.

To show that π is surjective, let $h_0 \in H_0$ be given. Choose a basepoint $x_0 \in S^1_0$ and define $y_0 = h_0(x_0)$. Since h_0 is freely homotopic to $\mathrm{id}_{S^1_0}$, choose such a homotopy and let α be the track of this homotopy at x_0; α is then a path from x_0 to y_0. It follows that given any loop ω at x_0, ω is homotopic to $\alpha \cdot (h_0\omega) \cdot \alpha^{-1}$ relative to x_0. Now choose $x \in S^1$ such that $q(x) = x_0$, and let $\widetilde{\alpha}$ be the lift of α starting at x. Define $y = \widetilde{\alpha}(1)$ and $y_0 = q(y) = \alpha(1)$.

We want to show that h_0 lifts to a homeomorphism $h \in H$ such that:

$$qh = h_0 q \quad \text{and} \quad h(x) = y\,. \tag{2.8}$$

We first show that $h_0 q \colon S^1 \to S^1_0$ lifts to a map $h \colon S^1 \to S^1$ satisfying (2.8). It follows then easily that $h \in H$.

By covering space theory, it is enough to show that if γ is any loop in S^1 at x then there is a loop σ in S^1 at y such that $h_0 q\gamma$ is homotopic to $q\sigma$ relative to y_0. Given γ, since $q\gamma$ is homotopic to $\alpha \cdot (h_0 q\gamma) \cdot \alpha^{-1}$ relative to x_0 and $q\gamma$ lifts to a loop at x, it follows that $\alpha \cdot (h_0 q\gamma) \cdot \alpha^{-1}$ lifts to a loop τ at x. Let σ be the lift of $h_0 q\gamma$ at y. We claim that σ is a loop. Indeed, if $\sigma(1) = g^m y$ for some integer m, then $\tau(1) = g^m x$. Hence $m = 0$ and σ is a loop, as claimed. Therefore h_0 lifts to an $h \in H$ satisfying (2.8).

It only remains to show that h commutes with g, i.e., $h \in Z_H(C)$. Let $x \in S^1$. Let $\rho(g) = \frac{p}{q}$ with $(p, q) = 1$. By Proposition 2.4(v) we know that the cyclic order of $Cx = \{x, gx, g^2x, \ldots, g^{q-1}x\}$ in S^1 is the same as that of $\{0, \frac{p}{q}, \frac{2p}{q}, \frac{(q-1)p}{q}\}$. Since h preserves cyclic order, it follows that the cyclic orders of

$$\begin{aligned} Ch(x) &= \{h(x), gh(x), g^2h(x), \ldots, g^{q-1}h(x)\}\,,\\ h(Cx) &= \{h(x), h(gx), h(g^2x), \ldots, h(g^{q-1}x)\}\,,\\ \text{and } &\{0, \tfrac{p}{q}, \tfrac{2p}{q}, \tfrac{(q-1)p}{q}\} \end{aligned}$$

are all the same. Since $qh = h_0 q$, h sends orbits to orbits, and therefore we have $Ch(x) = h(Cx)$, from which it follows that $gh(x) = h(gx)$.

Finally, assume that $C < T$. Since $Z_T C = T \cap Z_H C$ and $N_T C = T \cap N_H C$, it is now clear that $Z_T C = N_T C$. For all $h \in H$, since membership in $T(X, p)$ is a local property, $h \in T = T(S^1, u)$ if and only if $\pi(h) \in T_0 = T(S^1_0, qu)$, therefore (2.7) induces a short exact sequence

$$1 \longrightarrow C \longrightarrow Z_T C \xrightarrow{\pi|} T_0 \longrightarrow 1\,. \tag{2.9}$$

But as observed above, $T_0 \cong T$, and so the theorem is proved. □

3 Assembly maps and algebraic K-theory of T

In this last section we review assembly maps and isomorphism conjectures in algebraic K-theory, focusing on the rationalized case and referring the reader to [11, 10] for comprehensive surveys. Then we explain the main results of [12] and how they imply Theorem 1.1 as well as a generalization to higher algebraic K-theory.

Let G be a discrete group. The algebraic K-theory groups $K_n(\mathbb{Z}G)$ of the integral group ring of G play a central role in geometric topology, in particular in the classification of high-dimensional manifolds and their automorphisms. Arguably the most important K-theoretic invariant is the *Whitehead group* $Wh(G)$, which classifies high-dimensional h-cobordisms, and which is defined as the quotient of $K_1(\mathbb{Z}G) = \left(\bigcup_{k\in\mathbb{N}} GL_k(\mathbb{Z}G)\right)_{\mathrm{ab}}$ by the image of the 1-by-1 invertible matrices $(\pm g)$, $g \in G$. The following conjecture is one of the most well-known and consequential open problems in this area.

Conjecture 3.1. *If G is torsion-free, then $Wh(G) = 0$. If G has torsion, then the inclusions of finite subgroups H of G induce an injective homomorphism*

$$\operatorname*{colim}_{H\in \mathsf{Sub}_{\mathcal{F}\mathrm{in}}G} Wh(H) \otimes_{\mathbb{Z}} \mathbb{Q} \longrightarrow Wh(G) \otimes_{\mathbb{Z}} \mathbb{Q}. \tag{3.2}$$

The colimit in (3.2) is taken over the finite subgroup category $\mathsf{Sub}_{\mathcal{F}\mathrm{in}}G$, whose objects are the finite subgroups H of G and whose morphisms are defined as follows. Given subgroups H and K of G, let $\mathrm{conhom}_G(H, K)$ be the set all group homomorphisms $H \to K$ given by conjugation by an element of G. The group $\mathrm{inn}(K)$ of inner automorphisms of K acts on $\mathrm{conhom}_G(H, K)$ on the left by post-composition. The set of morphisms in $\mathsf{Sub}_{\mathcal{F}\mathrm{in}}G$ from H to K is then defined as the quotient $\mathrm{inn}(K)\backslash \mathrm{conhom}_G(H, K)$. For example, in the special case when G is abelian, then $\mathsf{Sub}_{\mathcal{F}\mathrm{in}}G$ is just the poset of finite subgroups of G ordered by inclusion. Equivalently, the colimit in (3.2) could be taken over the restricted orbit category $\mathsf{Or}_{\mathcal{F}\mathrm{in}}G$, which has as objects the homogeneous G-sets G/H for any finite subgroup H of G, and as morphisms the G-equivariant maps. The relation between $\mathsf{Sub}_{\mathcal{F}\mathrm{in}}G$ and $\mathsf{Or}_{\mathcal{F}\mathrm{in}}G$ and the equivalence of the two approaches is explained, for example, in [13, page 152, Lemma 3.11].

Conjecture 3.1 is known to be true for all Gromov hyperbolic groups [2] and all CAT(0)-groups [1], for example. One of the most interesting open cases of Conjecture 3.1 is Thompson's group F: is $Wh(F) = 0$? Our main result, Theorem 1.1, is that for Thompson's group T Conjecture 3.1 is true.

Before explaining this, we want to discuss how Conjecture 3.1 is a special case of the more general Farrell-Jones Conjecture in algebraic K-theory. This conjecture asserts that certain *assembly maps* are isomorphisms. The targets of the assembly maps are the algebraic K-theory groups $K_n(\mathbb{Z}G)$ that we are interested in. The sources are other groups that are easier to compute and homological in nature, and that only depend on the algebraic K-theory of relatively "small" subgroups of G. The construction of these assembly maps is rather technical, and we will not explain it here—see e.g. [11, 12] for details. But the picture simplifies after rationalizing, i.e., after tensoring with $\mathbb{Q}$, and we are going to focus on it now.

The *rationalized classical assembly map* for $K_n(\mathbb{Z}G)$, $n \in \mathbb{Z}$, is a homomorphism

$$\bigoplus_{\substack{s,t\geq 0\\ s+t=n}} H_s(BG;\mathbb{Q}) \underset{\mathbb{Q}}{\otimes} (K_t(\mathbb{Z}) \underset{\mathbb{Z}}{\otimes} \mathbb{Q}) \longrightarrow K_n(\mathbb{Z}G) \underset{\mathbb{Z}}{\otimes} \mathbb{Q}. \tag{3.3}$$

The *rationalized Farrell-Jones assembly map* for $K_n(\mathbb{Z}G)$, $n \in \mathbb{Z}$, is a homomorphism

$$\bigoplus_{C\in(\mathcal{F}\text{in}\mathcal{C}\text{yc})} \bigoplus_{\substack{s\geq 0,t\geq -1\\ s+t=n}} H_s(BZ_GC;\mathbb{Q}) \underset{\mathbb{Q}[W_GC]}{\otimes} \Theta_C(K_t(\mathbb{Z}C) \underset{\mathbb{Z}}{\otimes} \mathbb{Q}) \longrightarrow K_n(\mathbb{Z}G) \underset{\mathbb{Z}}{\otimes} \mathbb{Q}. \tag{3.4}$$

Here $(\mathcal{F}\text{in}\mathcal{C}\text{yc})$ denotes the set of conjugacy classes of finite cyclic subgroups in G, Z_GC denotes the centralizer in G of C, W_GC denotes the quotient N_GC/Z_GC of the normalizer modulo the centralizer, and $\Theta_C(K_t(\mathbb{Z}C) \otimes_{\mathbb{Z}} \mathbb{Q})$ is a direct summand of $K_t(\mathbb{Z}C) \otimes_{\mathbb{Z}} \mathbb{Q}$ naturally isomorphic to

$$\operatorname{coker}\left(\bigoplus_{D\lneqq C} K_t(\mathbb{Z}D) \underset{\mathbb{Z}}{\otimes} \mathbb{Q} \longrightarrow K_t(\mathbb{Z}C) \underset{\mathbb{Z}}{\otimes} \mathbb{Q} \right).$$

The dimensions of the $\mathbb{Q}$-vector spaces $\Theta_C(K_n(\mathbb{Z}C) \otimes_{\mathbb{Z}} \mathbb{Q})$ can be explicitly computed; see [18, Theorem on page 9].

Moreover, the summand in the source of (3.4) corresponding to $C = 1$ is the same as the source of (3.3), since $K_{-1}(\mathbb{Z}) = 0$. Therefore, if G is torsion-free, then the classical and the Farrell-Jones assembly maps are the same.

Conjecture 3.5 (Rationalized Farrell-Jones Conjecture). *For any group G and for any $n \in \mathbb{Z}$, the rationalized Farrell-Jones assembly map* (3.4) *is an isomorphism. In particular, if G is torsion-free, then the map* (3.3) *is an isomorphism for any $n \in \mathbb{Z}$.*

Conjecture 3.5, even in its much stronger integral version that we are not discussing here, is known to be true for all Gromov hyperbolic groups [2] and all CAT(0)-groups [1, 19], for example.

Theorem 1.3 and Corollary 1.5 have the following immediate consequence.

Corollary 3.6. *The source of the rationalized Farrell-Jones assembly map for Thompson's group T is isomorphic to*

$$\bigoplus_{k\geq 0}\bigoplus_{\substack{s\geq 0, t\geq -1\\ s+t=n}} H_s(BT;\mathbb{Q})\otimes_{\mathbb{Q}}\Theta_{C_k}(K_t(\mathbb{Z}C_k)\otimes_{\mathbb{Z}}\mathbb{Q}) \tag{3.7}$$

for any $n\in\mathbb{Z}$. In particular, if the Farrell-Jones conjecture is true for T, then $K_n(\mathbb{Z}T)\otimes_{\mathbb{Z}}\mathbb{Q}$ is isomorphic to (3.7) for any $n\in\mathbb{Z}$.

As we already remarked, thanks to theorems of Ghys-Sergiescu and Patronas, the dimension over $\mathbb{Q}$ of each individual summand in (3.7) is explicitly computable.

Now we recall a famous result about the injectivity of the rationalized classical assembly map.

Theorem 3.8 (Bökstedt-Hsiang-Madsen [3]). *Let G be any group, not necessarily torsion-free. Assume that for every $s\in\mathbb{N}$ the abelian group $H_s(BG;\mathbb{Z})$ is finitely generated. Then for every $n\in\mathbb{N}$ the rationalized classical assembly map (3.3) is injective.*

In particular, Theorem 3.8 applies to Thompson's group F, since any group of type F_∞ satisfies the assumption above. However, this injectivity result produces no information about $Wh(G)$.

In [12], Theorem 3.8 is generalized to the Farrell-Jones assembly map, yielding also information about $Wh(G)$.

Theorem 3.9. *([12, Main Theorem 1.13]). Let G be any group. Assume that for every finite cyclic subgroup C of G the following conditions hold:*

(i) for every $s\in\mathbb{N}$ the abelian group $H_s(BZ_GC;\mathbb{Z})$ is finitely generated;

(ii) let k be the order of C and let ζ_k be any primitive k^{th} root of unity; for every $t\in\mathbb{N}$ the natural homomorphism

$$K_t(\mathbb{Z}[\zeta_k])\longrightarrow\prod_{p\in\mathbb{P}}K_t\Big(\mathbb{Z}_p\otimes_{\mathbb{Z}}\mathbb{Z}[\zeta_k];\mathbb{Z}_p\Big)$$

is injective after tensoring with $\mathbb{Q}$, where $\mathbb{P}$ denotes the set of all primes and $\mathbb{Z}_p$ denotes the ring of p-adic integers for $p\in\mathbb{P}$.

Then the restriction of the rationalized Farrell-Jones assembly map (3.4) to the summands where $t\neq -1$ induces an injective homomorphism

$$\bigoplus_{C\in(\mathcal{F}\mathrm{in}\mathcal{C}\mathrm{yc})}\bigoplus_{\substack{s,t\geq 0\\ s+t=n}} H_s(BZ_GC;\mathbb{Q})\otimes_{\mathbb{Q}[W_GC]}\Theta_C(K_t(\mathbb{Z}C)\otimes_{\mathbb{Z}}\mathbb{Q})\longrightarrow K_n(\mathbb{Z}G)\otimes_{\mathbb{Z}}\mathbb{Q}$$

for every $n\geq 0$.

Corollary 3.10. *([12, Theorem 1.1]). Assume that a group G satisfies assumption (i) of Theorem 3.9. Then there is an injective homomorphism*

$$\operatorname*{colim}_{H \in \mathsf{Sub}_{\mathcal{F}\mathrm{in}} G} Wh(H) \underset{\mathbb{Z}}{\otimes} \mathbb{Q} \longrightarrow Wh(G) \underset{\mathbb{Z}}{\otimes} \mathbb{Q},$$

i.e., the second part of Conjecture 3.1 is true for G.

Some remarks are in order about assumption (ii) of Theorem 3.9. First of all, (ii) is true for all k if $t = 0, 1$, and for all t if $k = 1$. This explains why the assumption is absent from Theorem 3.8 and Corollary 3.10. Moreover, assumption (ii) is conjecturally always true, in the sense that it is automatically satisfied if a weak version of the Leopoldt-Schneider conjecture holds for cyclotomic fields; see [12, Section 2] for details.

Now Theorem 1.3 and its Corollaries 1.5 and 3.6, combined with Theorem 3.9 and Corollary 3.10, immediately imply our main result; cf. Theorem 1.1.

Corollary 3.11. *Conjecture 3.1 is true for Thompson's group T, and there is an injective homomorphism*

$$\operatorname*{colim}_{k \in \mathbb{N}} Wh(C_k) \underset{\mathbb{Z}}{\otimes} \mathbb{Q} \longrightarrow Wh(T) \underset{\mathbb{Z}}{\otimes} \mathbb{Q}.$$

In particular, $Wh(T) \otimes_{\mathbb{Z}} \mathbb{Q}$ is an infinite dimensional $\mathbb{Q}$-vector space. Moreover, if assumption (ii) of Theorem 3.9 holds for all $k, t \in \mathbb{N}$, then there is an injective homomorphism

$$\bigoplus_{k \geq 0} \bigoplus_{\substack{s,t \geq 0 \\ s+t=n}} H_s(BT; \mathbb{Q}) \underset{\mathbb{Q}}{\otimes} \Theta_{C_k}(K_t(\mathbb{Z}C_k) \underset{\mathbb{Z}}{\otimes} \mathbb{Q}) \longrightarrow K_n(\mathbb{Z}T) \underset{\mathbb{Z}}{\otimes} \mathbb{Q}$$

for all $n \in \mathbb{N}$.

Proof. The only step that remains to be explained is the identification

$$\operatorname*{colim}_{k \in \mathbb{N}} Wh(C_k) \cong \operatorname*{colim}_{H \in \mathsf{Sub}_{\mathcal{F}\mathrm{in}} T} Wh(H), \tag{3.12}$$

where on the left-hand side we have the colimit described right after Theorem 1.1. Recall that all finite subgroups of T are cyclic. Suppose that C and D are finite subgroups of T of orders k and ℓ, respectively, and assume that $k \mid \ell$. Then there is exactly one subgroup C' of D of order k, and C and C' are conjugate in T by Corollary 2.6. As explained in the proof of that Corollary, C has a unique generator with rotation number $\frac{1}{k}$, and the same is true for C'. Since rotation numbers are preserved by conjugation by Proposition 2.4(iii), we conclude that there is exactly one morphism in $\mathsf{Sub}_{\mathcal{F}\mathrm{in}} T$ from C to D. Now, identifying C_k with the cyclic subgroup $\langle \gamma_k \rangle$ of T generated by the pseudo-rotation of order k from Example 2.1, the isomorphism (3.12) follows by cofinality. □

References

[1] Arthur Bartels and Wolfgang Lück. The Borel conjecture for hyperbolic and CAT(0)-groups. *Ann. of Math. (2)*, 175(2):631–689, 2012.

[2] Arthur Bartels, Wolfgang Lück, and Holger Reich. The K-theoretic Farrell-Jones conjecture for hyperbolic groups. *Invent. Math.*, 172(1):29–70, 2008.

[3] Marcel Bökstedt, Wu Chung Hsiang, and Ib Madsen. The cyclotomic trace and algebraic K-theory of spaces. *Invent. Math.*, 111(3):465–539, 1993.

[4] Kenneth S. Brown. Finiteness properties of groups. *J. Pure Appl. Algebra*, 44(1-3):45–75, 1987. Proceedings of the Northwestern conference on cohomology of groups (Evanston, Ill., 1985).

[5] Kenneth S. Brown and Ross Geoghegan. An infinite-dimensional torsion-free FP_∞ group. *Invent. Math.*, 77(2):367–381, 1984.

[6] James W. Cannon, William J. Floyd, and Walter R. Parry. Introductory notes on Richard Thompson's groups. *Enseign. Math. (2)*, 42(3-4):215–256, 1996.

[7] Ross Geoghegan. *Topological methods in group theory*, volume 243 of *Graduate Texts in Mathematics*. Springer, 2008.

[8] Étienne Ghys and Vlad Sergiescu. Sur un groupe remarquable de difféomorphismes du cercle. *Comment. Math. Helv.*, 62(2):185–239, 1987.

[9] Anatole Katok and Boris Hasselblatt. *Introduction to the modern theory of dynamical systems*, volume 54 of *Encyclopedia of Mathematics and its Applications*. Cambridge University Press, 1995.

[10] Wolfgang Lück. K- and L-theory of group rings. In *Proceedings of the International Congress of Mathematicians vol. II*, volume II, pages 1071–1098. Hindustan Book Agency, 2010.

[11] Wolfgang Lück and Holger Reich. The Baum-Connes and the Farrell-Jones conjectures in K- and L-theory. In *Handbook of K-theory*, volume 2, pages 703–842. Springer, 2005.

[12] Wolfgang Lück, Holger Reich, John Rognes, and Marco Varisco. Algebraic K-theory of group rings and the cyclotomic trace map. *Adv. Math.*, 304:930–1020, 2017.

[13] Wolfgang Lück, Holger Reich, and Marco Varisco. Commuting homotopy limits and smash products. *K-Theory*, 30(2):137–165, 2003.

[14] Conchita Martínez-Pérez, Francesco Matucci, and Brita E.A. Nucinkis. Cohomological finiteness conditions and centralisers in generalisations of Thompson's group V. *Forum Math.*, 28(5):909–921, 2016.

[15] Conchita Martínez-Pérez and Brita E.A. Nucinkis. Bredon cohomological finiteness conditions for generalisations of Thompson groups. *Groups Geom. Dyn.*, 7(4):931–959, 2013.

[16] Francesco Matucci. *Algorithms and classification in groups of piecewise-linear homeomorphisms.* Ph.D. thesis, Cornell University, 2008.

[17] Robert Oliver. *Whitehead groups of finite groups*, volume 132 of *London Mathematical Society Lecture Note Series.* Cambridge University Press, Cambridge, 1988.

[18] Dimitrios Patronas. *The Artin defect in algebraic K-theory.* Ph.D. thesis, Freie Universität Berlin, 2014.

[19] Christian Wegner. The K-theoretic Farrell-Jones conjecture for CAT(0)-groups. *Proc. Amer. Math. Soc.*, 140(3):779–793, 2012.

Special cube complexes

Robert Kropholler

Abstract

We give an account on the programme of Wise for proving residual finiteness of hyperbolic groups with quasi-convex hierarchy. The key role is played by special cube complexes introduced by Haglund and Wise. We explain the result of Haglund and Wise saying that special cube complexes are invariant under Malnormal Amalgamation. We also suggest how cubical small cancellation leads to Wise's Malnormal Special Quotient Theorem. As a consequence, all closed hyperbolic 3-manifolds with a geometrically finite incompressible surface are virtually special. This article is based on a series of lectures given by Piotr Przytycki given at an LMS symposium in Durham.

1 Introduction

One of many interesting properties of groups is that of residual finiteness,

Definition 1.1. A group is *residually finite* if for all $g \neq e$ there exists a finite-index subgroup H such that $g \notin H$.

There are many examples of groups with this property such as, free groups and surface groups. One of the leading questions in geometric group theory is whether all hyperbolic groups are residually finite.

Wise [12] proved that the amalgam of two free groups over a malnormal subgroup is residually finite. This used the theory of clean square complexes which became the precursor of special cube complexes, the subject of this paper.

The category of special cube complexes introduced by Haglund and Wise [6], is designed to facilitate proofs of residual finiteness and similar properties such as separability, which for simplicity will not be discussed in this article.

Even though being special is a particular condition, the key phenomenon we will discuss is the stability of virtual specialness under amalgams (Section 3) and quotients (Section 4), under hyperbolicty conditions discussed later.

The programme has culminated in Agol [1] proving Wise's conjecture that all compact cube complexes with hyperbolic fundamental groups are virtually special, implying the virtual Haken conjecture. However, in this article we restrict to the state of affairs before Agol's result.

These notes will culminate in Wise's result that hyperbolic groups with a quasiconvex hierarchy are fundamental groups of compact virtually special cube complexes. Using Haken hierarchy this will be applicable to fundamental groups of closed hyperbolic 3-manifolds with a geometrically finite incompressible surface. One of the consequences, which we will not discuss in this article, is that any immersed incompressible surface in such a manifold is virtually embedded. Another consequence is that such a manifold virtually fibers.

Acknowledgements

I would like to thank Piotr Przytycki and Martin Bridson for carefully reading this article through its iterations and taking the time to give helpful feedback. I would also like to thank the referee for his suggestions.

2 Special Cube Complexes

Definition 2.1. A simplicial complex is called a *flag complex* if whenever there are n pairwise adjacent vertices, they span an (n-1) simplex

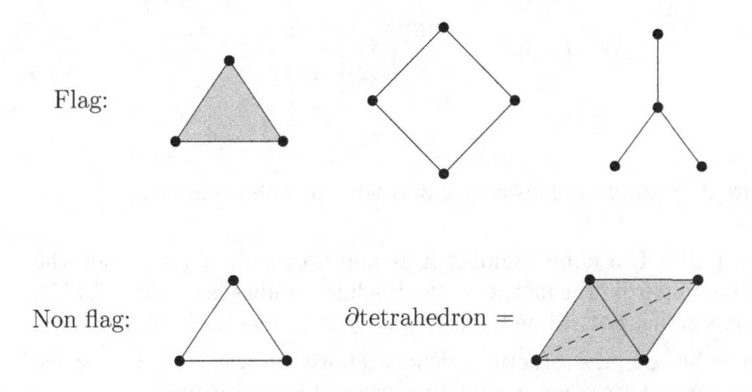

Figure 1: Some examples of flag and non flag complexes.

Another way to say this is that if we see the 1 skeleton of a higher dimensional simplex, then that simplex is included in the complex.

Definition 2.2. A *cube complex* is a complex which is built out of Euclidean cubes $[0, 1]^n$ where faces are identified via isometries.

We can give Euclidean n-space the structure of a cube complex by taking as vertices points in $\mathbb{Z}^n$ and joining any two vertices which are distance 1 apart by an edge, we now glue in an n-cube whenever we see its 1-skeleton.

If we take the quotient of this cube complex by the action of $\mathbb{Z}^n$ we have a cube complex structure on the torus T^n.

Definition 2.3. The *link of a vertex* in a cube complex is a complex built out of simplices whose n-simplices are corners of $(n+1)$-cubes adjacent to the vertex.

The link can be realised as the set of all points at distance $\frac{1}{4}$ from the vertex.

Definition 2.4. We say that a cube complex is *non positively curved (npc)* if the link of every vertex is a flag complex. A cube complex is CAT(0) if it is simply connected and the link of every vertex is flag.

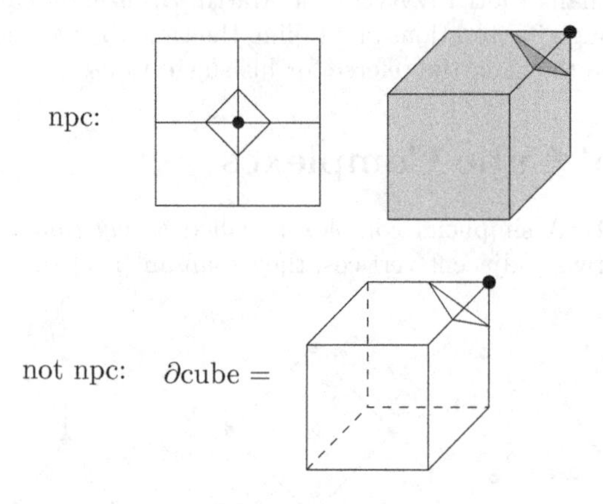

Figure 2: Some examples of npc and not npc cube complexes.

Remark. [3, p.210] If a cube complex X is non positively curved, then the metric on X coming from the metric on the Euclidean cubes is locally CAT(0). However in this article we will be taking a purely combinatorial point of view.

With the cube complex structures defined above, both Euclidean n-space $\mathbb{R}^n$ and the n-torus T^n are non-positively curved cube complexes.

Each individual cube is a copy of $[0,1]^n$ and as such is endowed with n projection maps to $[0,1]$

Definition 2.5. A *midcube* is the pre-image of $\frac{1}{2}$ under one of the coordinate projections.

Definition 2.6. A *hyperplane* is a connected subspace of a non positively curved cube complex that intersects each cube in a single midcube or $\emptyset$.

Proposition 2.7. *If X is a CAT(0) cube complex, then every midcube lies in a unique hyperplane.*

Sketch of Proof. We start with our midcube m in a cube c. We then add all midcubes m' from cubes c' adjacent to c such that $m' \cap m$ is a midcube of $c' \cap c$, call this new complex J. We then add any mid cube n from a cube e such that $n \cap J$ is a mid cube of $e \cap J$. We carry on this process until we have a stable subspace. We then check that this is a hyperplane. To do this we check that the complex does not cross itself, assume that there is a self crossing then we can find a smallest loop l in our hyperplane witnessing this self crossing. Using a technique known as bigon removal we can find a smaller loop thus contradicting the minimality of l. Hence there will be no self crossing and we have a hyperplane containing m. For full details see [13]. □

Definition 2.8. Let X be a non positively curved cube complex and $\tilde{X}$ its universal cover. An *immersed hyperplane* H in X is the space $H = \tilde{H}/Stab(H)$ where $\tilde{H}$ is a hyperplane in $\tilde{X}$ and $Stab(H)$ is the subgroup of $\pi_1(X)$ stabilising $\tilde{H}$. There is a natural map $H \to X$.

Since each midcube in $\tilde{X}$ defines a unique hyperplane, each midcube in X defines a unique immersed hyperplane.

An immersed hyperplane embeds if the natural map $H \to X$ is an embedding. i.e. is a hyperplane in the sense of Def 2.6

Definition 2.9. An edge is *dual* to a hyperplane if it intersects that hyperplane in a midcube.

Definition 2.10. A hyperplane is *2-sided* if we can give a consistent orientation to all of its dual edges.

Definition 2.11. A 2-sided hyperplane *directly self osculates*, if once given a consistent orientation, there are two edges dual to the hyperplane which share the same initial or terminal vertex.

Definition 2.12. A 2-sided hyperplane *indirectly self osculates*, if once given a consistent orientation, there are two edges dual to the hyperplane such that the terminal vertex of one is the initial vertex of the other.

Remark. We can get rid of all indirect self osculations by subdividing cubes.

Definition 2.13. Distinct hyperplanes H, H' *interosculate* if there are dual edges e_1, e_2 to H and f_1, f_2 to H' such that e_1, f_1 lie in a square and e_2, f_2 share a vertex but do not lie in a square.

Definition 2.14. We say that a non positively curved cube complex is *special* if the immersed hyperplanes.

1) embed (as such we will refer to them as hyperplanes from now on).
2) are 2-sided.
3) do not directly self-osculate.
4) do not interosculate.

See Figure 5.

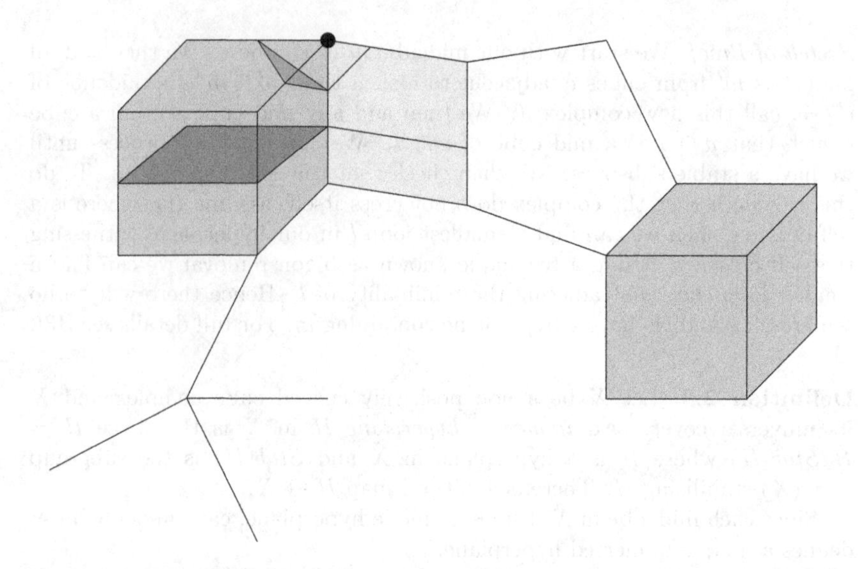

Figure 3: An example of a cube complex. The link of the bold vertex is the union of a triangle and an edge glued along a vertex. There is an example of a hyperplane which is the union of a square and an edge glued along a vertex.

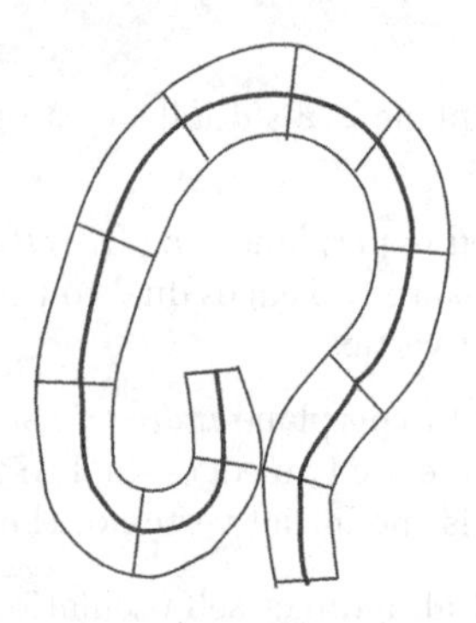

Figure 4: An indirect self osculation.

Some examples of special cube complexes include graphs (if a graph contains a loop of length 1, there is an indirect osculation, we can get rid of this by subdividing this edge.), surfaces and tori.

The properties 1)-4) in the definition of special cube complex fit together to give the following natural map into the Salvetti complex of a particular right angled Artin group, which we will define in the course of the proof.

Theorem 2.15. *[6] A special cube complex X with finitely many hyperplanes*

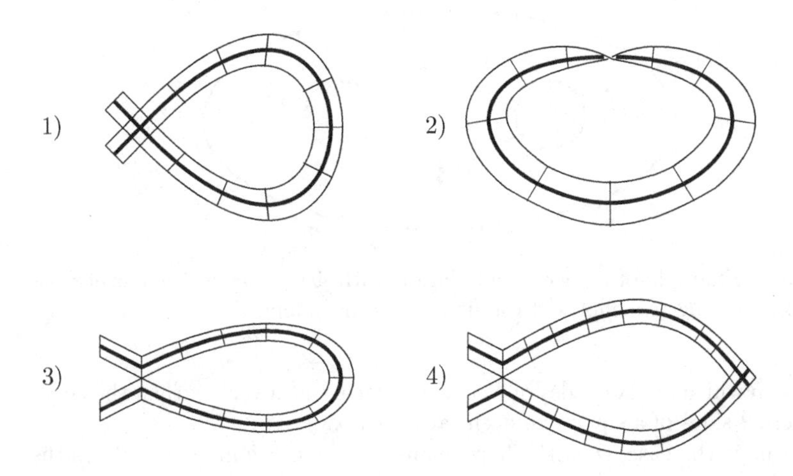

Figure 5: The 4 pathologies which are disallowed in special cube complexes.

(e.g. if X is compact) admits a local isometry into the Salvetti complex R of a finitely generated right angled Artin group (RAAG) F.

Proof. Let $h_1, \dots, h_n$ be the hyperplanes of X. Let F be the right angled Artin group $\langle s_1, \dots, s_n | s_i s_j = s_j s_i \text{ if } h_i \cap h_j \neq \emptyset \rangle$. Then $F = \pi_1(R)$ where $R \subset T^n = \prod_{i=1}^{n} S_i^1$ is the subcomplex consisting of cubes corresponding to pairwise commuting s_i.

We define the local isometry as follows:

$$\begin{aligned} \text{Vertices} &\mapsto \text{unique vertex of } R \\ \text{Edge dual to } h_i &\mapsto \text{edge labelled } s_i. \end{aligned}$$

We consider the dual edges oriented in a consistent way, which is possible by property 2), and then we send them in an orientation preserving way to the directed edge s_i.

This extends to the cubes by the definition of F and R.

We need to check that this is a local embedding. We can see this from properties 1) and 3) since if two edges which meet at a vertex are mapped down to the same s_i, then they are dual to the same hyperplane which implies that this hyperplane is either not embedded or self osculates.

To check that this is a local isometry we start with two edges a, b which are incident at the vertex v we must check that if these edges map forward to

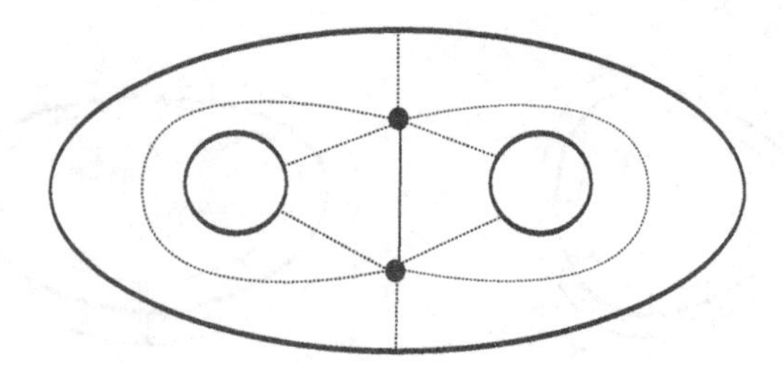

Figure 6: Example of a special cubulation with 4 vertices of the surface obtained by gluing two copies of the figure together along the bold circles.

edges $\bar{a}, \bar{b}$ which are two sides of a square adjacent at a vertex, then the edges a, b were 2 sides of a square adjacent at a vertex.

If this is the case, then the hyperplanes through $\bar{a}, \bar{b}$ intersect. Since the hyperplanes do not interosculate (property 4)), it follows that a, b must have been adjacent edges of a square. □

We now present the key construction of covers of special cube complexes, implying in particular residual finiteness of their fundamental groups without appealing to Theorem 2.15 and residual finiteness of linear groups. We will later see that this construction enjoys several functorial properties.

Theorem 2.16. *[6] If $f : Y \to X$ is a local isometry of a compact cube complex Y to a special cube complex X, then there is a finite cover of X, called the* canonical completion $C(Y \to X)$, *such that f lifts to an embedding $\hat{f}$ of Y and there exists a retraction called the* canonical retraction $r : C(Y \to X) \to Y$. *See Figure 7.*

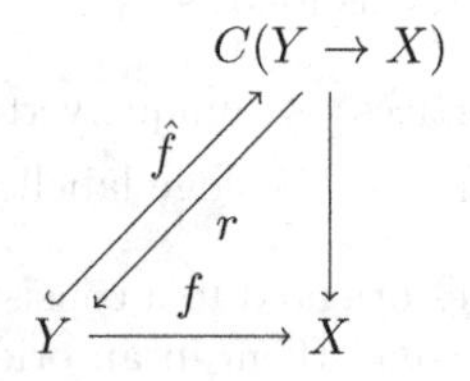

Figure 7: Diagrammatic scheme for Theorem 2.16

We will assume throughout the proof of Theorem 2.16 that X has no indirect self osculations.

Proof. We construct $C(Y \to X)$ as follows. The vertices of $C(Y \to X)$ are $\mathrm{Vert}(X) \times \mathrm{Vert}(Y)$.

For each edge (x, x') and vertex $x \times y \in \mathrm{Vert}(X) \times \mathrm{Vert}(Y)$, we will construct an edge of $C(Y \to X)$ starting at $x \times y$ and covering (x, x') by specifying its other endpoint $x' \times y'$. If possible, we choose y' so that (y, y') is an edge of Y and (x, x') and $f(y, y')$ are dual to the same hyperplane, if no such y' exists, then we choose $y = y'$.

To make sure that y' is unique we require both condition 3) and that there are no indirect self osculations. This finishes the construction of the 1-skeleton of $C(Y \to X)$. The retraction forgets the second coordinate and the edges where $y = y'$ are collapsed to vertices.

We use condition 4) to verify that the boundary of each square in X is covered, in the 1-skeleton of $C(Y \to X)$, by 4-cycles. We fill these with squares to form the 2-skeleton of $C(Y \to X)$. We similarly construct the higher dimensional cubes of $C(Y \to X)$. □

For examples see Figures 8 and 9.

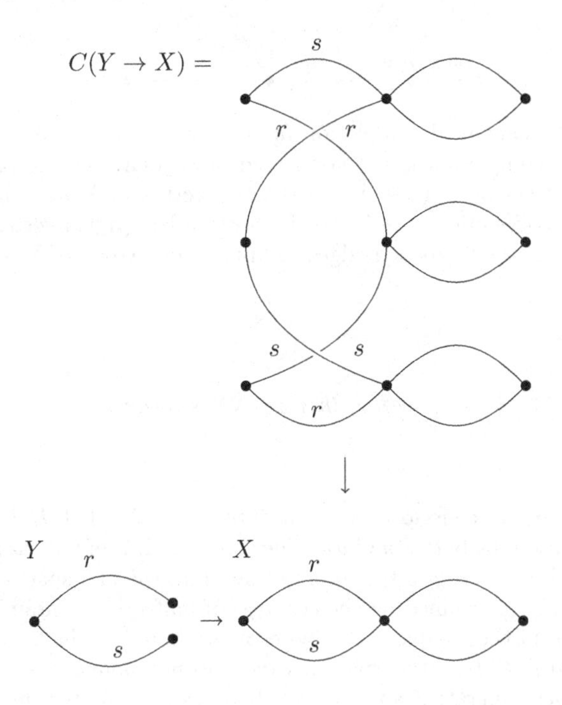

Figure 8: Example of canonical completion. Canonical retraction collapses all horizontal edges to points and maps diagonal edges labelled s and r to appropriate edges of Y.

This proof leads us to the first of many interesting consequences of being special.

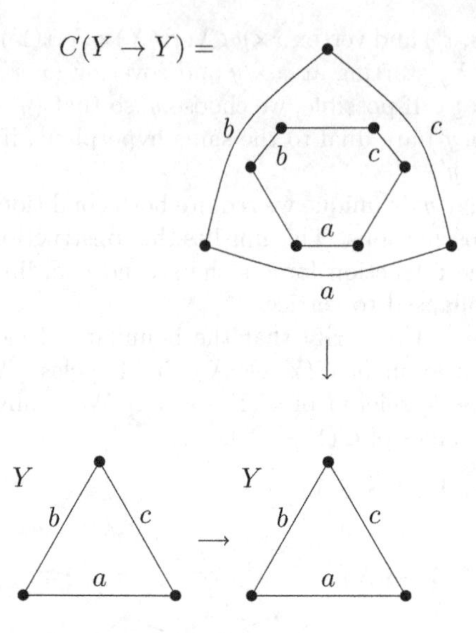

Figure 9: The canonical completion of the identity map. We can see that the canonical completion is not connected or regular. The retraction sends the top 3 vertices of $C(Y \to Y)$ to the top vertex of Y and the 3 bottom left (right) vertices of $C(Y \to Y)$ to the bottom left (right) vertex of Y. The retraction sends the 6 labelled edges to appropriate edges of Y and collapses the 3 remaining edges.

Theorem 2.17. *If X is special, then $\pi_1(X)$ is residually finite.*

Proof. Let γ be a geodesic path from $\tilde{x}$ to $g\tilde{x}$ in $\tilde{X}$. Let H be the set of halfspaces containing both $\tilde{x}$ and $g\tilde{x}$. The convex hull Y of γ is the subcomplex spanned by all vertices y in the intersection of all the half spaces in H. Each such vertex y is determined by choices $h(y)$ of halfspaces containing it for all hyperplanes h in the complex. For hyperplanes bounding halfspaces in H we much have $h(y) \in H$. Hyperplanes which do not bound a halfspace in H are those which seperate $\tilde{x}$ and $g\tilde{x}$ and there are only finitely many of these. So we have only finitely many choices for $h(y)$ and hence only finitely many vertices y, hence Y is compact.

The covering map gives a local isometry from Y to X. Since Y is compact we can use Theorem 2.16 to find a finite cover in which Y embeds. In this cover the path in X representing g lifts to a non closed path. $\square$

3 Malnormal Amalgamation

In this section we discuss the combination theorem for virtually special cube complexes due to Haglund and Wise [7].

Definition 3.1. We say that $H \leq G$ is *malnormal* if $g \notin H$ implies $g^{-1}Hg \cap H = \{e\}$.

Definition 3.2. We say that a cube complex is *virtually special* if there is a finite sheeted cover which is special.

Theorem 3.3. *[7] Let A, B be compact virtually special cube complexes with hyperbolic fundamental groups. Suppose that $M \to A$, $M \to B$ are local isometries of cube complexes such that $\pi_1(M)$ is malnormal in both $\pi_1(A)$ and $\pi_1(B)$. Then $X = A \cup_{M \times I} B$, formed by identifying $M \times 0$ and $M \times 1$ with their images in A and B respectively, is virtually special.*

Remark. $\pi_1(X)$ is hyperbolic by Bestvina-Feighn [2].

The proof of Theorem 3.3 generalises the proof of the particular case where A, B, M are graphs [12]. For simplicity, we restrict to this case from now on.

Lemma 3.4. *If $M \hookrightarrow A$ is a subgraph, then the preimage of M in $C(M \hookrightarrow A)$ is $C(M \xrightarrow{Id} M)$ and the canonical retraction to M on $C(M \hookrightarrow A)$ restricts to the retraction on $C(M \xrightarrow{Id} M)$*

Lemma 3.5. *If $M, M' \subset A$ are subgraphs such that $M \cap M'$ is a forest, then any loop in the preimage of M' in $C(M \hookrightarrow A)$ is homotopically trivial under the canonical retraction $r : C(M \to A) \to M$.*

Proof. Since r maps an edge covering an edge e in A either to e with an arbitrary orientation, or else to a vertex. r will map the cover of M' to $(M \cap M') \cup Vert(M)$ which is a forest. Thus the image under r of any loop in the cover of M' must be homotopically trivial. □

Proposition 3.6. *[7] For any finite covers $\hat{A}, \hat{B}$ of A, B there are further finite regular covers $\overset{\bowtie}{A}, \overset{\bowtie}{B}$ inducing the same cover $\overset{\bowtie}{M}$ of M.*

Remark. Proposition 3.6 is the key element in the proof of Theorem 3.3. Since we decided to keep this article simple, we will not give all the details of how to deduce that X in Theorem 3.3 is virtually special, but we will at least indicate how to prove that $\pi_1(X)$ is residually finite. In fact, we will only prove that for $e \neq g \in \pi_1(A)$, there is a finite cover $\overset{\bowtie}{X} \to X$ with $g \notin \pi_1(\overset{\bowtie}{X})$.

Proof of Remark. Take $e \neq g \in \pi_1(A)$, take $\hat{B} = B$ and using Theorem 2.17 take a finite cover $\hat{A}$ of A such that $g \notin \pi_1(\hat{A})$. Then using Proposition 3.6 we can take further finite covers $\overset{\bowtie}{A}, \overset{\bowtie}{B}$ of $\hat{A}, \hat{B}$ which induce the same cover $\overset{\bowtie}{M}$ of M.

We now glue together appropriate number of copies of $\overset{\bowtie}{A}$ and $\overset{\bowtie}{B}$ along $\overset{\bowtie}{M}$ to get a finite cover $\overset{\bowtie}{X}$ of X with $g \notin \pi_1(\overset{\bowtie}{X})$. □

Definition 3.7. For a map $M \to A$ and a cover $\overset{\bowtie}{A} \to A$, consider a component $\overset{\bowtie}{M}$ of the pullback of $\overset{\bowtie}{A}$ via $M \to A$. The map $\overset{\bowtie}{M} \to \overset{\bowtie}{A}$ is called an *elevation* of $M \to A$ to $\overset{\bowtie}{A}$. Note that $\overset{\bowtie}{M} \to M$ is a covering map; if it is of degree 1, then $M = \overset{\bowtie}{M} \to \overset{\bowtie}{A}$ is called a *lift*. An elevation is *based* if its domain contains the basepoint of the pullback, i.e. the point mapping to the basepoints of M and $\overset{\bowtie}{A}$.

Proof of Proposition 3.6. By passing to finite covers $\hat{A}, \hat{B}$ we can assume, by taking canonical completions, that

- all elevations of $M \to A, B$ to $\hat{A}, \hat{B}$ are embeddings
- the based elevations of M to $\hat{A}, \hat{B}$ have the same domain $\hat{M}$ see the diagram below.

Elevations of $M \to A$ to $\hat{A}$ correspond to intersections of conjugates of $\pi_1(M)$ with $\pi_1(\hat{A})$. For any $\hat{M}' \subset \hat{A}$ an elevation of $M \to A$ distinct from $\hat{M}$ we have that $\hat{M}' \cap \hat{M}$ is a forest, since $\pi_1(M)$ is malnormal in $\pi_1(A)$.

We look at canonical completions $C(\hat{M} \to \hat{A}), C(\hat{M} \to \hat{B}), C(\hat{M} \to \hat{M})$ in the third row of the diagram below. We have inclusions and commutativity of retractions by Lemma 3.4.

Let $\tilde{M} \to M$ be a finite cover factoring through the domains of all elevations of $M \to A, M \to B$ to $C(\hat{M} \to A), C(\hat{M} \to B)$. In particular, $\tilde{M}$ covers $\hat{M}$.

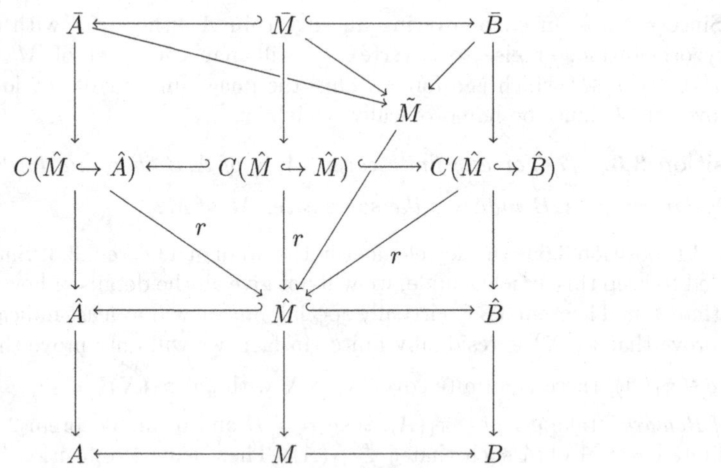

Let $\bar{A}$ be the pullback of the cover $\tilde{M} \to \hat{M}$ via the retraction $C(\hat{M} \hookrightarrow \hat{A}) \to \hat{M}$. Define $\bar{B}$ and $\bar{M}$ similarly.

Claim. The domains of all elevations of $M \to A$ to $\bar{A}$ are quotients of components of $\bar{M}$.

We treat the elevations in question depending on whether after descending to $\hat{A}$ they cover the elevation $\hat{M} \subset \hat{A}$ or a different elevation $\hat{M}' \subset \hat{A}$ of $M \to A$

- The elevations covering $\hat{M} \subset \hat{A}$ map to $C(\hat{M} \to \hat{M})$ by Lemma 3.4, hence they elevate to $\bar{M} \subset \bar{A}$.
- The elevations covering some $\hat{M}' \subset \hat{A}$ distinct from $\hat{M}$ map isomorphically into $C(\hat{M} \to \hat{A})$ by Lemma 3.5 and hence are involved in the definition of $\tilde{M}$.

Using the fact that $\hat{M}$ is a component of $C(\hat{M} \to \hat{M})$ on which r restricts to the identity, we see that a component of $\bar{M}$ is equal to $\tilde{M}$.

Now when we pass to the smallest covers $\overset{\bowtie}{A}, \overset{\bowtie}{B}$ of $\bar{A}, \bar{B}$ that are regular covers of A, B, they induce, as desired, the same cover $\overset{\bowtie}{M}$ of M, which is the smallest regular cover of M factoring through $\bar{M}$. □

4 Cubical Small Cancellation

In this section we discuss results of Wise describing to what extent the category of special cube complexes is invariant under filling, i.e. passing to quotients of π_1. The main tool is cubical small cancellation, which generalises classical small cancellation theory [9]. We will start by giving an overview of the classical setting and then talk about some of the key concepts in cubical small cancellation theory.

4.1 Classical Small Cancellation Theory

The idea of small cancellation theory is that if we have a group presentation $\langle X|R \rangle$ and the relators do not have much in common, then the group satisfies properties such as being word hyperbolic and the presentation complex is a classifying space.

Definition 4.1. Given a subset R of a free group $F(X)$ we define the *symmetric closure* R^* of R to be the set of cyclically reduced conjugates of elements of $R^{\pm 1}$.

Example. 1. $R = \{xyx^{-1}y^{-1}\}$, $R^* = \{xyx^{-1}y^{-1}, yx^{-1}y^{-1}x, x^{-1}y^{-1}xy,$ $y^{-1}xyx^{-1}, yxy^{-1}x^{-1}, xy^{-1}x^{-1}y, y^{-1}x^{-1}yx, x^{-1}yxy^{-1}\}$

2. $R = \{x^7\}$, $R^* = \{x^7, x^{-7}\}$

Definition 4.2. A *piece* in a presentation $\langle X|R \rangle$ is a word u such that there exist 2 distinct words $r_1, r_2 \in R^*$ such that $r_1 \equiv uv_1$ and $r_2 \equiv uv_2$

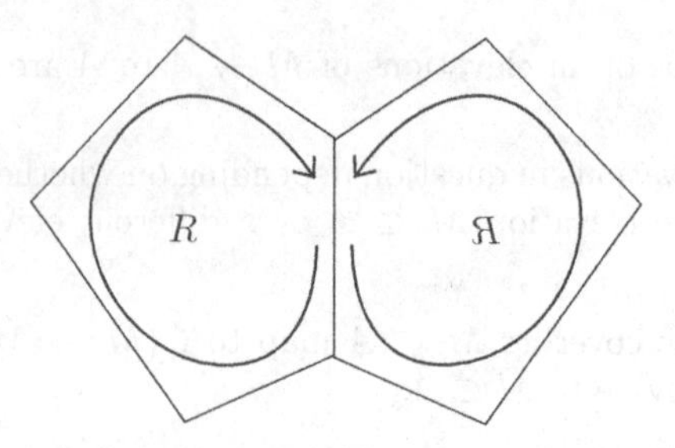

Figure 10: A non reduced van Kampen diagram

Definition 4.3. The presentation $\langle X|R\rangle$ satisfies the $C'(\lambda)$ *condition* if every piece u satisfies $|u| < \lambda r$ for each $r \in R^*$ with u a sub word of r.

The presentation $\langle X|R\rangle$ satisfies the $C(p)$ *condition* if no relator can be written as the concatenation of less than p pieces.

Remark. $C'(\frac{1}{p}) \Rightarrow C(p)$

Pieces can be restated in terms of van Kampen diagrams which are the key tool in small cancellation theory.

Definition 4.4. A *van Kampen diagram* is a contractible planar cell complex where the 1-cells are oriented and labelled by elements of $X^{\pm 1}$ and the label around the boundary of each 2-cell is an element of R^*.

The word read around the boundary of the diagram read from any base point is easily seen to be a product of conjugates of relators, we can also get the converse of this statement in an easy way.

Lemma 4.5. *Let w be a word in $X^{\pm 1}$. Then $w = e$ in the group presented by $\langle X|R\rangle$ if and only if there exists a van Kampen diagram with boundary label w.*

For a proof of the above Lemma see [4]

Definition 4.6. We call a van Kampen diagram *reduced* if there is no pair of adjacent faces with inverse labels. See figure 10

We can now consider pieces in terms of reduced diagrams. A piece is now a path in a van Kampen diagram which is a subset of a connected component of the intersection of two 2-cells. One can suppress vertices of valence 2 in a diagram and then the $C(p)$ condition says that the boundary of each 2-cell has combinatorial length at least p.

By studying reduced van Kampen diagrams we can see many desirable properties of our group. The first is a version of Greendlinger's Lemma.

Theorem 4.7. *If $G = \langle X|R\rangle$ satisfies the $C'(\frac{1}{6})$ condition and w is a reduced word in $F(X)$ representing the identity in G, then there is a subword s of w and a relator $r \in R^*$ such that:*

1. *s is a subword of r,*

2. *$|s| > \frac{|r|}{2}$.*

Using this we can easily verify that the group in question has a linear isoperimetric inequality and is therefore hyperbolic.

This is proved by looking at reduced van Kampen diagrams and using a combinatorial of the Gauss-Bonnet Theorem to prove that the diagram has negative curvature.

We will now look at how to transport these ideas to a cubical setting, where our generators will correspond to cube complexes and relators to locally isometric subcomplexes.

4.2 Cubical Small Cancellation Theory

Definition 4.8. Suppose that we have a non positively curved cube complex X, a family of cube complexes Y_i and local isometries $\phi_i : Y_i \to X$. This data forms the *cubical presentation* $\langle X|\{Y_i\}\rangle$.

We use these presentations to build a complex similar to the presentation 2 complex of a group presentation.

Definition 4.9. Let $CY_i = Y_i \times I/\sim$ where $(y, 0) \sim (y', 0)$ for all y and y' in Y_i. The *presentation complex* X^* is defined as $X \coprod CY_i/\sim$ where $\phi_i(y) \sim (y, 1)$ for all $y \in Y_i$ and all i.

This is a generalisation of classical presentations since if X is a rose with n petals and Y_i are cycles, then we get the same presentation 2 complex we would with the classical presentation $\langle x_1, \ldots, x_n|R\rangle$, where R is the set of words spelled out by the local isometries ϕ_i.

Example:

$\langle X|Y\rangle$ where

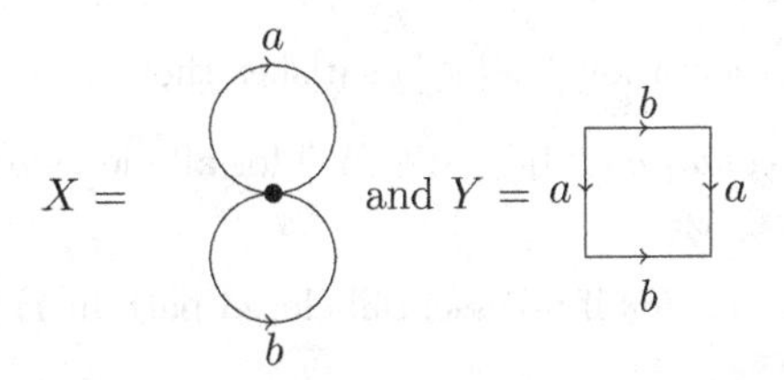

then $X^* =$

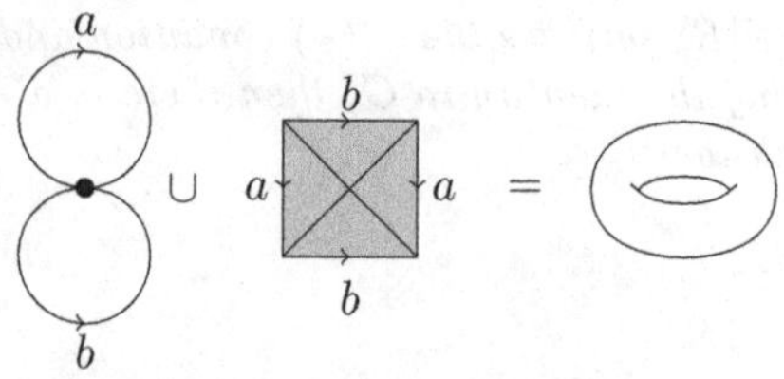

We are interested in studying the quotient $\pi_1(X)/\langle\langle\{\pi_1(Y_i)\}\rangle\rangle$ which concincides with $\pi_1(X^*)$.

We now have to define the analogue of pieces from classical small cancellation theory.

Definition 4.10. [10] An *abstract cone piece* in Y_i of Y_j is the intersection $P = \tilde{Y}_i \cap \tilde{Y}_j{}'$ of some elevations $\tilde{Y}_i, \tilde{Y}_j{}'$ of Y_i, Y_j to universal cover $\tilde{X}$ of X. In the case where $j = i$ we require the elevations are distinct in the send that for the projections $P \to Y_i, Y_j$ there is no automorphism $Y_i \to Y_j$ such that the following diagram commutes.

$$\begin{array}{ccc} P & \longrightarrow & Y_i \\ \downarrow & \swarrow & \downarrow \\ Y_j & \longrightarrow & X \end{array}$$

See Figure 11

Definition 4.11. The *carrier* $N(A)$ of a hyperplane A, in a CAT(0) cube complex, is the union of all the closed cubes which intersect the hyperplane A.

Definition 4.12. Let A be a hyperplane in $\tilde{X}$ disjoint form $\tilde{Y}_i$. An *abstract wall piece* is the intersection $\tilde{Y}_i \cap N(A)$.

Definition 4.13. An *abstract piece* is an abstract cone piece or an abstract wall piece.

A path $p : I \to Y_i$ is a *piece* in Y_i, if it lifts to $\tilde{Y}_i$ into an abstract piece in Y_i.

Definition 4.14. The *systole* $\|Y_i\|$ is the minimum combinatorial length of any essential closed path in Y_i.

We also use $|p|_{Y_i}$ to denote the minimum combinatorial length of a path homotopic to $p : I \to Y_i$ fixing $p(0)$ and $p(1)$

Definition 4.15. A cubical presentation $\langle X|\{Y_i\}\rangle$ satisfies the:

- $C'(\frac{1}{n})$ *small cancellation condition* if $|p|_{Y_i} < \frac{1}{n}\|Y_i\|$ for all pieces $p : I \to Y_i$ for all i
- $C(n)$ *small cancellation condition* if no essential closed path in Y_i is the concatenation of $< n$ pieces

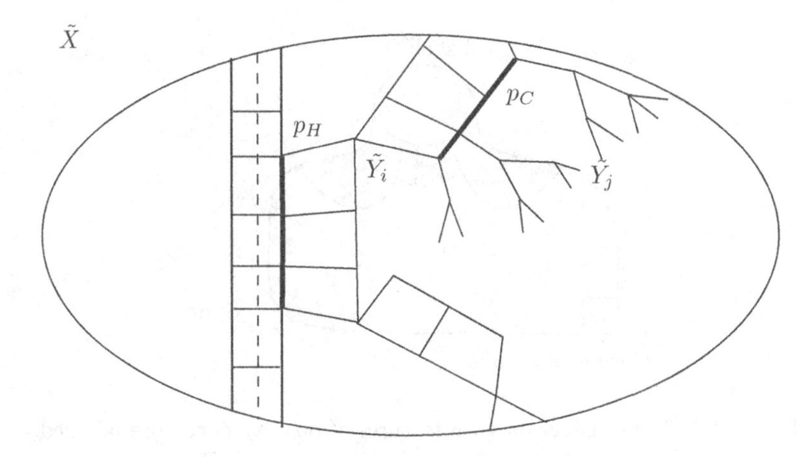

Figure 11: The two types of piece, p_H is an abstract hyperplane piece and p_C is an abstract cone piece.

Remark. $C'(\frac{1}{n}) \Rightarrow C(n+1)$

Definition 4.16. A *disc diagram* D is a combinatorial map from, a contractible, planar 2-complex $D \to X^*$ such that the boundary cycle $\partial D \cong S^1 \to X$. The 2-cells of D decompose into squares mapping to X and triangles mapping to $\text{Cone}(Y_i)$.

A *cone cell* in a disc diagram is a union of triangles such that they share a vertex mapping to a cone point in X^*.

Given a cone cell C which meets the boundary of the diagram D has a boundary path which can be decomposed into an *outer path* $C \cap \partial D$ and an *inner path* $\partial C \setminus \partial D$.

Complexity(D) = (# of cone cells, # of squares) ordered by the lexicographical ordering.

Disc diagrams are the analogue of Van Kampen diagrams in classical small cancellation theory.

Definition 4.17. The features of positive curvature in these diagrams are:

1. Corners of squares.

2. k-shells, which is a cone cell whose boundary is a concatenation of 2 paths Q, S. S is a subpath of ∂D, and Q is the inner path which can be decomposed into at most k pieces.

3. Spurs, which are edges one of whose endpoints does not lie in any other cell of the diagram.

See Figure 12 and for precise definitions see [8].

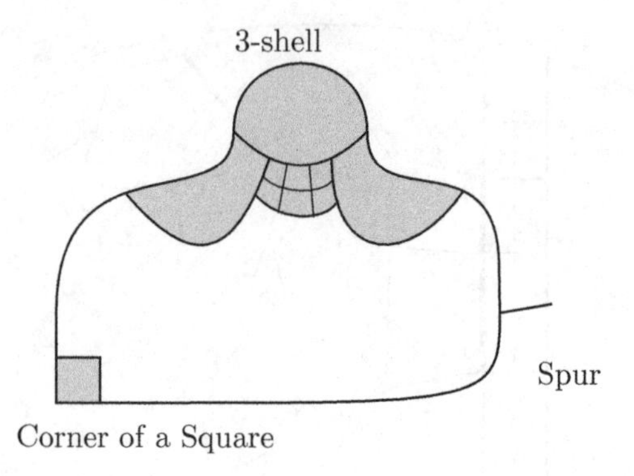

Figure 12: A Disc Diagram with features of positive curvature labelled

These are the features seen in our diagram that will provide some positive curvature on the boundary. The idea of the proofs is to redistribute this to the rest of the diagram through careful angle assignment.

Definition 4.18. A *pseudorectangle* is a square disc diagram R with $\partial R = e_1 \dots e_n f_1 \dots f_k e'_n \dots e'_1 g_l \dots g_1$ where $k, l \geq 0$ and $n > 0$ such that,

- e_i and e'_i are dual to the same hyperplane for each i,
- The hyperplane dual to e_i is disjoint from the one dual to e_j for $i \neq j$
- $e_n f_1 \dots f_k e'_n$ is a path in the carrier of the hyperplane dual to e_n and $e'_1 g_l \dots g_1 e_1$ is a path in the carrier of the hyperplane dual to e_1

The paths $e_1 \dots e_n$ and $e'_1 \dots e'_n$ are called the *sides* of the pseudorectangle.

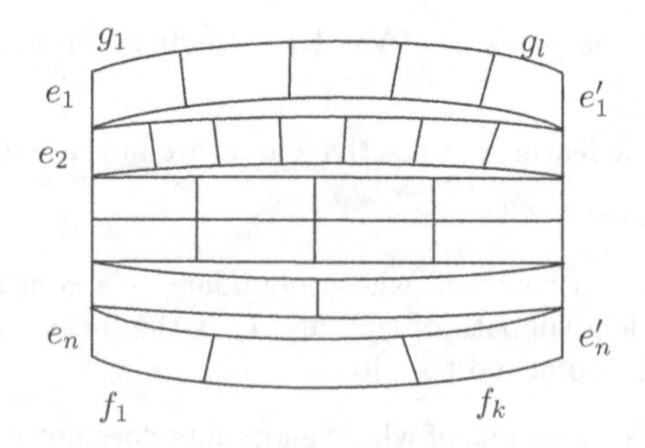

Figure 13: An example of a pseudorectangle.

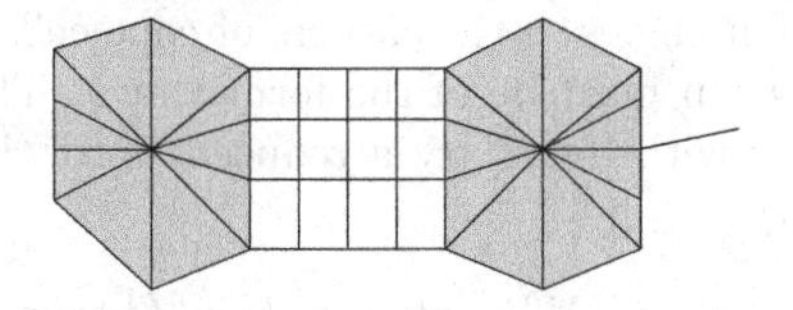

Figure 14: An example of a ladder.

Definition 4.19. A *ladder* is a disc diagram L consisting of a sequence $C_1, \ldots, C_n$ where each C_i is a cone-cell or a vertex, joined by square complexes in the following way.

- If $n = 2$ one of the following holds,
  - C_1 and C_2 are cone-cells glued together along a vertex.
  - C_1 and C_2 are joined by a single edge e which intersects each C_i in a vertex.
  - C_1 and C_2 are joined by a pseudorectangle with one side on each C_i
- If $n > 2$, then for each C_i, $i \neq 1, n$, $L - C_i$ has 2 components and each component union C_i is a ladder.

See Figure 14 for an example.

The following variant of Greendlinger's Lemma is Wise's fundamental theorem of cubical small cancellation, verified to hold under the $C(9)$ condition by Jankiewicz [8].

Theorem 4.20. *[8, 11] If $\langle X | \{Y_i\} \rangle$ satisfies the $C(9)$ small cancellation condition, then a minimal-complexity disc diagram with fixed boundary cycle is:*

- *A point or a single cone cell*

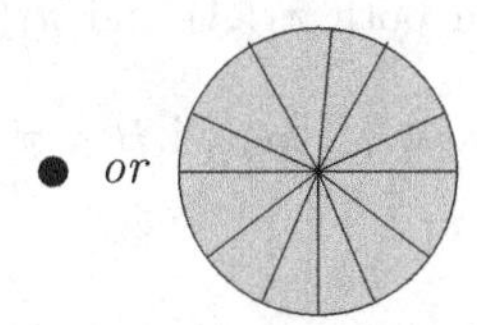

- *A ladder*
- *or else it has 3 or more features of positive curvature (k-shells $k \leq 4$, corners of squares or spurs)*

The idea is to assign angles to the corners of cone cells in such a way that we see a reduced diagram must be of the form above. This allows us to say things about diagrams with boundary a reduced word. For instance we get the following Corollary.

Corollary 4.21. *[11] If $\langle X|\{Y_i\}\rangle$ satisfies the $C'(\frac{1}{8})$ small cancellation condition and $\pi_1(X)$ is hyperbolic, then $\pi_1(X^*)$ is hyperbolic.*

Since we are interested in remaining in the category of special cube complexes, we not only want to know that $\pi_1(X^*)$ is hyperbolic, but also that it is *virtually compact special* i.e. that it has a finite index subgroup that is the fundamental group of a compact special cube complex. This leads to the following Malnormal Special Quotient Theorem due to Wise.

Definition 4.22. Let G be a hyperbolic group with fixed word metric. $H < G$ is *quasiconvex* if there is a constant K such that every geodesic in G connecting $h_1, h_2 \in H$ lies entirely in the K neighbourhood of H.

Definition 4.23. A collection of subgroups $\{H_1, \dots, H_n\}$ is a *malnormal collection*, if $g^{-1}H_ig \cap H_j = \{e\}$, unless $i = j$ and $g \in H_i$.

Theorem 4.24. *[11, 13] If G is hyperbolic virtually compact special and we have a malnormal collection of quasiconvex subgroups $\{H_1, \dots, H_n\}$, then there are finite index subgroups $\dot{H}_i < H_i$ such that for any further finite index subgroups $H_i' < \dot{H}_i$ the quotient $G/\langle\langle\{H_i'\}\rangle\rangle$ is hyperbolic and virtually compact special.*

Theorem 4.24 is an advanced corollary of Theorem 4.20 and the proof is outside the scope of the current article. We now indicate how Theorem 4.24 allows one to generalise Theorem 3.3.

Theorem 4.25. *[13] [11, Thm 13.1](see also [1]) Let A, B be compact virtually special and let $M \to A, B$ be local isometries. Let $X = A \cup_{M\times I} B$ and suppose $\pi_1(X)$ is hyperbolic. Then X is virtually special.*

Sketch of Proof. We will give a sketch that $G = \pi_1(X)$ is residually finite.

If $\pi_1(M)$ is malnormal in both $\pi_1(A)$ and $\pi_1(B)$, then we are done by Theorem 3.3.

Otherwise, given $g \in G$ we want to find $H < \pi_1(M)$ such that

- $g \notin \langle\langle H\rangle\rangle$
- H is malnormal in $\pi_1(A), \pi_1(B)$ so that the Malnormal Special Quotient Theorem 4.24 applies.
- $\pi_1(M)/\langle\langle H\rangle\rangle$ is simpler than $\pi_1(M)$, the notion of simplicity we will use is height.

Definition 4.26. The *height* of N is the maximal $n \in \mathbb{N}$ such that there are distinct cosets $g_1N, \ldots, g_nN \in G/N$ such that the intersection $g_1Ng_1^{-1} \cap \cdots \cap g_nNg_n^{-1}$ is infinite.

If N is malnormal, then it has height 1. It was proved in [5] that quasiconvex groups have finite height. We can also check that $\pi_1(M)$ is quasiconvex in $\pi_1(X)$.

Finding a subgroup H with the above properties is the main difficulty of the proof. The details of this are outside the scope of this article and will be omitted.

We will thus consider

$$\pi_1(A)/\langle\langle H\rangle\rangle *_{\pi_1(M)/\langle\langle H\rangle\rangle} \pi_1(B)/\langle\langle H\rangle\rangle$$

reducing the height of the amalgamating subgroup on this quotient. Using the Malnormal Special Quotient Theorem 4.24 the groups still satisfy the hypothesis of the theorem but we have reduced the height of the group which we are amalgamating over. We continue this process until we have height 1 and we can apply Theorem 3.3. At each stage the image of g was non trivial so in the quotient $\bar{g}$ will be non trivial so we can find a finite quotient such that $\bar{g}$ maps to a non trivial element which shows that the group is residually finite. □

Corollary 4.27. *[11], [13] The fundamental group of any closed hyperbolic 3 manifold containing a closed, geometrically finite, incompressible surface is virtually compact special.*

To prove this theorem we use the Haken hierarchy and repeatedly use theorem 4.25 used as in [13].

References

[1] Ian Agol. The virtual Haken conjecture. *Doc. Math.*, 18:1045–1087, 2013. With an appendix by Agol, Daniel Groves, and Jason Manning.

[2] M. Bestvina and M. Feighn. A combination theorem for negatively curved groups. *J. Differential Geom.*, 35(1):85–101, 1992.

[3] M. R. Bridson and A. Haefliger. *Metric spaces of non-positive curvature*, volume 319 of *Grundlehren der Mathematischen Wissenschaften [Fundamental Principles of Mathematical Sciences]*. Springer-Verlag, Berlin, 1999.

[4] Martin R. Bridson and Simon M. Salamon, editors. *Invitations to Geometry and Topology.* Oxford Graduate Texts in Mathematics 7, October 978.

[5] R. Gitik, M. Mitra, E. Rips, and M. Sageev. Widths of Subgroups. *Trans. Amer. Math. Soc.*, 350(1):321–329, 1998.

[6] F. Haglund and D. T. Wise. Special cube complexes. *Geom. Funct. Anal.*, 17(5):1551–1620, 2008.

[7] F. Haglund and D. T. Wise. A combination theorem for special cube complexes. *Ann. of Math. (2)*, 176(3):1427–1482, 2012.

[8] K. Jankiewicz. Greendlinger's Lemma in cubical small cancellation theory. *arXiv:1401.4995 [math]*, January 2014.

[9] Roger C. Lyndon and Paul E. Schupp. *Combinatorial group theory*. Classics in Mathematics. Springer-Verlag, Berlin, 2001. Reprint of the 1977 edition.

[10] P. Przytycki and D. Wise. Mixed 3-manifolds are virtually special. *arXiv:1205.6742 [math]*, May 2012. arXiv: 1205.6742.

[11] D. Wise. The Structure of Groups with a Quasiconvex Hierarchy. 2011.

[12] Daniel T. Wise. The residual finiteness of negatively curved polygons of finite groups. *Invent. Math.*, 149(3):579–617, 2002.

[13] Daniel T. Wise. *From riches to raags: 3-manifolds, right-angled Artin groups, and cubical geometry*, volume 117 of *CBMS Regional Conference Series in Mathematics*. Published for the Conference Board of the Mathematical Sciences, Washington, DC; by the American Mathematical Society, Providence, RI, 2012.

A hyperbolic group with a finitely presented subgroup that is not of type FP_3

Yash Lodha*

Abstract

Brady proved that there are hyperbolic groups with finitely presented subgroups that are not of type FP_3 (and hence not hyperbolic). We reprove Brady's theorem by presenting a new construction. Our construction uses Bestvina-Brady Morse theory, but does not involve branched coverings.

2010 Mathematics Subject Classification: 20F67
Key words and phrases: hyperbolic group, finiteness properties

1 Introduction

Hyperbolic groups were introduced by Gromov [10] as a generalization of fundamental groups of negatively curved manifolds and of finitely generated free groups. A finitely generated group is hyperbolic if its Cayley graph is hyperbolic as a metric space. Hyperbolic groups are finitely presented, and have solvable word and conjugacy problems.

The property of being hyperbolic is not inherited by subgroups. For instance finitely generated free groups have infinitely generated subgroups which cannot be hyperbolic since hyperbolic groups are finitely presentable. It is natural to then ask if the property of being hyperbolic is inherited by finitely generated subgroups. Rips constructed the first examples of finitely generated subgroups of hyperbolic groups that are not finitely presentable.[11]

Theorem 1. *(Rips) There are hyperbolic groups with subgroups that are finitely generated but not finitely presentable.*

So then one asks whether finitely presented subgroups inherit hyperbolicity. Gromov gave what appeared to be an example of a non-hyperbolic finitely presented subgroup of a hyperbolic group. But it was later discovered by Bestvina that the ambient group of this example is not hyperbolic [3]. In 1999 Noel Brady constructed the first example of a hyperbolic group with a

*This work was supported in part by NSF grant DMS-1262019

finitely presented subgroup that is not hyperbolic. In particular, he showed the following. (Theorem 6.1 in [3].)

Theorem 2. *(Brady) There exists a piecewise Euclidean cubical complex Y and a continuous map $f : Y \to S^1$ with the following properties.*

1. *The image of $\phi = f_* : \pi_1(Y) \to \pi_1(S^1)$ is of finite index in $\pi_1(S^1)$.*
2. *The universal cover $X = \widetilde{Y}$ is a* CAT(0) *metric space.*
3. *X has no embedded flat planes.*
4. *The map f lifts to a ϕ-equivariant Morse function $\tilde{f} : X \to \mathbb{R}$ whose ascending and descending links are all homeomorphic to S^2.*

The cube complex Y is constructed as a branched covering of a product of graphs. A consequence of this is the following theorem. (Theorem 1.1 in [3])

Theorem 3. *(Brady) There exists a short exact sequence of groups*

$$1 \to H \to G \to \mathbb{Z} \to 1$$

such that

1. *G is torsion free hyperbolic.*
2. *H is finitely presented.*
3. *H is* not *of type FP_3.*

In particular, H is not hyperbolic.

The group $G = \pi_1(Y)$ is the fundamental group of a nonpositively curved cube complex whose universal cover contains no flat planes. So by a theorem of Bridson, Eberlin and Gromov [4] $\widetilde{Y}$ admits a hyperbolic metric. Since G acts properly, cocompactly and by isometries on $\widetilde{Y}$, it is a hyperbolic group. The subgroup H emerges as the kernel of the induced homomorphism to $\mathbb{Z}$. Brady uses Bestvina–Brady Morse theory [2] to prove the finiteness properties of H. Brady asks if there are examples of hyperbolic groups with subgroups of type F_n but not type F_{n+1} for all $n \in \mathbb{N}$ [3].

We shall reprove the results of Rips and Brady using slightly different methods. In Section 3 we present an example of a finitely generated but not finitely presentable subgroup of a CAT(-1) group. Reading this construction will prepare the reader for the construction in Section 4. In Section 4, we shall construct a hyperbolic group with a finitely presented subgroup which is not of type FP_3. Our cube complexes will emerge as subcomplexes of products of graphs, and do not require branched coverings.

2 Preliminaries

2.1 Nonpositively curved metric spaces and groups.

Let G be a finitely generated group with a finite generating set A that is closed under inverses. Recall that the *cayley graph* $\Gamma_{G,A}$ of the group with respect to A is formed as follows: The vertices of the graph are elements of the group and for every $g \in G$ and $a \in A$, the vertices g, ga are connected by an edge. This forms a metric space by declaring that each edge is isometric to the unit interval and the metric on the graph is the induced path metric. This is a *geodesic* metric space, i.e. for any pair of points $x, y \in \Gamma_{G,A}$ there is an isometric embedding $\phi : [0, r] \to \Gamma_{G,A}$ such that $\phi(0) = x, \phi(r) = y$ and $r = d(x, y)$.

Given a geodesic metric space X, a triple $x, y, z \in X$ forms a *geodesic* triangle $\Delta(x, y, z)$ obtained by joining x, y, z pairwise by geodesics. Also let $\Delta_{\mathbb{E}^2}(x', y', z')$ be a Euclidean triangle such that the Euclidean distance between each pair $(x', y'), (x', z'), (y', z')$ is equal to the distance between the corresponding pair $(x, y), (x, z), (y, z)$. The triangle $\Delta_{\mathbb{E}^2}(x', y', z')$ is called a *comparison triangle* for $\Delta(x, y, z)$. If p is a point on the geodesic joining x, y, there is a point p' in the geodesic joining x', y' such that $d_X(x, p) = d_{\mathbb{E}^2}(x', p')$. This is called a *comparison point.*

X is said to be CAT(0) if for any geodesic triangle $\Delta(x, y, z)$ and a pair $p, q \in \Delta(x, y, z)$, the corresponding comparison points $p', q' \in \Delta_{\mathbb{E}^n}(x', y', z')$ have the property that $d_X(p, q) \leq d_{\mathbb{E}^2}(p', q')$.

Similarly we define CAT(-1) spaces by replacing Euclidean space in the above definition by a complete, simply connected, Riemannian 2-manifold of constant curvature -1. A group is said to be CAT(0) (or CAT(-1)) if it acts properly, cocompactly and by isometries on a CAT(0) (or respectively a CAT(-1)) space.

A path connected metric space X is said to be hyperbolic if there is a $\delta > 0$ such that for any geodesic triangle $\Delta(x, y, z)$ and a point p on the geodesic connecting x, y there is a point q in the union of the geodesics connecting y, z and x, z such that $d_X(p, q) < \delta$. This property is a quasi isometry invariant. A group is said to be hyperbolic if its cayley graph (with respect to any generating set) is hyperbolic as a metric space.

A comprehensive introduction and survey of CAT(0), CAT(-1) and hyperbolic spaces can be found in [5].

2.2 Cube Complexes and nonpositive curvature

By a *regular n-cube* $\square^n$ we mean a cube in $\mathbb{R}^n$ which is isometric to the cube $[0, 1]^n$ in $\mathbb{R}^n$. Informally a cube complex is a cell complex of regular Euclidean cubes glued along their faces by isometries. More formally, a cube complex is a cell complex X that satisfies the following conditions.

1. For each n-cell e in X there is an isometry $\chi_e : \square^n \to e$.

2. A map $f : \square^n \to e$ is an *admissible characteristic function* if it is χ_e precomposed with a partial isometry of $\mathbb{R}^n$. For any cell e in X the restriction of any χ_e to a face of $\square^n$ is an admissible characteristic function of a cell of C.

The metric on such cube complexes is the piecewise Euclidean metric (see [5]).

Definition 1. Given a face f of a regular cube $\square^n$, let x be the center of this face. The link $Lk(f, \square^n)$ is the set of unit tangent vectors at x that are orthogonal to f and point in $\square^n$. This is a subset of the unit sphere S^{n-1} which is homeomorphic to a simplex of dimension $n - \dim(f) - 1$. This admits a natural spherical metric, in which the dihedral angles are right angles.

Definition 2. Let f be a cell in X. Let

$$S = \{C \mid C \text{ is a cube in } X \text{ that contains } f \text{ as a face }\}.$$

The link $Lk(f, X) = \bigcup_{C \in S} Lk(f, C)$. This is a complex of spherical "all right" simplices glued along their faces by isometries. This admits a natural piecewise spherical metric.

Gromov gave the following characterizations of CAT(0) and CAT(−1) cube complexes by combinatorial conditions on the links of vertices in [10]. A nice survey of these results can be found in [7] and [8] (see Proposition I.6.8).

Definition 3. A simplicial complex is said to satisfy the *no $\square$-condition* if there are no 4-cycles in the 1-skeleton for which none of the pairs of opposite vertices of the cycle are connected by an edge. A simplicial complex Z is called a "flag" complex if any set $v_1, ..., v_n$ of vertices of Z that are pairwise connected by an edge span a simplex. This is also known as the "no empty triangles" condition.

Definition 4. A cube complex X is said to be nonpositively curved if the link of each vertex is a flag complex.

Theorem 4. *(Gromov) A cube complex X is* CAT(0) *if and only if it is non-positively curved and simply connected. Furthermore, X admits a $CAT(-1)$ metric if and only if it is* CAT(0) *and the link of each vertex satisfies the no $\square$-condition.*

The following is a characterization of CAT(0) metric spaces that are also hyperbolic.[4]

Theorem 5. *(Gromov, Eberlin, Bridson) A* CAT(0) *metric space with a cocompact group of isometries is hyperbolic if and only if it does not contain isometrically embedded flat planes.*

2.3 Topological finiteness properties of groups.

The classical finiteness properties of groups are that of being finitely generated and finitely presented. These notions were generalized by C.T.C. Wall [12]. In this paper we are concerned with the properties *type* F_n and *type* FP_n. These properties are quasi-isometry invariants of groups [1]. In order to discuss the property type F_n first we need to define Eilenberg-Maclane complexes.

Definition 5. An Eilenberg-Maclane complex for a group G, or a $K(G,1)$, is a connected CW-complex X such that $\pi_1(X) = G$ and $\widetilde{X}$ is contractible.

It is a fact that for any group G, there is an Eilenberg-Maclane complex X which is unique up to homotopy type. A group is said to be *of type* F_n if it has an Eilenberg Maclane complex with a finite n-skeleton. Clearly, a group is finitely generated if and only if it is of type F_1, and finitely presented if and only if it is of type F_2. (For more details see [9].)

Torsion-free hyperbolic groups are of type F_∞, which means that they are of type F_n for all $n \in \mathbb{N}$. This follows from the following result of Rips that appears in [11].

Theorem 6. *(Rips) Let H be a hyperbolic group. Then there exists a locally finite, simply connected, finite dimensional simplicial complex on which H acts faithfully, properly, simplicially and cocompactly. In particular, if H is torsion free, then the action is free and the quotient of this complex by H is a finite Eilenberg-Maclane complex $K(H,1)$.*

2.4 Homological finiteness properties of groups.

For a group G, consider the group ring $\mathbb{Z}G$. We view $\mathbb{Z}$ as a $\mathbb{Z}G$ module where the action of G is trivial, i.e. $g \cdot 1 = 1$ for every $g \in G$. A module is called *projective* if it is the direct summand of a free module. The group G is said to be of type FP_n if there is a projective $\mathbb{Z}G$-resolution (an exact sequence):

$$\ldots \to P_n \to P_{n-1} \to \ldots \to P_0 \to \mathbb{Z}$$

of the trivial $\mathbb{Z}G$ module $\mathbb{Z}$ such that for each $1 \leq i \leq n$ P_i is finitely generated as a $\mathbb{Z}G$ module.

A group is of type FP_1 if and only if it is finitely generated. However, there are examples of groups that are of type FP_2 but not finitely presented. If a group is of type F_n then it is of type FP_n. In general for $n > 1$ type FP_n does not imply type F_n. (There are examples due to Bestvina-Brady [2].) However, whenever a group is finitely presented and of type FP_n, it is also of type F_n. For a detailed exposition about homological finiteness properties, we refer the reader to [9].

2.5 Bestvina-Brady Morse theory

Here we shall sketch the main tool used in this paper. Bestvina–Brady Morse theory was introduced in [2] to study finiteness properties of subgroups of certain right angled Artin groups. Morse theory is defined more generally for affine cell complexes, but we shall only discuss the special case of piecewise euclidean cube complexes.

Let X be a piecewise Euclidean cube complex. Let G act freely, properly, cocompactly, cellularly and by isometries on X. Let $\mathbb{Z}$ act on $\mathbb{R}$ in the usual way. Let $\phi : G \to \mathbb{Z}$ be a homomorphism, and let $H = Ker(\phi)$. We fix these assumptions for the rest of this subsection.

Definition 6. A ϕ-equivariant Morse function is a map $f : X \to \mathbb{R}$ that satisfies,

1. For any cell F of X (with the characteristic function $\chi_F : \square^n \to F$), the composition $f \circ \chi_F$ is the restriction of a nonconstant affine map $\mathbb{R}^n \to \mathbb{R}$.

2. The image of the 0-skeleton is a discrete subset of $\mathbb{R}$.

3. f is G-equivariant, i.e for all $g \in G, x \in X$, $f(g \cdot x) = \phi(g) \cdot f(x)$.

One can think of a Morse function as a height function on X. The kernel $H = Ker(\phi)$ acts on the level sets of X, i.e. inverse images $f^{-1}(x)$ for $x \in \mathbb{R}$. Topological properties of level sets can be used to deduce the finiteness properties of H. In [2] it was shown that the topological properties of the level sets are determined by the topology of links of vertices. We make this precise below.

Definition 7. Given a vertex v of X, the ascending link $Lk^{\uparrow}(v, X)$ is defined as

$$Lk^{\uparrow}(v, X) = \bigcup\{Lk(v, F) \mid v \text{ is the minimum of } f \circ \chi_F\}$$

Similarly, the descending link $Lk^{\downarrow}(v, X)$ is defined as

$$Lk^{\downarrow}(v, X) = \bigcup\{Lk(v, F) \mid v \text{ is the maximum of } f \circ \chi_F\}$$

We now summarize the main result of Bestvina–Brady Morse theory. For the details of the proof see [2] or [3].

Theorem 7. *(Bestvina, Brady) Consider the situation described above. Then the following holds:*

1. *If for every vertex $v \in X$, $Lk^{\uparrow}(v, X)$ and $Lk^{\downarrow}(v, X)$ are simply connected, then H is finitely presented.*

2. *If for every vertex $v \in X$, and for each k with $0 \leq k \leq n-1$ or $k = n+1$, both $\widetilde{H}_k(Lk^{\uparrow}(v,X),\mathbb{Z}) = 0$ and $\widetilde{H}_k(Lk^{\downarrow}(v,X),\mathbb{Z}) = 0$ and also $\widetilde{H}_n(Lk^{\uparrow}(v,X),\mathbb{Z}) \neq 0$, $\widetilde{H}_n(Lk^{\downarrow}(v,X),\mathbb{Z}) \neq 0$, then H is of type FP_n but not of type FP_{n+1}.*

We remark that if a group is finitely presented and of type FP_n, then it is of type F_n. (See [9].)

3 A CAT(−1) example

In this section we produce a square complex X with a map $f : X \to S^1$ which lifts to an f_*-equivariant Morse function $\widetilde{f} : \widetilde{X} \to \mathbb{R}$ with the properties:

1. The link of each vertex is a finite graph with no 3 or 4 cycles,
2. The ascending and descending links of all vertices are homeomorphic to S^1.

First we define a bipartite graph Γ which will be an ingredient in both constructions.

Definition 8. Γ is the following graph:
The vertex set: $V(\Gamma) = A_+ \cup A_- \cup B_- \cup B_+$ where,

1. $A_+ = \{a_0^+, a_1^+, ..., a_{10}^+\}$.
2. $A_- = \{a_0^-, a_1^- ..., a_{10}^-\}$.
3. $B_+ = \{b_0^+, b_1^+, ..., b_{10}^+\}$.
4. $B_- = \{b_0^-, b_1^-, ..., b_{10}^-\}$.

The edge set: $E(\Gamma) = E_1 \cup E_2 \cup E_3$ where

1. E_1 consists of the edges $\{a_i^s, b_j^s\}$ for $s \in \{+,-\}$ and $i = j$ or $j = i + 1 (\mathrm{mod}\ 11)$.
2. E_2 consists of the edges $\{a_i^+, b_j^-\}$ for $j = i + 3 (\mathrm{mod}\ 11)$ or $j = i + 5 (\mathrm{mod}\ 11)$.
3. E_3 consists of the edges $\{a_i^-, b_j^+\}$ for $i = j$ or $j = i + 2 (\mathrm{mod}\ 11)$.

Lemma 8. Γ *satisfies the following:*

1. *The subgraphs spanned by the vertex sets $A_+ \cup A_-$ and $B_+ \cup B_-$ have no edges, and $A_+ \cup B_+$, $A_+ \cup B_-$, $A_- \cup B_+$, $A_- \cup B_-$ span subgraphs that are each a cycle.*

2. *There are no 3-cycles or 4-cycles.*

Proof. Γ is a bipartite graph so there are no 3-cycles. Property (1) in the statement of the lemma is easily verified. We claim that Γ has no 4-cycles. Assume there is a 4-cycle C. There are five cases to check. (We write a cycle as a set of edges.)

1. C lies in the subgraph spanned by $A_+ \cup A_- \cup B_-$.
   So C is of the form $\{\{a_i^+, b_j^-\}, \{a_k^-, b_j^-\}, \{a_k^-, b_l^-\}, \{a_i^+, b_l^-\}\}$.
2. C lies in the subgraph spanned by $A_+ \cup A_- \cup B_+$.
   So C is of the form $\{\{a_i^+, b_j^+\}, \{a_k^-, b_j^+\}, \{a_k^-, b_l^+\}, \{a_i^+, b_l^+\}\}$.
3. C lies in the subgraph spanned by $B_+ \cup B_- \cup A_-$.
   So C is of the form $\{\{a_j^-, b_i^+\}, \{a_j^-, b_k^-\}, \{a_l^-, b_k^-\}, \{a_l^-, b_i^+\}\}$.
4. C lies in the subgraph spanned by $B_+ \cup B_- \cup A_+$.
   So C is of the form $\{\{a_j^+, b_i^+\}, \{a_j^+, b_k^-\}, \{a_l^+, b_k^-\}, \{a_l^+, b_i^+\}\}$.
5. The vertices of C $\{v_1, ..., v_4\}$ satisfy
   $v_1 \in A_+, v_2 \in B_+, v_3 \in A_-, v_4 \in B_-$.
   So C is of the form $\{\{a_i^+, b_j^+\}, \{a_k^-, b_j^+\}, \{a_k^-, b_l^-\}, \{a_i^+, b_l^-\}\}$.

We treat cases (1) and (5). Cases (2), (3), (4) are similar to (1).

Case (1): Since the distinct vertices b_l^-, b_j^- share a neighbor a_k^-, we can assume without the loss of generality that $l = k$ and $j = k + 1(\text{mod } 11)$. So $|i - j| = \pm 1(\text{mod } 11)$. Now either $l = i + 3(\text{mod } 11)$ and $j = i + 5(\text{mod } 11)$, or $l = i + 5(\text{mod } 11)$ and $j = i + 3(\text{mod } 11)$. In either case we conclude that $|i - j| = \pm 2(\text{mod } 11)$. This is a contradiction. Therefore such a 4-cycle cannot exist.

Case (5): By construction we observe,

1. $j = i$ or $j = i + 1(\text{mod } 11)$.
2. $j = k$ or $j = k + 2(\text{mod } 11)$.
3. $l = k$ or $l = k + 1(\text{mod } 11)$.
4. $l = i + 3(\text{mod } 11)$ or $l = i + 5(\text{mod } 11)$.

From (1), (2), (3) above we deduce that if $|l - i| \cong n(\text{mod } 11)$, then $n \in \{-2, -1, 0, 1, 2\}$. From (4) on the other hand $|l - i| \cong 3$ or $5(\text{mod } 11)$. This is a contradiction. Therefore we conclude that there are no 4-cycles in Γ. □

We remark that in the above construction each set A_+, A_-, B_+, B_- has cardinality 11, but for any prime greater than 11 we can obtain similar constructions. However, 11 is the smallest number for which we are able to do such a construction. Now we will define our square complex X. Let Θ_1 be a graph with vertices v_1, v_2 and the edge set $E(\Theta_1) = A_- \cup A_+ \subseteq V(\Gamma)$, such that each edge meets v_1 and v_2. The A_- edges are oriented in the direction of v_2 (i.e. the arrow points to v_2) and the A_+ edges are oriented in the direction of v_1. Similarly, let Θ_2 be a graph with vertices v_3, v_4, and the edge set $E(\Theta_2) = B_- \cup B_+ \subseteq V(\Gamma)$. The B_- edges are oriented in the direction of v_4 and the B_+ edges are oriented in the direction of v_3.

Consider the 2-complex $J = \Theta_1 \times \Theta_2$. The squares of J are ordered pairs (a, b) where $a \in A_+ \cup A_-, b \in B_+ \cup B_-$. We now define a subcomplex X of J in the following manner. Let $X^{(1)} = J^{(1)}$. Glue a 2-face along the boundary of every square in $X^{(1)}$ that has the property that the corresponding pair $\{a, b\}$ is an edge in Γ. The resulting square complex is X.

Identify S^1 with $\mathbb{R}/\mathbb{Z}$. The orientations on the edges of Θ_1, Θ_2 determine maps $l_i : \Theta_i \to S^1$. Under this map each vertex maps to $[0]$, and the map on each edge is as follows: Identify the edge with the unit interval $[0, 1]$ in a way that the edge is oriented towards the vertex identified with 1. Then define the map l_i on an edge as $x \mapsto [x]$. Now identify each square C of X isometrically with the unit square $[0, 1]^2$, where the edges $\{v\} \times [0, 1]$ and $[0, 1] \times \{v\}$ are oriented towards $(v, 1)$ and $(1, v)$ respectively for each $v \in \{0, 1\}$. The map f on C is now defined as $(x, y) \mapsto [x + y]$.

The map $f_* : \pi_1(X) \to \pi_1(S^1)$ is the induced map on the fundamental groups, and $\pi_1(X)$ acts on the universal cover $\widetilde{X}$ by deck transformations. f lifts to a map $\widetilde{f} : \widetilde{X} \to \mathbb{Z}$, which is a f_*-equivariant Morse function on $\widetilde{X}$. Conditions (1), (2), (3) of the definition of a Morse function are apparent, and condition (4) follows from the definition of the lift $\widetilde{f}$.

Theorem 9. *The square complex X has the following properties.*

1. *$\widetilde{X}$ admits a* CAT(-1) *metric.*

2. $\mathrm{Ker}(f_* : \pi_1(X) \to \pi_1(S^1))$ *is finitely generated but not finitely presented.*

Proof. By construction, the link of each vertex is homeomorphic to the graph Γ. Since Γ is a flag simplicial complex with no empty-squares, (1) follows from Theorem 4. By construction the ascending and descending links of each vertex are homeomorphic to S^1. Therefore (2) follows from Theorem 7. □

4 A subgroup of type F_2 but not type F_3

We shall construct a three-dimensional cube complex Δ and a piecewise linear function $g : \Delta \to S^1$ that lifts to a g_* equivariant Morse function $\widetilde{g} : \widetilde{\Delta} \to \mathbb{R}$

satisfying the hypothesis of Theorem 7 with $n = 2$. It will be shown that $\widetilde{\Delta}$ is a hyperbolic metric space. The cube complex Δ will be constructed as a subcomplex of a product of finite graphs.

We first define graphs U, V, W, each of which is isomorphic to $K_{22,22}$, the complete bipartite graph with 22 vertices in each "part". Let the parts of U, V, W be U_1, U_2, V_1, V_2 and W_1, W_2 respectively. The vertices of the parts U_1, V_1, W_1 are $\{a_0^+, ..., a_{10}^+, a_0^-, ..., a_{10}^-\}$ and the vertices of the parts U_2, V_2, W_2 are $\{b_0^+, ..., b_{10}^+, b_0^-, ..., b_{10}^-\}$.

For each of the graphs U, V, W, we fix the following orientations on edges. Given an edge $\{a_n^s, b_m^t\}$ for $0 \leq m, n \leq 10, s, t \in \{+, -\}$, if $s = t$ then the edge is oriented toward a_n^s otherwise the edge is oriented toward b_m^t. So any vertex of U, V, W has 11 incoming and 11 outgoing edges. Declare each edge of U, V, W to be isometrically identified with the unit interval $[0, 1]$ in such a way that the edge is oriented towards the vertex identified with 1. Let $U \times V \times W$ be the product cube complex. We define a cube subcomplex Δ as follows.

Definition 9. Let (u, v, w) be a vertex in $U \times V \times W$. Then $(u, v, w) \in \Delta$ if one of the following holds. (Recall that Γ is the graph defined in section 3.)

1. For some $i \in \{1, 2\}$, $u \in U_i, v \in V_i, w \in W_i$.
2. $u \in U_1, v \in V_2$ and $\{u, v\}$ is an edge in Γ.
3. $v \in V_1, w \in W_2$ and $\{v, w\}$ is an edge in Γ.
4. $u \in U_2, w \in W_1$ and $\{u, w\}$ is an edge in Γ.

We declare a cell of $U \times V \times W$ to be in Δ if all its incident vertices are in Δ.

It follows immediately that Δ is a piecewise Euclidean cube complex. Vertices in $U \times V \times W$ that satisfy (1) above are said to be *type 1* vertices, and vertices that satisfy either (2), (3) or (4) are said to be *type 2* vertices. It is an easy exercise to show that any two vertices in Δ are connected by a path in $\Delta^{(1)}$. (Check this for type 1 vertices first.) We conclude that Δ is connected.

The graph Ω of Figure 1 serves as a tool for determining when a given vertex is in Δ. The edges of the graph encode the conditions of the definition above as follows: Given a vertex $\tau = (u, v, w)$ such that $u \in U_i, v \in V_j, w \in W_k$, either τ is a type 1 vertex (and hence is in Δ), or there is an edge connecting two of the three nodes U_i, V_j, W_k in the graph Ω. Then $\tau \in \Delta$ if and only if the corresponding pair from u, v, w forms an edge in Γ.

One observes that the map $U \times V \times W \to U \times V \times W; (u, v, w) \mapsto (w, u, v)$ and its iterates are cell permuting isometries whose restriction to Δ induces a map $\Delta \to \Delta$ that is a cell permuting isometry. These symmetries of our construction will be invoked in arguments that follow.

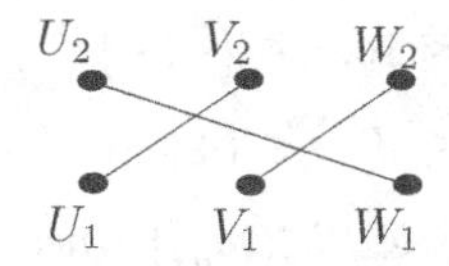

Figure 1: The graph Ω

The orientations on the edges induce a PL function $f : \Delta \to S^1$, which is $x \to [x]$ on each edge, and on the product it is defined as $(x, y, z) \to [x+y+z]$. (Recall the identification of each edge with $[0, 1]$ that was described above, and the identification of S^1 with $\mathbb{R}/\mathbb{Z}$.) The map f lifts as a PL Morse function between the universal covers $\widetilde{f} : \widetilde{\Delta} \to \mathbb{R}$. Now we prove two key lemmas, the first of which examines the links of vertices.

Lemma 10. *Let $\tau \in \Delta$ be a vertex. Then the following holds. (Here $\star$ denotes the topological join and Υ is a discrete set of four points.)*

1. *If τ is a type 1 vertex then $Lk(\tau, \Delta)$ is homeomorphic to $\Upsilon \star \Upsilon \star \Upsilon$.*

2. *If τ is a type 2 vertex then $Lk(\tau, \Delta)$ is homeomorphic to $\Gamma \star \Upsilon$.*

In particular, Δ is nonpositively curved. Furthermore, in each case $Lk^{\uparrow}(\tau, \widetilde{\Delta})$ and $Lk^{\downarrow}(\tau, \widetilde{\Delta})$ are both homeomorphic to S^2.

Proof. Let $\tau = (u, v, w)$ be a type 1 vertex. Assume that $u \in U_1$, $v \in V_1$, and $w \in W_1$. Recall that $u, v, w \in \{a_0^+, ..., a_{10}^+, a_0^-, ..., a_{10}^-\} \subseteq V(\Gamma)$. Let the neighbors of u, v, w in Γ be

$$\{b_{k_1}^+, b_{k_2}^+, b_{k_3}^-, b_{k_4}^-\}, \{b_{l_1}^+, b_{l_2}^+, b_{l_3}^-, b_{l_4}^-\}, \{b_{j_1}^+, b_{j_2}^+, b_{j_3}^-, b_{j_4}^-\}$$

respectively. The following are the 1-cells adjacent to τ in Δ:

1. $[u, b_{j_i}^+] \times v \times w$ for $1 \leq i \leq 2$ and $[u, b_{j_i}^-] \times v \times w$ for $3 \leq i \leq 4$.

2. $u \times [v, b_{k_i}^+] \times w$ for $1 \leq i \leq 2$ and $u \times [v, b_{k_i}^-] \times w$ for $3 \leq i \leq 4$.

3. $u \times v \times [w, b_{l_i}^+]$ for $1 \leq i \leq 2$ and $u \times v \times [w, b_{l_i}^-]$ for $3 \leq i \leq 4$.

It follows that $Lk(\tau, \Delta \cap (U \times v \times w))$, $Lk(\tau, \Delta \cap (u \times V \times w))$, $Lk(\tau, \Delta \cap (u \times v \times W))$ are all discrete sets of four points each. Furthermore, it follows from the definition of Δ that $[u, p] \times [v, q] \times [w, r]$ is a cube in Δ for each $p \in \{b_{j_1}^+, b_{j_2}^+, b_{j_3}^-, b_{j_4}^-\}, q \in \{b_{k_1}^+, b_{k_2}^+, b_{k_3}^-, b_{k_4}^-\}, r \in \{b_{l_1}^+, b_{l_2}^+, b_{l_3}^-, b_{l_4}^-\}$. So $Lk(\tau, \Delta)$ is the topological join of these sets. Observe that exactly two of the four 1-cells in each of (1), (2), (3) are oriented away from τ and the remaining are oriented towards τ. So it follows immediately that $Lk^{\uparrow}(\tau, \widetilde{\Delta}), Lk^{\downarrow}(\tau, \widetilde{\Delta})$ are homeomorphic to S^2. The case where $u \in U_2, v \in V_2, w \in W_2$ is similar.

Now consider the case where $\tau = (u, v, w) \in \Delta$ is a type 2 vertex. Assume that $u \in U_2, v \in V_1, w \in W_1$. The 1-cells incident to τ in Δ are:

1. $[u, a_i^s] \times v \times w$ for $0 \leq i \leq 10$, $s \in \{+, -\}$.
2. $u \times [v, b_i^s] \times w$ for $0 \leq i \leq 10$, $s \in \{+, -\}$.
3. $u \times v \times [w, p]$, for $p \in \{b_{n_1}^+, b_{n_2}^+, b_{n_3}^-, b_{n_4}^-\}$ where $\{b_{n_1}^+, b_{n_2}^+, b_{n_3}^-, b_{n_4}^-\}$ are the four neighbors of v in Γ.

Now given 1-cells $[u, a_i^s] \times v \times w$, $u \times [v, b_j^t] \times w$, observe that there is a square $[u, a_i^s] \times [v, b_j^t] \times w$ in Δ if and only if a_i^s, b_j^t are connected in Γ by an edge. This means that $Lk(\tau, \Delta \cap (U \times V \times w)) \cong \Gamma$. Furthermore, the definition of Δ implies that $[u, a_i^s] \times [v, b_j^t] \times [w, b_k^r]$ is a cube in Δ if and only if $[u, a_i^s] \times [v, b_j^t] \times w$ is a square in Δ and $u \times v \times [w, b_k^r]$ is a 1-cell in Δ. This means that $Lk(\tau, \Delta)$ is the topological join of Γ and the discrete set of four points, $Lk(\tau, \Delta \cap (u \times v \times W))$.

Now $Lk^{\uparrow}(\tau, \Delta \cap (U \times V \times w))$, $Lk^{\downarrow}(\tau, \Delta \cap (U \times V \times w))$ are both cycles, and $Lk^{\uparrow}(\tau, \Delta \cap (u \times v \times W))$, $Lk^{\downarrow}(\tau, \Delta \cap (u \times v \times W))$ are both discrete sets of two points each. This means that $Lk^{\uparrow}(\tau, \widetilde{\Delta}) \cong S^2$ and $Lk^{\downarrow}(\tau, \widetilde{\Delta}) \cong S^2$. For any other vertex τ of type 2, the analysis is similar by symmetry of the construction. □

So far we have shown the following.

1. Δ is a nonpositively curved cube complex.
2. By Lemma 10 and Theorem 7 it follows that $Ker(f_* : \pi_1(\Delta) \to \pi_1(S^1))$ is finitely presented but not of type F_3.

Now we will show that $\widetilde{\Delta}$ is a hyperbolic metric space and hence $\pi_1(\Delta)$ is a hyperbolic group. We have already established that $\widetilde{\Delta}$ is a $CAT(0)$ space, and so by Theorem 5 it suffices to show that $\widetilde{\Delta}$ does not contain isometrically embedded flat planes.

Let $\tau = (u, v, w)$ be a type 2 vertex in Δ. From the previous lemma $Lk(\tau, \Delta) \cong \Gamma \star \Upsilon$, where Υ is a discrete set of four points. So $Lk(\tau, \Delta^{(1)})$ is naturally identified with $V(\Gamma) \cup \Upsilon$.

Lemma 11. *Let τ, τ' be type 2 vertices in Δ such that $[\tau, \tau']$ is a 1-cell in Δ. The $Lk(\tau, [\tau, \tau']) \in \Upsilon$ if and only if $Lk(\tau', [\tau, \tau']) \in \Upsilon$.*

Proof. We assume that $\tau = (u, v, w), \tau' = (u', v, w)$. Let $u \in U_i, u' \in U_j$ where $\{i, j\} = \{1, 2\}$. Also, let $v \in V_k, w \in W_l$. Assume that $Lk(\tau, [\tau, \tau']) \in \Upsilon$. It follows that $Lk(\tau, \Delta \cap (U \times v \times w)) = \Upsilon$. So U_j is connected by an edge in Ω with either V_k or W_l. We will show that U_i is connected by an edge with either V_k or W_l in Ω, and hence $Lk(\tau', [\tau, \tau']) \in \Upsilon$.

Assume that this is not the case. Then since τ' is a type 2 vertex it must be the case that V_k, W_l are connected by an edge in Ω. This cannot be true since U_j is connected by an edge with either V_k or W_l. This proves our assertion. By symmetry of our construction this follows for any arbitrary type 2 vertex in Δ. □

Definition 10. A 1-cell $[\tau, \tau']$ in Δ satisfying the statement of Lemma 11 i.e., $Lk(\tau, [\tau, \tau']) \in \Upsilon$ and $Lk(\tau', [\tau, \tau']) \in \Upsilon$ is called a *special* 1*-cell.* Denote the union of all special 1-cells in Δ by L. A lift of a special 1-cell in $\widetilde{\Delta}$ is a special 1-cell in $\widetilde{\Delta}$ and $\widetilde{L}$ is the union of all special 1-cells in $\widetilde{\Delta}$.

Figure 2 depicts a cube in Δ. The three bold 1-cells are the special 1-cells, the vertices τ_1, τ_2 are the type 1 vertices and the remaining vertices are of type 2.

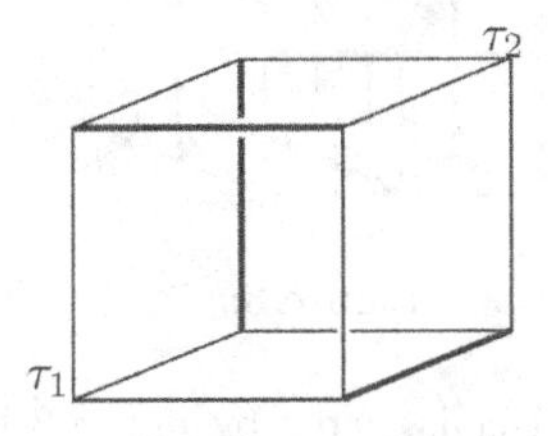

Figure 2:

Lemma 12. *$\widetilde{\Delta}$ does not contain isometrically embedded flat planes, and hence is a hyperbolic metric space.*

Proof. The proof is similar to the proof of 6.1(3) in [3], and we claim no originality here. We adapt that argument to our construction. Let us assume that there is an isometric embedding $i : \mathbb{R}^2 \to \widetilde{\Delta}$. We say that a point $x \in i(\mathbb{R}^2)$ is a *transverse intersection point* if there is a neighborhood U of x in $i(\mathbb{R}^2)$ such that the intersection of U with $\widetilde{L}$ is the point x. Following [3], our proof is divided into two steps. In *step* 1 we will show that $i(\mathbb{R}^2)$ has a transverse intersection point. In *step* 2 we will show that the angle around the transverse intersection point in $i(\mathbb{R}^2)$ is greater than 2π, contradicting the fact that this is an isometric embedding.

Step 1: Let C be a cube in $\widetilde{\Delta}$ such that $i(\mathbb{R}^2) \cap C$ is nonempty and two dimensional. (Such cubes must exist in $\widetilde{\Delta}$ since $i(\mathbb{R}^2)$ is an isometric embedding.)There are four cases to consider.

In the first case, $i(\mathbb{R}^2)$ intersects a special 1-cell e of C in a vertex p. Now $i(\mathbb{R}^2)$ must also intersect a neighboring cube C' of C that shares a 2-face with C and contains a special 1-cell e' incident to p. Then either $i(\mathbb{R}^2)$ contains e' or p is a transverse intersection point. Since $i(\mathbb{R}^2)$ is an isometric embedding, it cannot contain e' or else it would also contain e. In the second case, $i(\mathbb{R}^2)$ intersects a special 1-cell of C in an interior point, in which case it is clear that this is a transverse intersection point. In the third case, $i(\mathbb{R}^2)$ contains a special 1-cell of C. In this case it transversely intersects a different special 1-cell of C. (Recall that $i(\mathbb{R}^2) \cap C$ is two dimensional and see Figure 2).

Finally, consider the case where $i(\mathbb{R}^2)$ does not intersect a special 1-cell of C. Then $i(\mathbb{R}^2)$ intersects a 1-cell incident to a type 1 vertex τ in C. Let

J be the subcomplex of $\widetilde{\Delta}$ consisting of cubes in $\widetilde{\Delta}$ that have a nonempty intersection with $i(\mathbb{R}^2)$. Recall that $Lk(\tau, \widetilde{\Delta}) \cong \Upsilon \star \Upsilon \star \Upsilon$ where Υ is a discrete set of four points. The set of cubes incident to τ in J is a subcomplex of a stack of 8 cubes depicted in Figure 3. (C is one of the 8 cubes.) Here the

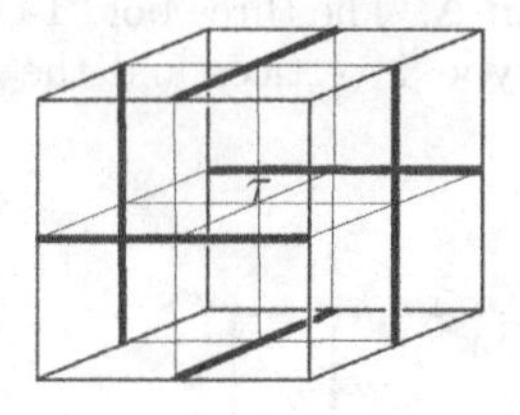

Figure 3:

bold 1-cells are the special 1-cells. Now by figure 3 it must be the case that $i(\mathbb{R}^2)$ transversely intersects a special 1-cell in a neighboring cube of C.

Step 2: Now we demonstrate a contradiction to our assumption that $i(\mathbb{R}^2)$ is an isometrically embedded flat plane in $\widetilde{\Delta}$. This involves computing the angle in $i(\mathbb{R}^2)$ around a transverse intersection point p.

Let Y be the subcomplex consisting of cubes in $\widetilde{\Delta}$ that contain p and for which $C \cap i(\mathbb{R}^2)$ is 2-dimensional. For each cube C in Y, $i^{-1}(i(\mathbb{R}^2) \cap C)$ is a polygon, and $i^{-1}(i(\mathbb{R}^2) \cap Y)$ is a union of polygons in $\mathbb{R}^2$ that are incident to $i^{-1}(p)$ such that the sum of the angles at $i^{-1}(p)$ in each polygon is 2π.

Let C, C' be cubes in Y such that $i^{-1}(i(\mathbb{R}^2) \cap C), i^{-1}(i(\mathbb{R}^2) \cap C')$ are adjacent polygons. Then $i(\mathbb{R}^2) \cap (C \cup C')$ is locally the intersection of $i(\mathbb{R}^2)$ with a Euclidean half space. This means that the angle sum of any two consecutive polygons in $i^{-1}(i(\mathbb{R}^2) \cap Y)$ is π. To establish a contradiction, it suffices to show that the number of polygons in $i^{-1}(i(\mathbb{R}^2) \cap Y)$ is greater than 4. Let $e = [\tau, \tau']$ be a special 1-cell containing p. Recall that since τ is a type 2 vertex, $Lk(\tau, \widetilde{\Delta})$ is naturally identified with $\Gamma \star \Upsilon$. Now $Lk(\tau, Y)$ is a subcomplex of $\Gamma \star \Upsilon$ and by Lemma 11 we know that $Lk(\tau, [\tau, \tau']) \in \Upsilon$. So there is a natural bijection between the aforementioned set of polygons and the set of edges of a cycle in Γ. Since all cycles in Γ have more than four edges, we have established that the angle around p in $i(\mathbb{R}^2)$ is greater than 2π contradicting the fact that this is an isometric embedding. □

5 Concluding remarks

At this point we do not have a concrete way of distinguishing the groups in Section 4 from Brady's groups in [3], other than the method of construction. (Our complexes are different from Brady's complexes since the links are different.) In fact, there is a striking similarity between the two examples, even though the methods of construction are entirely different. Nevertheless, we do believe that our approach is less abstract.

This construction does not seem to have a natural generalization in higher dimensions. It seems likely that any natural generalization in dimensions four and higher always produces flat planes in the universal cover. As a result, it is not clear whether such an approach can be used to construct hyperbolic groups with subgroups that are of type F_n but not of type F_{n+1} for $n > 2$.

References

[1] J. Alonso, 'Finiteness conditions on Groups and Quasi-isometries.', *Journal of Pure and Applied Algebra* 95 (1994) 121-129

[2] M. Bestvina and N. Brady, 'Morse theory and finiteness properties of groups', *Invent Math.* 129 (1997) 445-40.

[3] N. Brady, 'Branched coverings of cubical complexes and subgroups of Hyperbolic groups', *J. London Math. Soc.* (2) 60 (1999) 461-480

[4] M.R. Bridson, 'On the existence of flat planes in spaces of nonpositive curvature', *Proc. Amer Math. Soc.* 123 (1995) 223-235.

[5] M.R. Bridson and A. Haefliger *Metric spaces of non-positive curvature*, Springer-Verlag, Berlin, 1999.

[6] K. Brown, *Cohomology of groups*, Graduate Texts in Mathematics 87 (Springer, New York, 1982).

[7] M. Davis, 'Nonpositive curvature and reflection groups', Proceedings of the Eleventh Annual Workshop in Geometric Group Theory, Park City, Utah (1994).

[8] M. Davis, 'The Geometry and Topology of Coxeter groups', Princeton University press, 2008.

[9] R. Geoghegan, 'Topological Methods in Group Theory', Graduate Texts in Mathematics (Volume 243, Springer 2008)

[10] M. Gromov, 'Hyperbolic groups', *Essays in group theory*, Mathematical Sciences Research Institute Publications 8 (ed S.M. Gersten, Spriger, New York, 1987).

[11] E. Rips, 'Subgroups of small cancellation groups', *Bull. London Math. Soc.* 14 (1982) 45-47.

[12] C.T.C. Wall, 'Finiteness Conditions for C.W. Complexes', *The Annals of Mathematics*, 2nd Ser., Vol 81, No. 1. (Jan., 1965), 56-69.

The structure of euclidean Artin groups

Jon McCammond

Abstract

The Coxeter groups that act geometrically on euclidean space have long been classified and presentations for the irreducible ones are encoded in the well-known extended Dynkin diagrams. The corresponding Artin groups are called euclidean Artin groups and, despite what one might naively expect, most of them have remained fundamentally mysterious for more than forty years. Recently, my coauthors and I have resolved several long-standing conjectures about these groups, proving for the first time that every irreducible euclidean Artin group is a torsion-free centerless group with a decidable word problem and a finite-dimensional classifying space. This article surveys our results and the techniques we use to prove them.

2010 Mathematics Subject Classification: 20F36, 20F55
Key words and phrases: euclidean Coxeter groups, euclidean Artin groups, Garside structures, dual presentations.

The reflection groups that act geometrically on spheres and euclidean spaces are all described by presentations of an exceptionally simple form and general Coxeter groups are defined by analogy. These spherical and euclidean Coxeter groups have long been classified and their presentations are encoded in the well-known Dynkin diagrams and extended Dynkin diagrams, respectively. Artin groups are defined by modified versions of these Coxeter presentations, and they were initially introduced to describe the fundamental group of a space constructed from the complement of the hyperplanes in a complexified version of the reflection arrangement for the corresponding spherical or euclidean Coxeter group. The most basic example of a Coxeter group is the symmetric group and the corresponding Artin group is the braid group, the fundamental group of a quotient of the complement of a complex hyperplane arrangement called the braid arrangement.

The spherical Artin groups, that is the Artin groups corresponding to the Coxeter groups acting geometrically on spheres, have been well understood ever since Artin groups themselves were introduced by Pierre Deligne [Del72] and by Brieskorn and Saito [BS72] in adjacent articles in the *Inventiones* in 1972. One might have expected the euclidean Artin groups to be the next class of Artin groups whose structure was well-understood, but this was not to be. Despite the centrality of euclidean Coxeter groups in Coxeter theory and Lie theory more generally, euclidean Artin groups have remained fundamentally mysterious, with a few minor exceptions, for the past forty years.

In this survey, I describe recent significant progress in the study of these groups. In particular, my coauthors and I have succeeded in clarifying the structure of all euclidean Artin groups. We do this by showing that each of these groups is isomorphic to a subgroup of a new class of Garside groups that we believe to be of independent interest. The results discussed are contained in the following papers: "Factoring euclidean isometries" with Noel Brady [BM15], "Dual euclidean Artin groups and the failure of the lattice property" [McC15], and "Artin groups of euclidean type" with Robert Sulway [MS]. The first two are foundational in nature; the third establishes the main results. The structure of this survey follows that of the talks I gave in Durham. The first part corresponds to my first talk and the second part corresponds to my second talk.

Part I. Factoring euclidean isometries

I begin with a brief sketch of some elementary facts and known results about Coxeter groups and Artin groups in order to establish a context for our results. The discussion then shifts to a seemingly unrelated topic: the structure of the poset of all minimum length reflection factorizations of an arbitrary euclidean isometry. The connection between these two disparate topics is rather indirect and its description is postponed until the second part of the article.

1 Coxeter groups

Recall that a group is said to act *geometrically* when it acts properly discontinuously and cocompactly by isometries, and an action on euclidean space is *irreducible* if there does not exist a nontrivial orthogonal decomposition of the underlying space so that the group is a product of subgroups acting on these subspaces.

Definition 1.1 (Spherical Coxeter groups). The irreducible *spherical Coxeter groups* are those groups generated by reflections that act geometrically and irreducibly on a sphere in some euclidean space fixing its center. The classification of such groups is classical and their presentations are encoded in the well-known Dynkin diagrams. The *type* of a Dynkin diagram is its name in the Cartan-Killing classification and it is *crystallographic* or *non-crystalographic* depending on whether or not it extends to a euclidean Coxeter group. The crystallographic types consist of three infinite families (A_n, $B_n = C_n$, and D_n) and five sporadic examples (G_2, F_4, E_6, E_7, and E_8). The non-crystallographic types are H_3, H_4 and $I_2(m)$ for $m \neq 3, 4, 6$. The

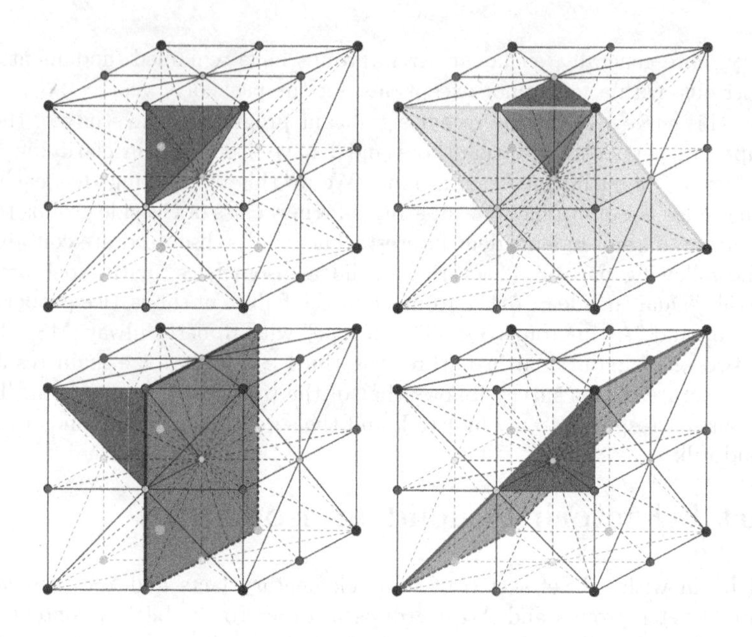

Figure 1: The spherical Coxeter group $\text{Cox}(B_3)$.

subscript is the dimension of the euclidean space containing the sphere on which it acts.

Example 1.2 (Simplices and cubes). The spherical Coxeter groups of types A and B are the best known and represent the symmetry groups of regular simplices and high-dimensional cubes, respectively. As groups they are the *symmetric groups* and extensions of symmetric groups by elementary 2-groups called *signed symmetric groups*. For example, the group $\text{Cox}(A_3) \cong \text{Sym}_4$ is the symmetric group of a regular tetrahedron and the group $\text{Cox}(B_3) \cong (\mathbb{Z}_2)^3 \rtimes \text{Sym}_3$ and is the group of symmetries of the 3-cube shown in Figure 1.

Definition 1.3 (Euclidean Coxeter groups). The irreducible *euclidean Coxeter groups* are the groups generated by reflections that act geometrically and irreducibly on euclidean space. The classification of such groups is also classical and their presentations are encoded in the extended Dynkin diagrams shown in Figure 2. There are four infinite families ($\widetilde{A}_n$, $\widetilde{B}_n$, $\widetilde{C}_n$ and $\widetilde{D}_n$) and and five sporadic examples ($\widetilde{G}_2$, $\widetilde{F}_4$, $\widetilde{E}_6$, $\widetilde{E}_7$, and $\widetilde{E}_6$). The subscript is the dimension of the euclidean space on which it acts. Removing the white dot and the attached dashed edge or edges from the extended Dynkin diagram $\widetilde{X}_n$ produces the corresponding Dynkin diagram X_n.

These extended Dynkin diagrams index many different objects including

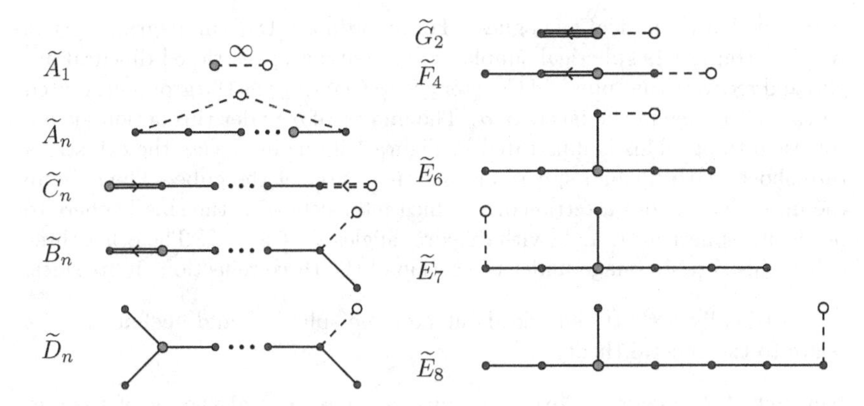

Figure 2: Four infinite families and five sporadic examples.

the Artin groups that are our primary focus, but in the present context, it is more relevant that they index euclidean simplices with restricted dihedral angles.

Definition 1.4 (Euclidean Coxeter simplices). Every extended Dynkin diagram encodes a simplex in euclidean space, unique up to rescaling, with the following properties: the vertices of the diagram are in bijection with the facets of the simplex, i.e. its codimension one faces, and vertices s and t in the diagram are connected with 0, 1, 2, or 3 edges iff the corresponding facets intersect with a dihedral angle of $\frac{\pi}{2}$, $\frac{\pi}{3}$, $\frac{\pi}{4}$, or $\frac{\pi}{6}$, respectively. These conventions are sufficient to describe the simplices associated to each diagram with one exception: the diagram $\widetilde{A}_1$ corresponds to a 1-simplex in $\mathbb{R}^1$ whose facets are its endpoints. These do not intersect and this is indicated by the infinity label on its unique edge. The extended Dynkin diagrams form a complete list of those euclidean simplices where every dihedral angle is of the form $\frac{\pi}{m}$ for some integer $m > 1$. We call these *euclidean Coxeter simplices.*

From these euclidean Coxeter simplices we can recover the corresponding euclidean Coxeter groups and an associated euclidean tiling.

Definition 1.5 (Euclidean tilings). Let $\widetilde{X}_n$ be an extended Dynkin diagram and let σ be the corresponding euclidean n-simplex described above. The group generated by the collection of $n+1$ reflections which fix some facet of σ is the corresponding euclidean Coxeter group $W = \text{Cox}(\widetilde{X}_n)$ and the images of σ under the action of W group tile euclidean n-space. As an illustration, consider the extended Dynkin diagram $\widetilde{G}_2$. It represents a euclidean triangle with dihedral angles $\frac{\pi}{3}$, $\frac{\pi}{6}$ and $\frac{\pi}{2}$ and the euclidean Coxeter group $\text{Cox}(\widetilde{G}_2)$ generated by the reflections in its sides is associated with the tiling of $\mathbb{R}^2$ by congruent 30-60-90 triangles shown in Figure 10.

Remark 1.6 (Spherical analogues). For an ordinary Dynkin diagram of type X_n, one constructs spherical simplex σ with similarly restricted dihedral angles and recovers the spherical Coxeter group $\text{COX}(X_n)$ as the group generated by the reflections in the facets of σ. The images of σ under this action yield a spherical tiling. This is illustrated in Figure 1 if one intersects the cell structure shown with a small sphere around the center of the cube. The cube in the upper left shades a tetrahedron which intersects with the small sphere to produce a spherical triangle with dihedral angles $\frac{\pi}{3}$, $\frac{\pi}{4}$ and $\frac{\pi}{2}$. The other three cubes illustrate its image under the action of the three reflections in its sides.

And finally a short remark about how the spherical and euclidean cases relate to the general theory.

Remark 1.7 (General Coxeter groups). The general theory of Coxeter groups was pioneered by Jacques Tits in the early 1960s and the spherical and euclidean Coxeter groups are key examples that motivate their introduction. Coxeter groups are defined by simple presentations and in that first unpublished paper, Tits proved that every Coxeter group has a faithful linear representation preserving a symmetric bilinear form and thus has a solvable word problem. Irreducible Coxeter groups can be coarsely classified by the signature of the symmetric bilinear forms they preserve and the irreducible spherical and euclidean groups are those that preserve positive definite and positive semi-definite forms, respectively.

2 Artin groups

As mentioned in the introduction, Artin groups first appear in print in 1972 in a pair of articles by Pierre Deligne [Del72] and by Brieskorn and Saito [BS72]. Both articles focus on spherical Artin groups as fundamental groups of spaces constructed from complements of complex hyperplane arrangements and successfully analyze their struture using different techniques. The resulting presentations resemble Artin's standard presentation for the braid groups, which is, of course, the most prominent example of a spherical Artin group. In the spherical and euclidean context these presentations are extremely easy to describe.

Definition 2.1 (Euclidean Artin groups). Let $\widetilde{X}_n$ be an extended Dynkin diagram. The standard presentation for the Artin group of type $\widetilde{X}_n$ has a generator for each vertex and at most one relation for each pair of vertices. More precisely, if s and t are vertices connected by 0, 1, 2, or 3 edges, then the presentation contains the relation $st = ts$, $sts = tst$, $stst = tsts$ or $ststst = tststs$ respectively. And finally, in the case of $\widetilde{A}_1$, the edge labeled ∞ indicates that there is no relation corresponding to this pair of vertices. As an illustration, Figure 3 shows the extended Dynkin diagram of type $\widetilde{B}_3$

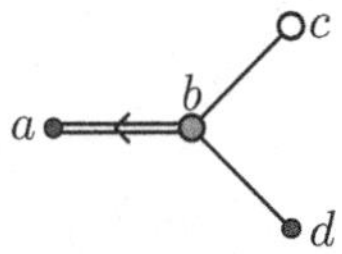

$$\text{ART}(\widetilde{B}_3) = \left\langle a, b, c, d \;\middle|\; \begin{array}{ll} abab = baba & cd = dc \\ bcb = cbc & ad = da \\ bdb = dbd & ac = ca \end{array} \right\rangle$$

Figure 3: The $\widetilde{B}_3$ diagram and the presentation for the corresponding euclidean Artin group.

along with the explicit presentation for the corresponding euclidean Artin group $\text{ART}(\widetilde{B}_3)$.

General Artin groups are defined by similarly simple presentations encoded in the same diagrams as general Coxeter groups and then coarsely classified in the same way. Given the centrality of euclidean Coxeter groups and the elegance of their structure, one might have expected euclidean Artin groups to be well understood shortly after being introduced. It is now 40 years later and these groups are still revealing their secrets.

Definition 2.2 (Four conjectures)**.** In their recent survey article, Eddy Godelle and Luis Paris highlight how little we know about general Artin groups by highlighting four basic conjectures that remain open [GP12]. Their four conjectures are:

(A) All Artin groups are torsion-free.

(B) Every non-spherical irreducible Artin group has a trivial center.

(C) Every Artin group has a solvable word problem.

(D) All Artin groups satisfy the $K(\pi, 1)$ conjecture.

Godelle and Paris also remark that these conjectures remain open and are a "challenging question" even in the case of the euclidean Artin groups. These are precisely the conjectures that my collaborators and I set out to resolve.

There are a few euclidean Artin groups with a well-understood structure. The earliest results are by Craig Squier.

Example 2.3 (planar Artin groups)**.** In his 1987 article, Squier successfully analyzes the structure of the three irreducible euclidean Artin groups $\text{ART}(\widetilde{A}_2)$, $\text{ART}(\widetilde{C}_2)$, and $\text{ART}(\widetilde{G}_2)$ that correspond to the three irreducible euclidean Coxeter groups which act geometrically on the euclidean plane [Squ87]. He works directly with the presentations and analyzes them as amalgamated products and HNN extensions of well-known groups. This technique does not appear to generalize to other euclidean Artin groups.

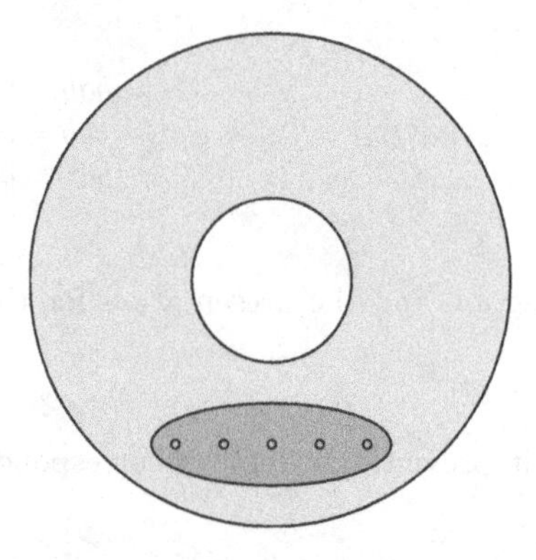

Figure 4: Five punctures in a disk inside an annulus.

A second result is the consequence of an unusual embedding of a euclidean Artin group into a spherical Artin group.

Example 2.4 (Annular braids). It has been repeatedly observed that the euclidean Artin group $\textsc{Art}(\widetilde{A}_n)$ embeds as a subgroup in the spherical Artin group $\textsc{Art}(B_{n+1})$ and is, in fact, part of a short exact sequence

$$\textsc{Art}(\widetilde{A}_n) \hookrightarrow \textsc{Art}(B_{n+1}) \twoheadrightarrow \mathbb{Z}$$

which greatly clarifies its structure [tD98, All02, KP02, CP03]. The group $\textsc{Art}(B_n)$ is sometimes called the *annular braid group* because it can be interpreted as the braid group of the annulus [Bir74]. If one selects a disk in the annulus containing all the punctures, as shown in Figure 4, then the path traced by each puncture, viewed as a path that starts and ends in the disk, has a winding number. The sum of these individual winding numbers is a global winding number for each element of $\textsc{Art}(B_{n+1})$, and this assignment of a global winding number is a group homomorphism onto $\mathbb{Z}$ with $\textsc{Art}(\widetilde{A}_n)$ as its kernel. In other words, the group $\textsc{Art}(\widetilde{A}_n)$ is the subgroup of annular braids with global winding number 0.

And finally, there are two recent results due to François Digne.

Example 2.5 (Garside structures). Digne showed that the Artin groups $\textsc{Art}(\widetilde{A}_n)$ and $\textsc{Art}(\widetilde{C}_n)$ have infinite-type Garside structures [Dig06, Dig12]. In the first article Digne uses the embedding $\textsc{Art}(\widetilde{A}_n) \hookrightarrow \textsc{Art}(B_{n+1})$ to show that the euclidean Artin groups of type $\widetilde{A}_n$ have infinite-type Garside structures and in the second he uses a delicate analysis of the some maps relating type C and type A to show that the euclidean Artin groups of type $\widetilde{C}_n$ also has an infinite-type Garside structure. Our approach to arbitrary euclidean

Artin groups is closely related to Digne's work and the second part of the article contains a more detailed description of Garside structures and their uses.

To my knowledge, these euclidean Artin groups, i.e. the ones of type $\widetilde{A}_n$, $\widetilde{C}_n$, and $\widetilde{G}_2$, are the only ones whose structure was previously fully understood. In fact, one of the main frustrations in the area is the stark contrast between the utter simplicity of the presentations involved and the fact that we typically know very little about the groups they define.

For example, all four conjectures identified by Godelle and Paris were open for the group $\textsc{Art}(\widetilde{B}_3)$ shown in Figure 3 – including a solution to its word problem – until 2010 when my Ph.D. student Robert Sulway analyzed its structure as part of his dissertation [Sul10]. As an extension of Sulway's work, he and I are now able to give positive solutions to Conjectures (A), (B) and (C) for all euclidean Artin groups and we also make some progress on Conjecture (D). We prove, in particular, that every irreducible euclidean Artin group $\textsc{Art}(\widetilde{X}_n)$ is a torsion-free centerless group with a solvable word problem and a finite-dimensional classifying space. Our proofs rely heavily on the structure of intervals in euclidean Coxeter groups and other euclidean groups generated by reflections, and so we now shift our attention to structural aspects of the set of all factorizations of a euclidean isometry into reflections.

3 Isometries

Every euclidean isometry can be built out of reflections and the Cayley graph of the euclidean isometry group with respect to this natural reflection generating set has bounded diammeter. This follows from a fact that most mathematicians learn early on in their education: every isometry of n-dimensional euclidean space is a product of at most $n+1$ reflections. The goal of the next few sections is to describe in some detail the structure of the portion of this Cayley graph between the identity and a fixed euclidean isometry. We begin with a coarse classification of euclidean isometries and their basic invariants following the approach taken in [BM15]. The first step is elementary but important for conceptual clarity: we make a sharp distinction between points and vectors.

Definition 3.1 (Points and vectors). Let V be a vector space with a simple transitive action on a set E as shown in Figure 5. We call E an *affine space*, the elements of E are called *points* and the elements of V are called *vectors*. The main difference between V and E is that E is essentially a vector space with no distinguished point identified as its origin.

Both V and E have a natural collections of subspaces which are used to defines the basic invariants of euclidean isometries.

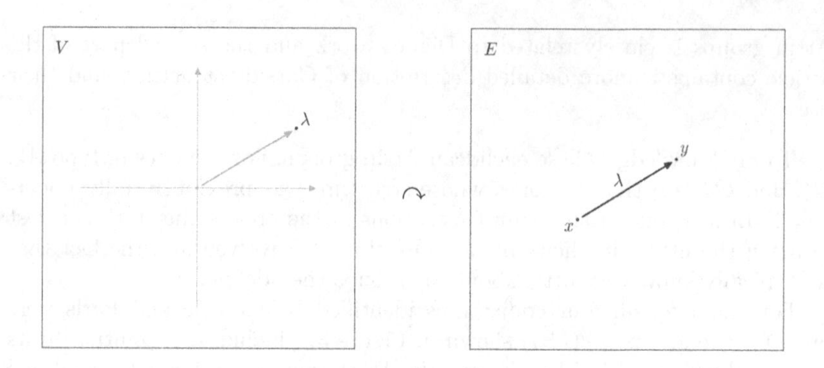

Figure 5: Vectors acting on points.

Definition 3.2 (Subspaces). A subset of V is *linear* if it is closed under linear combination. A subset of V or E is *affine* if for every pair of distinct elements in the subset, the line through these elements is also in the subset. Thus the vector space V has *linear subspaces* through the origin and other *affine subspaces* not through the origin. The affine space E only has affine subspaces. For any affine subspace $B \subset E$, vectors between points in B form a linear subspace $\text{DIR}(B) \subset V$ called its *space of directions*.

Posets are obtained by ordering these natural subspaces by inclusion.

Definition 3.3 (Poset structure). The linear subspaces of V ordered by inclusion define a poset $\text{LIN}(V)$ which is a graded, bounded, self-dual lattice. The affine subspaces of E ordered by inclusion define a poset $\text{AFF}(E)$ which is graded and bounded above, but not bounded below, not self-dual and not a lattice. Also note that there is a well-defined rank-preserving map $\text{AFF}(E) \twoheadrightarrow \text{LIN}(V)$ that sends B to $\text{DIR}(B)$.

If one equips V with a positive definite inner product, this induces a euclidean metric on E and a corresponding set of euclidean isometries that preserve this metric.

Definition 3.4 (Basic invariants). Let w be an isometry of the euclidean space E. Its *move-set* is the subset $\text{MOV}(w) \subset V$ of all the motions that its points undergo. In symbols,

$$\text{MOV}(w) = \{w(x) - x \in V \mid x \in E\}$$

and it is easy to show that $\text{MOV}(w)$ is an affine subspace of V. As an affine subspace, $\text{MOV}(w)$ is a translation of a linear subspace. If U denotes the unique linear subspace of V which differs from $\text{MOV}(w)$ by a translation and μ is the unique vector in $\text{MOV}(w)$ closest to the origin, then we call $U + \mu$ the

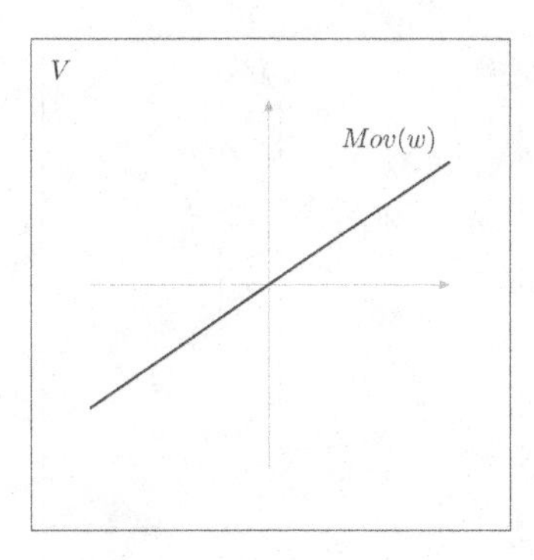

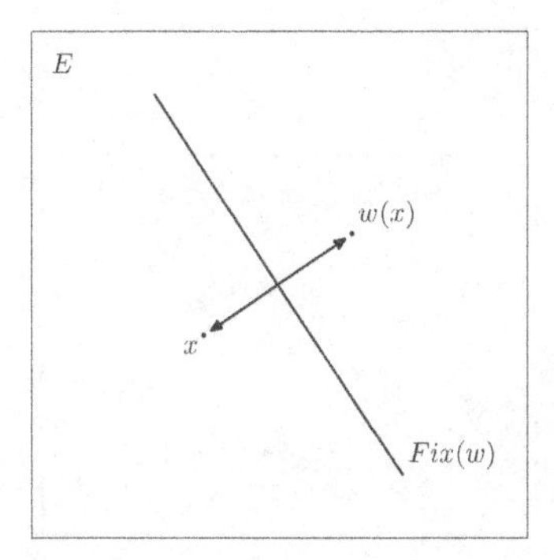

Figure 6: Basic invariants of a reflection.

standard form of $\text{MOV}(w)$. The points in E that undergo the motion μ are a subset $\text{MIN}(w) \subset E$ called the *min-set* of w and it is also easy to show that $\text{MIN}(w)$ is an affine subspace of E. We call these the *basic invariants* of w.

Euclidean isometries naturally divide into two types.

Definition 3.5 (Elliptic and hyperbolic)**.** Let w be a euclidean isometry. If its move-set $\text{MOV}(w)$ includes the origin, then μ is trivial, and its min-set $\text{MIN}(w)$ is also its fix-set $\text{FIX}(w)$. Under these equivalent conditions w is called *elliptic*. Otherwise, w is called *hyperbolic*.

The simplest euclidean isometries are reflections and translations.

Definition 3.6 (Translations)**.** For each vector $\lambda \in V$ there is a *translation* isometry t_λ whose min-set is all of E and whose move-set is the single point $\{\lambda\}$. So long as λ is nontrivial, t_λ is a hyperbolic isometry.

Definition 3.7 (Reflections)**.** For each *hyperplane* H in E (an affine subspace of codimension 1) there is a unique nontrivial isometry r fixing H called a *reflection*. It is elliptic with fix-set H and its move-set is a line through the origin in V. We call any nontrivial vector α in this line a *root* of r. The basic invariants of a typical reflection are shown in Figure 6.

A more interesting example which better illustrates these ideas is given by a glide reflection. The move-set of a glide reflection such as the one shown in Figure 7 is a non-linear affine line in V. It has a unique point μ closest to the origin and the points in E which undergo the motion μ are those on its min-set, also known as its glide axis.

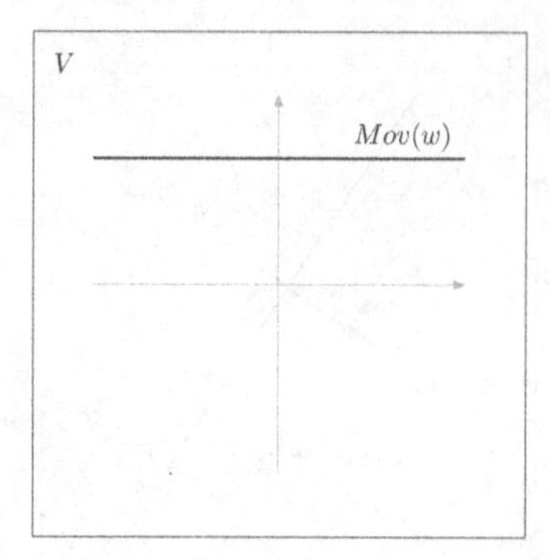

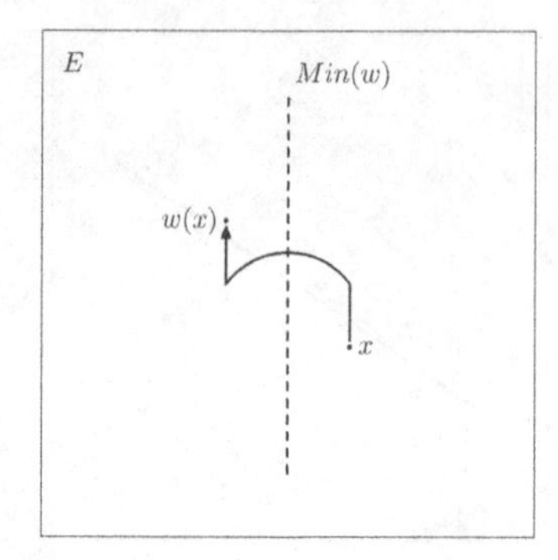

Figure 7: Basic invariants of a glide reflection.

4 Intervals

Let V be an n-dimensional vector space over $\mathbb{R}$, let E be an n-dimensional euclidean space on which it acts, and let $L = \text{ISOM}(E)$ be the Lie group of euclidean isometries of E. The structure I want to describe is the portion of the Cayley graph of L generated by its reflections between the identity element and a fixed euclidean isometry w. We begin by recalling that in any metric space there is a notion of "betweenness" which can used to construct intervals that are posets.

Definition 4.1 (Intervals in metric spaces)**.** In any metric space a point z is said to be *between* points x and y when the triangle inequality becomes an equality, that is when $d(x,z)+d(z,y) = d(x,y)$. The set of all points between x and y form the *interval* $[x,y]$ and this set can be given a partial ordering by defining $z \leq w$ if and only if $d(x,z) + d(z,w) + d(w,y) = d(x,y)$.

As an illustration, consider the unit 2-sphere with standard angle metric. If x and y are not antipodal, the only points between x and y are those along the unique geodesic connecting them with the usual ordering of an interval in $\mathbb{R}$. If, however, x and y are antipodal, say x is the south pole and y is the north pole, then all points on $\mathbf{S}^2$ are between x and y, the interval $[x,y]$ is all of $\mathbf{S}^2$ and its ordering is that $z < w$ iff they lie on a common longitude line with the latitude of z below that of w. See Figure 8. When Cayley graphs are viewed as metric spaces, they can be used to construct intervals.

Definition 4.2 (Intervals in groups)**.** Let G be a group with a fixed symmetric generating set S. If we assign the elements of S positive weights and the set of all possible weights is a discrete subset of the reals, then G can be viewed as a metric space where the distance $d(g,h)$ is calculated as the minimum total length of a path in the Cayley graph from v_g to v_h. One convention is to assign

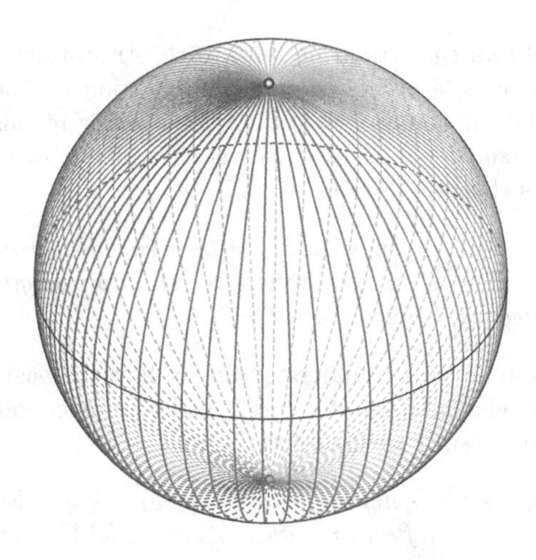

Figure 8: The interval between antipodal points on a 2-sphere.

every generator a weight of 1 and for finite generating sets the discreteness condition is always true, but for infinite generating sets with varying weights, some condition is needed so that the infimum of distances between two points is achieved by some actual path in the Cayley graph. Let $[g,h]^G$ denote the portion of the Cayley graph between g and h, by which I mean the union of all the minimum length directed paths from v_g to v_h. This is an edge-labeled directed graph which also encodes the poset structure.

Since Cayley graphs are homogeneous, $d(g,h) = d(1, g^{-1}h)$ and the interval $[g,h]^G$ is isomorphic as an edge-labeled directed graph to the interval $[1, g^{-1}h]^G$. Thus it is sufficient to restrict our attention to distances from the identity and intervals of the form $[1,g]^G$. Note that the formulas given above are for right Cayley graphs with their natural left group action; for left Cayley graphs $[g,h]^G \cong [1, hg^{-1}]^G$.

Question 4.3 (Euclidean intervals)**.** In this language our goal is to describe the poset structure of intervals in the Lie group $L = \textsc{Isom}(E)$ of all euclidean isometries generated by its full set of reflections with each reflection given unit weight. Questions one might ask include: What are the possible poset structures for these intervals $[1,w]^L$? To what extent is this poset structure independent of w? Are these posets lattices? i.e. do well-defined meets and joins always exist? A good test case for these questions is when w is a loxodromic "corkscrew" isometry of $\mathbb{R}^3$, to which we return at the end of Section 5.

In [BM15] Noel Brady and I answer these questions by completely characterizing the poset structure of all euclidean intervals. Our motivation was to create a technical tool that could be used to construct dual presentations of euclidean Artin groups, a process described in Section 7. The first step is to understand how far an isometry w is from the identity in this Cayley graph, i.e. its *reflection length* $\ell_R(w)$, and this is the content of a classical result known as Scherk's theorem [Sch50].

Theorem 4.4 (Reflection length). *Let w be a euclidean isometry with a k-dimensional move-set. If w is elliptic, its reflection length is k. If w is hyperbolic, its reflection length is $k+2$.*

From there we build up an understanding of how the basic invariants of a euclidean isometry change when it is multiplied by a reflection. The following lemma is one of the results we establish.

Lemma 4.5. *Suppose w is hyperbolic with $\ell_R(w) = k$ and $\textsc{Mov}(w) = U + \mu$ in standard form, r is a reflection with root α and let U_α denote the span of $U \cup \{\alpha\}$.*
- *If $\alpha \in U$ then rw is hyperbolic with $\ell_R(rw) = k-1$.*
- *If $\alpha \notin U$ and $\mu \in U_\alpha$ then rw is elliptic with $\ell_R(rw) = k-1$.*
- *If $\alpha \notin U$ and $\mu \notin U_\alpha$ then rw is hyperbolic with $\ell_R(rw) = k+1$.*

What I would like the reader to notice is that the geometric relationships between the basic invariants of w and the basic invariants of r combine to determine key properties of the basic invariants of rw. This type of detailed information makes it possible to prove results such as the following:

Proposition 4.6 (Elliptic intervals). *Let w be an elliptic isometry such that $\textsc{Mov}(w) = U \subset V$. The map $u \mapsto \textsc{Mov}(u)$ creates a poset isomorphism $[1,w]^L \cong \textsc{Lin}(U)$. In particular, $[1,w]^L$ is a lattice.*

Alternatively the map $u \mapsto \textsc{Fix}(u)$ gives a poset isomorphism with the affine subspaces containing $\textsc{Fix}(w)$ under reverse inclusion. Proofs of this result can be found [Sch50], [BW02b] or [BM15]. The most remarkable aspect of this proposition is that the structure of the interval $[1,w]^L$ only depends on the fact that w is elliptic and the dimension of its move-set (or equivalently the codimension of its fix-set); it is otherwise independent of w itself. In other words, the fix-set of w completely determines the order structure of the interval.

5 Models

The main new result established in [BM15] is an analysis of the structure of euclidean intervals for hyperbolic isometries. To describe these intervals, we first define an abstract poset which mimics the basic invariants of euclidean isometries.

Definition 5.1 (Global Poset). Let E be an n-dimensional euclidean space and let V be the n-dimensional vector space that acts on it. We construct a poset P called the *global poset* with two types of elements: it has an element we call h^M for each nonlinear affine subspace $M \subset V$ and an element we call e^B for each affine subspace $B \subset E$. The ordering of these elements is defined as follows:

$$\begin{array}{l} h^M \geq h^{M'} \text{ iff } M \supset M' \\ e^B \geq e^{B'} \text{ iff } B \subset B' \\ h^M > e^B \text{ iff } M^\perp \subset \textsc{Dir}(B) \\ \text{no } e^B \text{ is ever above } h^M \end{array}$$

Next, we define a map from the Lie group $L = \textsc{Isom}(E)$ to the global poset P.

Definition 5.2 (Invariant map). For each euclidean isometry w, the *invariant map* assigns an element of P based on its type and its basic invariants. More precisely, the invariant map $\textsc{inv} : L \to P$ is defined by setting $\textsc{inv}(u) = h^{\textsc{Mov}(u)}$ when u is hyperbolic and $\textsc{inv}(u) = e^{\textsc{Fix}(u)}$ when u is elliptic.

One reason to introduce the poset P and the map $\textsc{inv}$ is that there is a way to use distance from the identity to turn the Lie group L into a poset and under this ordering the invariant map is a rank-preserving poset map. It is, however, far from injective as can be seen from the fact that all rotations which fix the same subspace are sent to the same element of P. Because $\textsc{inv} : L \to P$ is a well-defined map between posets, it sends the elements below w to the elements below $\textsc{inv}(w)$. The former are the intervals $[1, w]^L$. The latter are what we called model posets.

Definition 5.3 (Model posets). For each affine subspace $B \subset E$, let P^B denote the poset of elements below e^B in the global poset P. Similarly, for each nonlinear affine subspace $M \subset V$, let P^M denote the poset of elements below h^M in global poset P. We call these our *model posets*. Finally, let $P(w)$ be the model poset of elements below $\textsc{inv}(w)$ in P.

As noted above, the invariant map sends elements in the interval $[1, w]^L$ to elements in the model poset $P(w)$. In fact, one of the main results in [BM15] is that these restrictions of the invariant map are poset isomorphisms.

Theorem 5.4 (Models for euclidean intervals). *For each isometry $w \in L$, the map $u \mapsto \textsc{inv}(u)$ is a poset isomorphism between the interval $[1, w]^L$ and the model poset $P(w)$.*

When w is elliptic, this reduces to the previously known Proposition 4.6, but when w is hyperbolic this result is new. Theorem 5.4 allows attention to shift away from the isometries themselves and to focus instead on these model posets defined purely in terms of the affine subspaces of V and E. In particular, we are able to understand the structure of euclidean intervals well enough that we can determine when meets and joins exist.

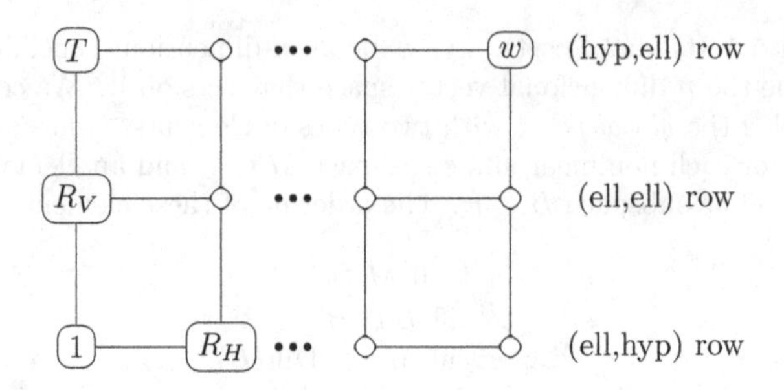

Figure 9: Coarse structure for a maximal hyperbolic isometry.

Corollary 5.5 (Lattice failure). *Let $w \in L$ be a euclidean isometry. The interval $[1,w]^L$ is not a lattice iff w is a hyperbolic isometry and its move-set has dimension at least* 2. *All other intervals are lattices.*

In [BM15] we give an explicit characterization of where these failures occur. For the application to euclidean Artin groups it is sufficient to describe these failures when w is a hyperbolic isometry of maximal reflection length. For such a w, its min-set is a line in E and its move-set is a nonlinear affine hyperplane (i.e. an affine subspace of codimension 1) in V. We call the direction of its min-set *vertical* and all of the orthogonal directions *horizontal*. More generally we call any motion with a non-trivial vertical component vertical. One consequence of Theorem 5.4 is that there is exactly one elliptic isometry in $[1,w]^L$ for each affine subspace $M \subset E$ and exactly one hyperbolic isometry for each affine subspace of $\textsc{Mov}(w) \subset V$. Using the model poset structure as a guide we coarsely partition the elements in the interval $[1,w]^L$ into a grid with three rows.

Definition 5.6 (Coarse structure). Let $w \in L$ be a hyperbolic euclidean isometry of maximal reflection length. For every $u \in [1,w]^L$ there is a unique v such that $uv = w$ and we coarsely partition the elements of $[1,w]^L$ into 3 rows based on the types of u and v and into columns based on the dimensions of their basic invariants. See Figure 9. When u or v is hyperbolic, it turns out that the other must be an elliptic isometry where every point undergoes a motion that is purely horizontal. In particular, it is an elliptic isometry whose fix-set is invariant under vertical translation, i.e translation in the direction of the line which is the min-set of w. When both u and v are elliptic, it turns out that both motions must have non-trivial vertical components and thus neither of their fix-sets is invariant under vertical translation. Within each row we grade based on the dimensions of the basic invariants. In the bottom row, the dimension of the fix-set of u decreases and the dimension of the move-set of v increases as we move from left to right. In the middle row, the dimension of the fix-set of u decreases and the dimension of the fix-set of v increases as we move from left to right. And in the top row, the dimension of the move-set of u increases and the dimension of the fix-set of v increases as we move from left to right.

The only element in the lower left-hand box is the identity element corresponding to the factorization $1 \cdot w = w$ and the only element in the upper right-hand box is the element w corresponding to the factorization $w \cdot 1 = w$. All other boxes in this grid contain infinitely many elements. Descending in the poset order involves moving to elements in boxes down and/or to the left and covering relations involve elements in boxes that are adjacent either vertically or horizontally. As a consequence, the box an element is placed in determines its reflection length: its length equals the number of steps its box is from the lower left-hand corner.

Definition 5.7 (Three special boxes)**.** There are three particular boxes in this grid that merit additional description. The elements placed in the upper left-hand corner are hyperbolic isometries in $[1,w]^L$ of reflection length 2 which means that they are produced by multiplying a pair of reflections fixing parallel hyperplanes. In other words they are pure translations t_λ, and by construction of the model poset $P(w)$, the pure translations T which occur in the interval are precisely those where the translation vector λ is a element of the non-linear affine subspace $\textsc{Mov}(w) \subset V$. The elements in the second box in the bottom row have reflection length 1, i.e. they are themselves reflections and since they are in the bottom row, they have fixed hyperplanes invariant under vertical translation. We call this set R_H the *horizontal reflections* since they move points in a horizontal direction. All such reflections occur in the interval $[1,w]^L$. And finally the first box in the middle row contains reflections whose fixed hyperplane is not invariant under vertical translation. We call this set R_V the *vertical reflections* because the motions they produce contain a nontrivial vertical component. All such reflections also occur in the interval $[1,w]^L$.

We conclude our discussion of intervals in the full euclidean isometry group by describing where in the grid meets and joins fail to exist.

Example 5.8 (Lattice failure)**.** The simplest euclidean isometry whose interval fails to be a lattice is a loxodromic "corkscrew" motion w in $\mathbb{R}^3$. This isometry has reflection length 4 and it has a coarse structure with three rows and three columns. If we consider any pair of hyperbolic isometries from the middle box of the top row whose min-sets are parallel vertically invariant planes and a pair of horizontal reflections in the middle box of the bottom row whose fixed planes are parallel to each other and to the min-sets of the chosen hyperbolic isometries, then it is straight-forward to check that these hyperbolic isometries are distinct minimal upper bounds for this pair of elliptic isometries and these elliptic isometries are distinct minimal lower bounds for these hyperbolic isometries. In [BM15] we call this situation a *bowtie*.

Part II. Crystallographic Garside groups

In this second part of the article I describe how knowing the structure of intervals in the full euclidean isometry group leads to an understanding of similar intervals inside a euclidean Coxeter group, and how these Coxeter intervals provide the technical foundation at the heart of our successful attempt to understand euclidean Artin groups using infinite-type Garside structures.

6 Coxeter elements

Let $W = \text{Cox}(\widetilde{X}_n)$ be an irreducible euclidean Coxeter group acting geometrically on an n-dimensional euclidean space E. The Coxeter group W is discrete subsgroup of the Lie group $L = \text{Isom}(E)$ and if we continue to view L as a group generated by all reflections and we view W as the subgroup generated by those reflections which occur in W, then one might naturally expect there to be a close relationship between the interval $[1, w]^W$ and the interval $[1, w]^L$ for each $w \in W$. For generic elements the connection is not as close as one might hope. In fact, even the distance to the origin might be different in the two contexts, which makes the sets of minimal length reflection factorizations completely disjoint, as Kyle Petersen and I explored in [MP11]. There is, however, a close connection when w is a Coxeter element of W.

Definition 6.1 (Coxeter elements). Let $W = \text{Cox}(\widetilde{X}_n)$ be an irreducible euclidean Coxeter group with Coxeter generating set S. A *Coxeter element* $w \in W$ is obtained by multiplying the elements of S in some order. This produces many different Coxeter elements depending on the order in which these elements are multiplied, but so long as the diagram $\widetilde{X}_n$ is a tree, as it is in all cases except for $\widetilde{A}_n$, all Coxeter elements in W belong to the same conjugacy class and act on the corresponding euclidean tiling in the exact same way [McC15, Proposition 7.5]. Thus it makes sense to talk about *the* Coxeter element in most irreducible euclidean contexts.

Coxeter elements of irreducible euclidean Coxeter groups are hyperbolic euclidean isometries whose geometric invariants play a large role in the our understanding of the structure of the interval $[1, w]^W$.

Definition 6.2 (Axial features). Let w be a Coxeter element for an irreducible euclidean Coxeter group $W = \text{Cox}(\widetilde{X}_n)$. It is a hyperbolic isometry whose reflection length is $n+1$ when measured in either W or L. In L this reflection length is the maximum possible, its min-set $\text{Min}(w)$ is a line in E called its *axis* and its move-set $\text{Mov}(w)$ is a non-linear affine hyperplane in V. The

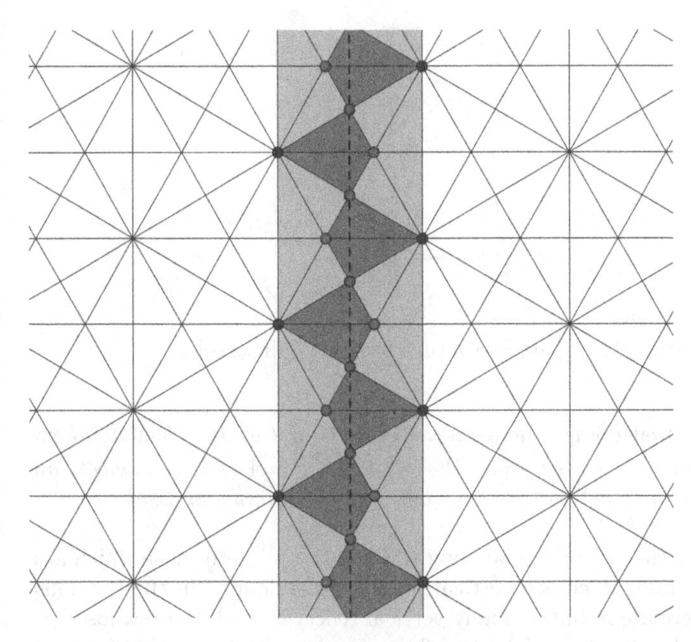

Figure 10: The euclidean Coxeter Group $\text{COX}(\widetilde{G}_2)$.

top-dimensional simplices whose interior nontrivially intersects the axis are called *axial simplices* and the vertices of these simplices are *axial vertices.*

Example 6.3 ($\text{COX}(\widetilde{G}_2)$)**.** Figure 10 illustrates these ideas for the Coxeter group $\text{COX}(\widetilde{G}_2)$. Its Coxeter element is a glide reflection whose glide axis, i.e. its min-set, is shown as a dashed line. The corresponding axial simplices are heavily shaded and their axial vertices are shown as enlarged dots.

For Coxeter elements, the interval $[1,w]^W$ is a restriction of edge-labeled poset $[1,w]^L$ to the union of minimum length paths in the Cayley graph of L from v_1 to v_w where every edge is labeled by a reflection in $W = \text{COX}(\widetilde{X}_n)$. The original product of elements in the Coxeter generating set S which produces w as a Coxeter element is one such minimal length path. As such, the elements of this *Coxeter interval* $[1,w]^W$ have a coarse structure as described in Definition 5.6. The first difference we find is that whereas every reflection in L labels some edge in the interval $[1,w]^L$, in the Coxeter interval $[1,w]^W$ only a proper subset of the reflections in W actually label edges in the interval. The unused reflections are those that do not occur in a minimum length factorization of w where every reflection must belong to W. In [McC15, Theorem 9.6] the reflections that do occur as edge labels in the interval are precisely characterized as follows.

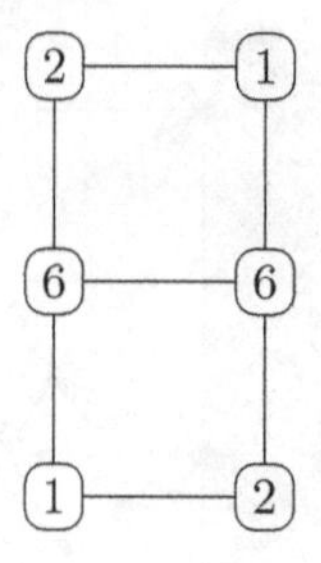

Figure 11: Coarse structure for the $\widetilde{G}_2$ interval.

Theorem 6.4 (Reflection generators). *Let w be a Coxeter element of an irreducible euclidean Coxeter group $W = \textsc{Cox}(\widetilde{X}_n)$. A reflection labels an edge in the interval $[1, w]^W$ iff its fixed hyperplane contains an axial vertex.*

If we separate the reflections labeling edges in $[1, w]^W$ into those which are horizontal and those which are vertical, in the sense defined in the previous section, then there are infinitely many vertical reflections (all those whose hyperplanes cross the Coxeter axis) and a finite number of horizontal reflections (those whose hyperplanes bound the convex hull of the axial vertices). More generally, the coarse structure of the interval $[1, w]^W$ will have only finitely many elements in each box along the top row and finitely many elements in each box along the bottom row. The boxes in the middle row, on the other hand, have infinitely many elements in each. Nevertheless, there is a periodicity to the convex hull of the axial simplices and this means that the infinitely many elements in each box in the middle row falls into a finite number of infinitely repeating patterns. We illustrate this with the $\widetilde{G}_2$ Coxeter group where one can view the entire euclidean tiling.

Example 6.5 (Coarse structure of the $\widetilde{G}_2$ interval). As can be seen in Figure 10, the euclidean Coxeter group $W = \textsc{Cox}(\widetilde{G}_2)$ has exactly 2 horizontal reflections, the ones with vertical fixed lines which bound the lightly shaded region and this is indicated by the 2 in the second box in the bottom row of the coarse structure for the interval $[1, w]^W$ shown schematically in Figure 11. On the other hand, there are 6 essentially different ways that a fixed line can cross the glide axis and the corresponding 6 infinite families of vertical reflections below w are indicated by the 6 in the first box of the middle row. Similarly, there are 6 infinite families of rotations fixing an axial vertex represented by the 6 in the second box of the middle row and exactly two pure translations below w indicated by the 2 in the first box of the top row. Finally, although it is not immmediately obvious, it turns out that this bounded, graded, self-dual poset with finite height and an infinite number of elements is a lattice.

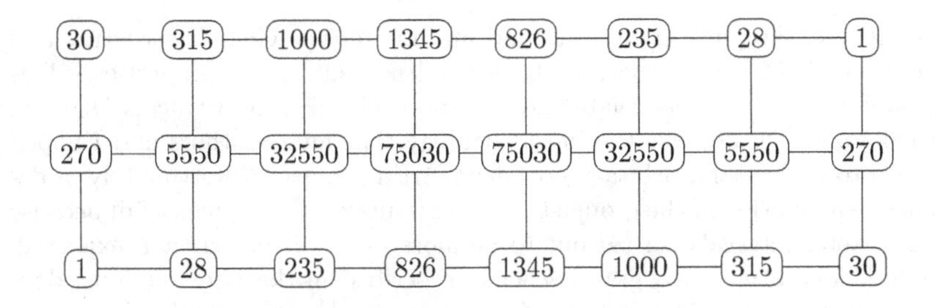

Figure 12: Coarse structure for the $\widetilde{E}_8$ interval.

For a more interesting example, consider the largest sporadic irreducible euclidean Coxeter group $\text{Cox}(\widetilde{E}_8)$.

Example 6.6 (Coarse structure of the $\widetilde{E}_8$ interval)**.** Let w be a Coxeter element for the Coxeter group $W = \text{Cox}(\widetilde{E}_8)$. The coarse structure of the interval $[1, w]^W$ is shown schematically in Figure 12. As in the case of $\text{Cox}(\widetilde{G}_2)$, the numbers listed in the top row and in bottom row indicate the actual number of elements in each box, but the numbers in the middle row only indicate the number of infinite families of such elements. The equivalence relation used is that two middle row elements below w belong to the same family iff they differ by (conjugation by) a translation of the tiling in the direction of the Coxeter axis. Thus, from the coarse structure we see that the interval $[1, w]^W$ has 28 horizontal reflections (second box in the bottom row), 30 translations (first box in the top row), 270 infinite families of vertical reflections (first box in the middle row) and 5550 infinite families of elements that rotation around a 6-dimensional fix-set that is not invariant under vertical translation (second box in the middle row), and so on. Representatives of each family were computed using a program I wrote called `euclid.sage` that is available from my webpage. The software also checks whether this bounded, graded, self-dual poset of finite height with an infinite number of elements is a lattice and in this case the answer is "No".

7 Dual Artin groups

Now that the technical foundations are in place, it is time to shift our attention to the irreducible euclidean Artin groups themselves, to finally explain why intervals in euclidean Coxeter groups are relevant and why we are interested in whether or not these intervals are lattices. The answers are relatively straightforward. First, intervals in irreducible euclidean Coxeter groups can be used to give alternative, so called "dual" presentations for irreducible euclidean Artin groups as I show in [McC15]. Next, when these Coxeter intervals are lattices, the dual Artin group has an infinite-type Garside structure and groups with Garside structures have good computational properties. This "grand scheme", closely related to the approach taken by François Digne in [Dig06] and [Dig12], was the initial strategy that my coauthors and I hoped to use to understand arbitary euclidean Artin groups. Unfortunately, a detailed examination of the groups themselves caused this scheme to fail because the Coxeter intervals turned out to be more poorly behaved than expected. Nonetheless, a modified grand scheme, described in the later sections, does eventually succeed. The first step is to understand how intervals lead to presentations of new groups.

Definition 7.1 (Interval groups)**.** Let G be a group with a fixed symmetric discretely weighted generating set and let $[1,g]^G$ be an interval in G, viewed as an edge-labeled directed graph sitting inside the Cayley graph of G. The *interval group* G_g is a new group generated by the labels of edges in the interval subject to the set of all relations that are visible in the interval. Since there is a natural function from the generators of G_g to G and since the set of relations used to define G_g is a subset of the relations which hold in G, this function extends to a group homomorphism $G_g \to G$. If, moreover, the labels on the edges in the interval $[1,g]^G$ include a generating set for G then this natural map is onto.

To see how this works in practice, consider the following example.

Example 7.2 (Noncrossing partitions)**.** Let $G = \text{SYM}_n$ be the symmetric group on n elements, let g be the n-cycle $(1,2,\ldots,n)$ and fix the full set of tranpositions as its generating set. It turns out in this case that the poset structure of the interval $[1,g]^G$ is a well-known combinatorial object called the *noncrossing partition lattice* defined as follows. Start with a convex regular n-gon in the plane whose vertices are labeled 1 through n in a clockwise fashion. A partitioning of its vertex set $[n] = \{1,2,\ldots,n\}$ is called *noncrossing* if the convex hulls of the blocks of the partition are pairwise disjoint. For example, the partition $\{\{1,3\},\{2,4\}\}$ of $[4]$ is not a noncrossing partition because the convex hulls are two line segments that intersect. Noncrossing partitions can be ordered by declaring $\sigma < \tau$ when every block of σ is a subset of a block of τ and a noncrossing partition can be converted into a permutation by clockwise permuting the vertices in the boundary of the convex hull of each block. This function defines a poset isomorphism between the noncrossing partition lattice NC_n and the Coxeter interval $[1,g]^G$. See [McC06] for an elementary discussion of these ideas. When $n = 3$, there are only 3 transpositions, the noncrossing parititon lattice is particularly simple and the presentation for the corresponding interval group G_g is $\langle a,b,c \mid ab = bc = ca\rangle$, which is an alternate presentation for the 3-string braid group.

This example, and its generalization to all spherical Coxeter groups described below, leads to the following general definition.

Definition 7.3 (Dual Artin groups). Let $W = \text{Cox}(\Gamma)$ be an arbitrary Coxeter group viewed as a group generated by its full set of reflections and let w be one of its Coxter elements. The interval group W_w defined by the interval $[1,w]^W$ is called a *dual Artin group* and denoted $\text{Art}^*(\Gamma, w)$. The notation is meant to highlight the fact that in general Coxeter groups there are geometrically distinct Coxeter elements and thus there are distinct dual presentations which heavily depend on the choice of Coxeter element.

The study of dual presentations in general is motivated by the work of Davis Besis [Bes03] and Tom Brady and Colum Watt [BW02a] on spherical Artin groups. Here are their main results translated into this terminology.

Theorem 7.4 (Dual spherical Artin groups). *Let $W = \text{Cox}(X_n)$ be a spherical Coxeter group generated by its reflections, and let w be a Coxeter element of W. In this case the Coxeter interval $[1,w]^W$ is isomorphic to the W-noncrossing partition lattice and the interval group $W_w = \text{Art}^*(X_n, w)$ is isomorphic to the corresponding Artin group $\text{Art}(X_n)$.*

The terminology "dual Artin group" was introduced in [McC15] because in general it is not known whether or not Artin groups and dual Artin groups are isomorphic. Fortunately, I was able to establish that they are isomorphic in the euclidean case [McC15].

Theorem 7.5 (Dual euclidean Artin groups). *If $W = \text{Cox}(\widetilde{X}_n)$ is an irreducible euclidean Coxeter group generated by its reflections, and w is a Coxeter element, then the dual Artin group $W_w = \text{Art}^*(\widetilde{X}_n, w)$ is naturally isomorphic to $\text{Art}(\widetilde{X}_n)$.*

The proof uses a result from quiver representation theory to greatly simplify the dual presentations for dual euclidean Artin groups and it is then relatively straight-forward to establish that these simplified presentations define the same groups as do the standard euclidean Artin presentations [McC15]. In other words, the interval $[1,w]^W$ gives a new presentation (with infinitely many generators and infinitely many relations) of the corresponding Artin group in the irreducible euclidean case. The interest in whether or not these Coxeter intervals are lattices has to do with the following result essentially due to Jean Michel.

Theorem 7.6 (Sufficient conditions). *Let G be a group with a fixed symmetric discretely weighted generating set closed under conjugation. For each $g \in G$, if the interval $[1,g]^G$ is a lattice, then G_g is a Garside group, possibly of infinite-type.*

In this article, Garside structures are treated as a black box. For a more detailed discussion see [Bes03], [McC15], and particularly the book [DDG+]. Interval groups appear in [DDG+] in Chapter VI as the "germ derived from a groupoid". The main idea is that if there is a portion of the Cayley graph which contains a generating set and has well-defined meets and joins (plus a few more technical conditions) then this local lattice structure can be used to systematically construct normal forms for all group elements, thereby solving the word problem for the group and allowing one to construct a finite dimensional classifying space.

Theorem 7.7 (Garside consequences). *If an interval group G_g is a Garside group, possibly of infinite-type, then G_g is a torsion-free group with a solvable word problem and a finite dimensional classifying space.*

Remark 7.8 (Artin groups with Garside structures). By 2000 or so, many dual Artin groups were known to have Garside structures. We have already mentioned that the dual Artin group $\text{ART}(\Gamma)$ is Garside when $\text{COX}(\Gamma)$ is spherical (due to Bessis [Bes03] and Brady-Watt [BW02a] independently) and when Γ is an extended Dynkin diagram of type $\widetilde{A}_n$ or $\widetilde{C}_n$ (due to Digne [Dig06, Dig12]). David Bessis also proved this for the free group, thought of as the Artin group where every $m_{ij} = \infty$ [Bes06]. In addition, there are unpublished results due to myself and John Crisp which show this to be true for all Artin groups with at most 3 standard generators and also for all Artin groups defined by a diagram in which every m_{ij} is at least 6. Note that the 3 standard generator result means that the three planar Artin groups investigated by Craig Squier in [Squ87] have dual presentations which are Garside structures (as do all of the 3-generator Artin groups whose Coxeter groups naturally act on the hyperbolic plane). In particular, $\text{ART}(\widetilde{G}_2)$ has an infinite-type Garside structure.

This list of results naturally lead one to conjecture that Coxeter intervals using Coxeter elements are always lattices and that all the corresponding dual Artin groups are Garside groups. Unfortunately, as we have already seen in Example 6.6, this natural conjecture turns out to be too optimistic and false even for some of the irreducible euclidean Coxeter groups such as the group $W = \text{COX}(\widetilde{E}_8)$.

8 Horizontal roots

Several years ago John Crisp and I systematically investigated whether every irreducible euclidean Coxeter group has a Coxeter interval which is a lattice, and what we found was not what we expected to find. The only irreducible euclidean Coxeter groups whose Coxeter intervals are lattices are those of type $\widetilde{A}_n$, $\widetilde{C}_n$, and $\widetilde{G}_2$. In other words, the only dual Artin groups with Garside structures are those where the group structure was already well understood by Craig Squier and/or François Digne. Further investigation revealed that the reason why a euclidean Coxeter group might have a Coxeter interval which

Type	Horizontal root system
A_n	$\Phi_{A_{p-1}} \cup \Phi_{A_{q-1}}$
C_n	$\Phi_{A_{n-1}}$
B_n	$\Phi_{A_1} \cup \Phi_{A_{n-2}}$
D_n	$\Phi_{A_1} \cup \Phi_{A_1} \cup \Phi_{A_{n-3}}$
G_2	Φ_{A_1}
F_4	$\Phi_{A_1} \cup \Phi_{A_2}$
E_6	$\Phi_{A_1} \cup \Phi_{A_2} \cup \Phi_{A_2}$
E_7	$\Phi_{A_1} \cup \Phi_{A_2} \cup \Phi_{A_3}$
E_8	$\Phi_{A_1} \cup \Phi_{A_2} \cup \Phi_{A_4}$

Table 1: Horizontal root systems by type.

failed to be a lattice is closely related to the structure of what we call its horizontal root system.

Definition 8.1 (Horizontal root system). Let $W = \text{Cox}(\widetilde{X}_n)$ be an irreducible euclidean Coxeter group and let w be a Coxeter element of W. The horizontal reflections in the interval $[1, w]^W$ are those whose roots are orthogonal to the direction of the Coxeter axis of w. The set of all such roots form a subroot system of the full root system of W that we call the *horizontal root system.*

It turns out that the horizontal root system is easy to describe as a subdiagram of the original extended Dynkin diagram.

Remark 8.2 (Finding horizontal roots). Let $W = \text{Cox}(\widetilde{X}_n)$ be an irreducible euclidean Coxeter group with Coxeter element w. The horizontal root system with respect to w is itself a root system for a spherical Coxeter group whose Dynkin diagram is obtained by removing two dots from the extended Dynkin diagram $\widetilde{X}_n$ or one dot from Dynkin diagram X_n. In Figure 2 the dots to be removed are slightly larger than the others. Removing the large white dot produces the Dynkin diagram of type X_n. Also removing the large shaded dot produces the diagram for the horizontal root system. In all cases, the shaded dot is the long end of a multiple bond or the branch point if either exists in X_n. The only case where neither exists is in type $\widetilde{A}_n$, where the shaded dot might be any of the remaining dots and different choices arise from the geometrically different choices for the Coxeter element in this case.

The types of the horizontal root systems are listed in Table 1. The key property turns out ot be whether or not the remaining diagram is connected, or equivalently, whether or not the horizontal root system is reducible. The horizontal root systems for types C and G are irreducible, the horizontal root systems for types B, D, E and F are reducible and for type A it depends on the choice of Coxeter element. The following theorem is a restatement of and explanation for the computational result originally discovered in collaboration with John Crisp. It is proved in [McC15], an article which is, morally speaking, the result of a collaboration with John Crisp even though it was only written up after he left research mathematics.

Theorem 8.3 (Failure of the lattice property). *The interval $[1, w]^W$ is a lattice iff the horizontal root system is irreducible. In particular, types C and G are lattices, types B, D, E and F are not, and for type A the answer depends on the choice of Coxeter element.*

As a consequence of this theorem, it is clear which irreducible euclidean Artin groups have dual presentations that lead to Garside structures.

Corollary 8.4. *The dual euclidean Artin group $\mathrm{ART}^*(\widetilde{X}_n, w)$ is Garside when X is C or G and it is not Garside when X is B, D, E or F. When the group has type A there are distinct dual presentations and the one investigated by Digne is the only one that is Garside.*

At this point, the grand scheme has failed and no additional euclidean Artin groups have been understood. There is, however, a ray of hope. The explicit nature of the euclidean model posets enables an explicit examination of the pairs of elements which fail to have well-defined meets or well-defined joins. It turns out that pairs of elements with no well-defined join must occur in the bottom row of the coarse structure and pairs of elements with no well-defined meet must occur in the top row of the coarse structure. Since the top and bottom rows only contain finitely many elements, this means that out of the infinitely many pairs of elements in the infinite interval $[1, w]^W$, only finitely pairs fail to be well-behaved. This leaves open the possibility that one can systematically fix these finitely many failures.

9 New groups

Let $W = \mathrm{COX}(\widetilde{X}_n)$ be an irreducible euclidean Coxeter group with Coxeter element w. As remarked above, the finitely pairs of elements in the Coxeter interval $[1, w]^W$ which cause the lattice property to fail all occur in the top and bottom rows of its coarse structure. Thus it makes sense to focus on the subgroup corresponding to this portion of the interval. Its structure is closely related to an elementary group which does not appear to have a standard name in the literature. In fact, we have not found any references to it in the literature so far.

Definition 9.1 (Middle groups). Consider the group of isometries of $\mathbb{Z}^n$ in $\mathbb{R}^n$ generated by coordinate permutations and integral translations. We call

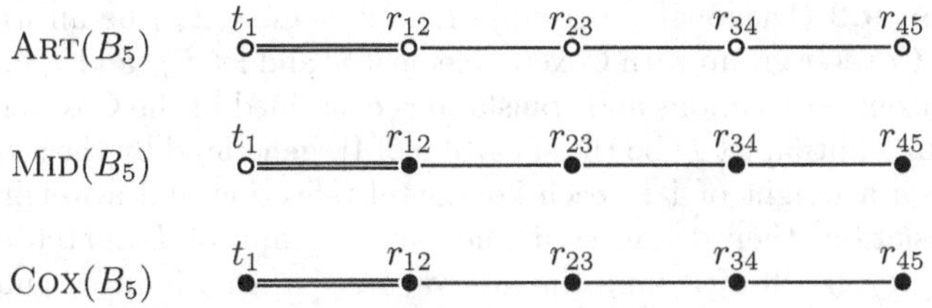

$\text{ART}(B_5)$: t_1, r_{12}, r_{23}, r_{34}, r_{45}

$\text{MID}(B_5)$: t_1, r_{12}, r_{23}, r_{34}, r_{45}

$\text{COX}(B_5)$: t_1, r_{12}, r_{23}, r_{34}, r_{45}

Figure 13: Presentation diagrams for the Artin, middle and Coxeter groups of type B_5.

$$\begin{array}{ccccc}
\text{ART}(\widetilde{A}_{n-1}) & \hookrightarrow & \text{ART}(B_n) & \twoheadrightarrow & \mathbb{Z} \\
\downarrow & & \downarrow & & \downarrow \\
\text{COX}(\widetilde{A}_{n-1}) & \hookrightarrow & \text{MID}(B_n) & \twoheadrightarrow & \mathbb{Z} \\
 & & \downarrow & & \\
 & & \text{COX}(B_n) & &
\end{array}$$

Figure 14: Relatives of middle groups.

this group the *middle group* and denote it $\text{MID}(B_n)$. It is generated by the reflections r_{ij} that switch coordinates i and j and the translations t_i that adds 1 to the i-th coordinate and it is a semidirect product $\mathbb{Z}^n \rtimes \text{SYM}_n$ with the translations t_i generating the normal free abelian subgroup. A standard minimal generating set for $\text{MID}(B_n)$ is the set $\{t_1\} \cup \{r_{12}, r_{23}, \ldots, r_{n-1n}\}$ and it has a presentation similar to $\text{ART}(B_n)$ and $\text{COX}(B_n)$ [MS]. See Figure 13. A solid dot means the corresponding generator has order 2 and an empty dot means the corresponding generator has infinite order.

If we consider $\text{MID}(B_n)$ as a group generated by the full set of translations and reflections, then the factorizations of $w = t_1 r_{12} r_{23} \cdots r_{n-1n}$ form an interval isomorphic to the type B noncrossing partition lattice, exactly the same poset as the Coxeter interval in the spherical Coxeter group $\text{COX}(B_n)$. This explains the use of B_n in the notation. The name "middle group" is suggested by its location in the center of a diagram that shows its relation to several closely related Coxeter groups and Artin groups. See Figure 14. The top row is the short exact sequence that is often used to understand $\text{ART}(\widetilde{A}_{n-1})$. Geometrically middle groups are easy to recognize as a symmetric group generated by reflections and a translation with a component out of this subspace. Middle groups are introduced in [MS] in order to succinctly describe the structure of the diagonal subgroup build from the top and bottom rows of the coarse structure of a euclidean Coxeter interval.

Definition 9.2 (Diagonal subgroup). Let $W = \text{Cox}(\widetilde{X}_n)$ be an irreducible euclidean Coxeter group with Coxeter element w and let R_H and T denote the set of horizontal reflections and translations contained in the Coxeter interval $[1,w]^W$. In addition let D be the subgroup of W generated by the set $R_H \cup T$. If we assign a weight of 1 to each horizontal reflection and a weight of 2 to each translation, then distances in the Cayley graph of D match distances in the Cayley graph of W and the interval $[1,w]^D$ is an induced subposet of the Coxeter interval $[1,w]^W$ consisting of only the top and bottom rows. We write D_w to denote the interval group defined by this restricted interval.

The interal $[1,w]^D$ is almost a direct product of posets and the group D is almost a direct product of middle groups. More precisely, the poset $[1,w]^D$ is almost a direct product of type B noncrossing partitions lattices and the missing elements are added if we factor the translations of D.

Definition 9.3 (Factored translations). Each pure translation t in $[1,w]^D$ projects nontrivially onto the Coxeter axis and onto each of the irreducible components of the horizontal root system of the corresponding Coxeter group $W = \text{Cox}(\widetilde{X}_n)$. Let $t^{(i)}$ be the translation which agrees with t on the i-th component and contains $1/k$ of the translation in the Coxeter direction where k is the number of irreducible components of the horizontal root system. We call each $t^{(i)}$ a *factored translation* and let T_F denote the set of all such factored translations.

We use the factored translations to introduce a slightly larger group.

Definition 9.4 (Factorable groups). The *factorable group* F is defined as the group of euclidean isometries generated by $R_H \cup T_F$. It is crystallographic in the sense that it acts geometrically on euclidean space but it is not a Coxeter group in general because it is not generated by reflections. If we assign a weight of $2/k$ to each factored translation then distances in the Cayley graph of F agree with those in D and the interval $[1,w]^D$ is an induced subinterval of $[1,w]^F$. The main advantage of the interval $[1,w]^F$ is that it factors as a direct product of type B noncrossing partition lattices with one factor for each irreducible component of the horizontal root system. The edge labels in the i-th factor poset correspond to the factored translations which project nontrivially onto i-th component of the horizontal root system together with the horizontal reflections whose roots belong to this component. Moreover, the isometries that occur as labels in any one factor generator a group isomorphic to a middle group acting on the subspace whose directions are spanned by a component of the horizontal root system and the direction of the Coxeter axis. The structure of the factorable group F is not quite as a clean since each of the component middle groups contain central elements which are pure translations in the direction of the Coxeter axis. Thus F is merely a central product of the associated middle groups.

Name	Symbol	Generating set
Horizontal	H	R_H
Diagonal	D	$R_H \cup T$
Coxeter	W	$R_H \cup R_V$ $(\cup\, T)$
Factorable	F	$R_H \cup T_F$ $(\cup\, T)$
Crystallographic	C	$R_H \cup R_V \cup T_F$ $(\cup\, T)$

Table 2: Five generating sets.

In addtion to the groups W, D, and F, we introduce several other groups that help to clarify the structure of the corresponding euclidean Artin group.

Definition 9.5 (Ten groups). Let $W = \text{Cox}(\widetilde{X}_n)$ be an irreducible euclidean Coxeter group with Coxeter element w and let D and F be the diagonal and factorable groups acting on n-dimensional euclidean space as defined above. There are two other euclidean isometry groups we need to define. Let H denote the subgroup of W generated by horizontal reflections R_H that label edges in the interval $[1,w]^W$ and let C be the group generated by the set $R_H \cup R_V \cup T_F (\cup T)$ of all generating isometries considered so far. The set T of translations is in parentheses because its elements are products of other generators. Note that C, like F, is crystallographic in that it acts geometrically on euclidean space but it is not in general a Coxeter group since we have added translation generators which are not products of reflections contained in W. The five generating sets introduced are listed in Table 2 along with the euclidean isometry groups they generate. Of these five groups H and W are Coxeter groups, while D, F and C are merely crystallographic.

Next we construct five groups defined by presentations. Let D_w, F_w, W_w and C_w denote the interval groups defined by the interval $[1,w]$ in each of the four contexts, but note that we write $A = W_w$ and $G = C_w$ since these turn out to be the corresponding Artin group and a previously unstudied Garside group, respectively. A final group H_w is defined by a presentation with R_H as its generators and subject to the relations among these generators which are visible in the interval $[1,w]^W$. This is not quite an interval group since there is no interval of the form $[1,w]^H$. This is because the element w is not in the subgroup H as it requires a vertical motion in order to be constructed. Some of the maps between these ten groups are shown in Figure 15. The maps in the bottom level are the natural inclusions among these five euclidean isometry groups, the vertical arrows are the projections from the five groups defined by presentations to the groups from which they were constructed, and the maps in the top level are the natural homomorphisms that extend the inclusions on their generating sets.

A review of some of the various posets and groups associated with the euclidean Coxeter group of type $\widetilde{E}_8$ might help to clarify these definitions.

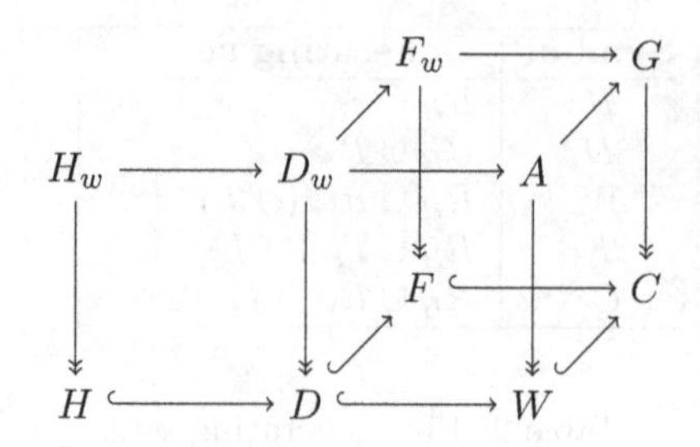

Figure 15: Ten groups.

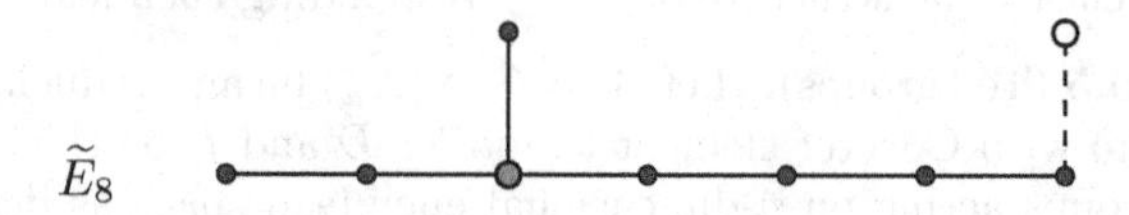

Figure 16: The $\widetilde{E}_8$ diagram.

Example 9.6 (Groups of type $\widetilde{E}_8$). When W is the irreducible euclidean Coxeter group of type $\widetilde{E}_8$, its horizontal root system decomposes as $\Phi_{A_1} \cup \Phi_{A_2} \cup \Phi_{A_4}$. See Figure 16. The factorable group F is a central product of middle groups $\text{Mid}(B_2)$, $\text{Mid}(B_3)$ and $\text{Mid}(B_5)$. The interval $[1,w]^F$ is isomorphic to the direct product $NC_{B_2} \times NC_{B_3} \times NC_{B_5}$ of type B noncrossing partition lattices, and the interval $[1,w]^F$ defines an interval group F_w which is a direct product $\text{Art}(B_2) \times \text{Art}(B_3) \times \text{Art}(B_5)$ of type B spherical Artin groups. The horizontal reflections in the interval $[1,w]^W$ generate a group H isomorphic to a direct product $\text{Cox}(\widetilde{A}_1) \times \text{Cox}(\widetilde{A}_2) \times \text{Cox}(\widetilde{A}_4)$ of type A euclidean Coxeter groups, and the relations among these reflections visible in the interval $[1,w]^W$ define a group H_w which is isomorphic to a direct product $\text{Art}(\widetilde{A}_1) \times \text{Art}(\widetilde{A}_2) \times \text{Art}(\widetilde{A}_4)$ of type A euclidean Artin groups.

10 Structural results

The addition of the factored translations as extra generators completely solves the lattice problem and makes it possible to to prove the three main structural results established by myself and Rob Sulway in [MS]. The first establishes the existence of a new class of Garside groups based on intervals in the crystallographic groups C introduced in the previous section.

Theorem A (Crystallographic Garside groups). *If $C = \text{Cryst}(\widetilde{X}_n, w)$ is the crystallographic group obtained by adding the factored translations to the* *generating set of the irreducible euclidean Coxeter group* $W = \textsc{Cox}(\widetilde{X}_n)$*, then the interval* $[1,w]^C$ *in the Cayley graph of* C *is a lattice. As a consequence, this interval defines a group* $G = C_w$ *with an infinite-type Garside structure.*

As is typical, the most difficult step in the entire article is the proof that these augmented intervals are lattices. Using the program `euclid.sage`, we verify that this is the case for all irreducible euclidean Coxeter groups up to rank 9, which includes all of the sporadic examples. Then special properties of the infinite families are used to complete the proof. It would, of course, be more desirable to have case-free proof of the lattice property, but that project has not yet been completed.

The second main result of [MS] establishes that the crystallographic Garside group G has the structure of an amalgamated free product, and as a consequence, the natural map from corresponding euclidean Artin group A to the crystallographic Garside group G is injective.

Theorem B (Artin groups as subgroups). *For each irreducible euclidean Coxeter group* $W = \textsc{Cox}(\widetilde{X}_n)$ *and for each choice of Coxeter element* w*, the crystallographic Garside group* $G = \textsc{Gar}(\widetilde{X}_n, w)$ *is an amalgamated free product of explicit groups with the euclidean Artin group* $A = \textsc{Art}(\widetilde{X}_n)$ *as one of its factors. In particular, the euclidean Artin group* A *injects into the crystallographic Garside group* G*.*

In terms of the groups defined in the previous section the crystallographic Garside group G is an amalgamated product of F_w and A over D_w. This also means that all of the group homomorphisms on the top level of Figure 15 are injective. Injectivity is a consequence of our structural analysis and not something that was immediately obvious from the definitions of the maps. The final result of [MS] uses these embeddings of euclidean Artin groups into crystallographic Garside groups to elucidate their structure.

Theorem C (Structure of euclidean Artin groups). *Every irreducible euclidean Artin group* $A = \textsc{Art}(\widetilde{X}_n)$ *is a torsion-free centerless group with a solvable word problem and a finite-dimensional classifying space.*

Most of these structural results follow immediately from Theorem B. The only aspect that required a bit more work is the center. The Garside structure on G, the product structure on F_w, and the fact that we are amalgamating over D_w are all used in the proof that shows the center of A is trivial. See [MS] for details.

To conclude this survey, I would like to highlight some of the questions that these results raise. Now that we understand the word problem for the euclidean types, can we devise an Artin group intrinsic solution that avoids the introduction of the crystallographic Garside groups? Or perhaps the crystallographic Garside groups we define are merely the first instance of a natural geometric completion process? What about hyperbolic Artin groups and beyond? Are there similar procedures that work in these more general contexts?

Acknowledgements: I would like to thank the organizers of the 2013 Durham symposium on geometric and cohomological group theory for their invitation to speak about these results and also the organizers of the 2003 Durham symposium since it was during that earlier conference that this project was initially conceived. I thank the anonymous referree for their careful reading of the manuscript and for pointing out a misattribution in an earlier version. And finally, my long-suffering collaborators, Noel Brady, John Crisp and Rob Sulway, deserve a special note of thanks for putting up with me as I refused to let this project die a quiet death during its darkest days.

References

[All02] Daniel Allcock, *Braid pictures for Artin groups*, Trans. Amer. Math. Soc. **354** (2002), no. 9, 3455–3474 (electronic). MR 1911508 (2003f:20053)

[Bes03] David Bessis, *The dual braid monoid*, Ann. Sci. École Norm. Sup. (4) **36** (2003), no. 5, 647–683. MR MR2032983 (2004m:20071)

[Bes06] ______, *A dual braid monoid for the free group*, J. Algebra **302** (2006), no. 1, 55–69. MR MR2236594 (2007i:20061)

[Bir74] Joan S. Birman, *Braids, links, and mapping class groups*, Princeton University Press, Princeton, N.J., 1974, Annals of Mathematics Studies, No. 82. MR 0375281 (51 #11477)

[BM15] Noel Brady and Jon McCammond, *Factoring Euclidean isometries*, Internat. J. Algebra Comput. **25** (2015), no. 1-2, 325–347. MR 3325886

[BS72] Egbert Brieskorn and Kyoji Saito, *Artin-Gruppen und Coxeter-Gruppen*, Invent. Math. **17** (1972), 245–271. MR 48 #2263

[BW02a] Thomas Brady and Colum Watt, *$K(\pi,1)$'s for Artin groups of finite type*, Proceedings of the Conference on Geometric and Combinatorial Group Theory, Part I (Haifa, 2000), vol. 94, 2002, pp. 225–250. MR 1 950 880

[BW02b] ______, *A partial order on the orthogonal group*, Comm. Algebra **30** (2002), no. 8, 3749–3754. MR MR1922309 (2003h:20083)

[CP03] Ruth Charney and David Peifer, *The $K(\pi,1)$-conjecture for the affine braid groups*, Comment. Math. Helv. **78** (2003), no. 3, 584–600. MR MR1998395 (2004f:20067)

[DDG+] Patrick Dehornoy, François Digne, Eddy Godelle, Daan Krammer, and Jean Michel, *Foundations of garside theory*, Available at `arXiv:1309.0796`.

[Del72] Pierre Deligne, *Les immeubles des groupes de tresses généralisés*, Invent. Math. **17** (1972), 273–302. MR 0422673 (54 #10659)

[Dig06] F. Digne, *Présentations duales des groupes de tresses de type affine* $\widetilde{A}$, Comment. Math. Helv. **81** (2006), no. 1, 23–47. MR 2208796 (2006k:20075)

[Dig12] ———, *A Garside presentation for Artin-Tits groups of type* $\widetilde{C}_n$, Ann. Inst. Fourier (Grenoble) **62** (2012), no. 2, 641–666. MR 2985512

[GP12] Eddy Godelle and Luis Paris, *Basic questions on Artin-Tits groups*, Configuration spaces, CRM Series, vol. 14, Ed. Norm., Pisa, 2012, pp. 299–311. MR 3203644

[KP02] Richard P. Kent, IV and David Peifer, *A geometric and algebraic description of annular braid groups*, Internat. J. Algebra Comput. **12** (2002), no. 1-2, 85–97, International Conference on Geometric and Combinatorial Methods in Group Theory and Semigroup Theory (Lincoln, NE, 2000). MR 1902362 (2003f:20056)

[McC06] Jon McCammond, *Noncrossing partitions in surprising locations*, Amer. Math. Monthly **113** (2006), no. 7, 598–610. MR MR2252931 (2007c:05015)

[McC15] ———, *Dual euclidean Artin groups and the failure of the lattice property*, J. Algebra **437** (2015), 308–343. MR 3351966

[MP11] Jon McCammond and T. Kyle Petersen, *Bounding reflection length in an affine Coxeter group*, J. Algebraic Combin. **34** (2011), no. 4, 711–719. MR 2842917 (2012h:20089)

[MS] Jon McCammond and Robert Sulway, *Artin groups of euclidean type*, Available at `arXiv:1312.7770 [math.GR]`, to appear in Invent. Math.

[Sch50] Peter Scherk, *On the decomposition of orthogonalities into symmetries*, Proc. Amer. Math. Soc. **1** (1950), 481–491. MR 12,157c

[Squ87] Craig C. Squier, *On certain* 3*-generator Artin groups*, Trans. Amer. Math. Soc. **302** (1987), no. 1, 117–124. MR 887500 (88g:20069)

[Sul10] Robert Sulway, *Braided versions of crystallographic groups*, Ph.D. thesis, University of California, Santa Barbara, 2010.

[tD98] Tammo tom Dieck, *Categories of rooted cylinder ribbons and their representations*, J. Reine Angew. Math. **494** (1998), 35–63, Dedicated to Martin Kneser on the occasion of his 70th birthday. MR 1604452 (99h:18010)

Finitely presented groups associated with expanding maps

Volodymyr Nekrashevych

Abstract

We associate with every locally expanding self-covering $f : \mathcal{M} \longrightarrow \mathcal{M}$ of a compact path connected metric space a finitely presented group $\mathcal{V}_f$. We prove that this group is a complete invariant of the dynamical system: two groups $\mathcal{V}_{f_1}$ and $\mathcal{V}_{f_2}$ are isomorphic as abstract groups if and only if the corresponding dynamical systems are topologically conjugate. We also show that the commutator subgroup of $\mathcal{V}_f$ is simple, and give a topological interpretation of $\mathcal{V}_f/\mathcal{V}'_f$.

Contents

1 Introduction

A dynamical system is *finitely presented* if it can be represented as the factor of a shift of finite type by an equivalence relation that is also a shift of finite type, see [Fri87, CP93]. If we think of finite type as analogous to finite generation (of a group or of a normal subgroup), then the notion of a finitely presented dynamical system becomes analogous to the notion of a finitely presented group. But the relation is deeper than just a superficial analogy.

The condition of being finitely presented for a dynamical system is very closely related to dynamical hyperbolicity, see [Fri87]. For example, if $f : \mathcal{J} \longrightarrow \mathcal{J}$ is a locally expanding self-covering of a compact metric space, then f is finitely presented. Dynamical hyperbolicity is very closely related to Gromov hyperbolicity for groups, see [Gro87, CP93, Nek15], and finite presentation is a prominent property of hyperbolic groups.

The aim of this paper is to show a new connection between expanding maps and finite presentations. We naturally associate with every finite degree self-covering $f : \mathcal{M} \longrightarrow \mathcal{M}$ of a path-connected space $\mathcal{M}$ a group $\mathcal{V}_f$ with the following property (see Theorem 5.9 and Theorem 7.1).

Theorem 1.1. *If $f : \mathcal{M} \longrightarrow \mathcal{M}$ is a locally expanding self-covering of a compact path connected metric space then $\mathcal{V}_f$ is finitely presented.*

If $f_i : \mathcal{M}_i \longrightarrow \mathcal{M}_i$ are as above, then the groups $\mathcal{V}_{f_1}$ and $\mathcal{V}_{f_2}$ are isomorphic if and only if f_1 and f_2 are topologically conjugate, i.e., there exists a homeomorphism $\phi : \mathcal{M}_1 \longrightarrow \mathcal{M}_2$ such that $f_1 = \phi^{-1} \circ f_2 \circ \phi$.

We also show that the commutator subgroup of $\mathcal{V}_f$ is simple, and give a dynamical interpretation of the abelianization $\mathcal{V}_f/\mathcal{V}_f'$, see Theorem 4.7 and Proposition 5.7.

The groups $\mathcal{V}_f$ are defined in the following way. Let $f : \mathcal{M} \longrightarrow \mathcal{M}$ be a finite degree covering map, and suppose that $\mathcal{M}$ is path connected. Choose $t \in \mathcal{M}$, and consider the tree T_t of preimages of t under the iterations of f. Its set of vertices is $\bigsqcup_{n \geq 0} f^{-n}(t)$, and a vertex $v \in f^{-n}(t)$ is connected to the vertex $f(v) \in f^{-(n-1)}(t)$. Let ∂T_t be its boundary, which can be defined as the inverse limit of the discrete sets $f^{-n}(t)$ with respect to the maps $f : f^{-(n+1)}(t) \longrightarrow f^{-n}(t)$.

Let γ be a path in $\mathcal{M}$ connecting a vertex $v \in f^{-n}(t)$ to a vertex $u \in f^{-m}(t)$ of the tree T_t. Considering lifts of γ by the coverings $f^k : \mathcal{M} \longrightarrow \mathcal{M}$, we get an isomorphism $S_\gamma : T_v \longrightarrow T_u$ between subtrees T_v, T_u of T_t. Namely, if γ_z is a lift of γ starting at $z \in f^{-k}(v)$, then $S_\gamma(z) \in f^{-k}(u)$ is the end of γ_z. Denote by the same symbol S_γ the induced homeomorphism $\partial T_v \longrightarrow \partial T_u$ of the boundaries of the subtrees, seen as clopen subsets of ∂T_t.

Definition 1.1. The group $\mathcal{V}_f$ is the group of all homeomorphisms $\partial T_t \longrightarrow \partial T_t$ locally equal to homeomorphisms of the form $S_\gamma : \partial T_v \longrightarrow \partial T_u$.

The group $\mathcal{V}_f$ is generated by the Higman-Thompson group $G_{\deg f,1}$ (as in [Hig74]) acting on ∂T_t and the *iterated monodromy group* IMG (f) of $f : \mathcal{M} \longrightarrow \mathcal{M}$. The iterated monodromy group can be defined as the subgroup of $\mathcal{V}_f$ consisting of homeomorphism $S_\gamma : \partial T_t \longrightarrow \partial T_t$, where γ is a loop starting and ending at the basepoint t. It is an invariant of the topological conjugacy class of $f : \mathcal{M} \longrightarrow \mathcal{M}$, and it becomes a complete invariant (in the expanding case), if we consider it as a *self-similar group.* Self-similarity is an additional structure on a group, and it can be defined using one of several equivalent approaches: virtual endomorphisms, wreath recursions, bisets, or structures of an automaton group. The fact that self-similar iterated monodromy group is a complete invariant of an expanding self-covering is one of the main topics of [Nek05].

From the point of view of group theory, $\mathcal{V}_f$ seems to be a "cleaner" object, since no additional structure is needed to make it a complete invariant of a dynamical system. Besides, it is finitely presented, unlike the iterated monodromy groups, which are typically infinite presented. However, iterated monodromy groups have better functorial properties than the groups $\mathcal{V}_f$, see [Nek08].

A group $\mathcal{V}_G$, analogous to $\mathcal{V}_f$, can be defined for any self-similar group G, so that $\mathcal{V}_f = \mathcal{V}_{\mathrm{IMG}(f)}$. Groups of this type were for the first time studied by C. Röver [Röv99, Röv02]. In particular, he showed that if G is the Grigorchuk group [Gri80], then $\mathcal{V}_G$ is finitely presented, simple, and is isomorphic to the abstract commensurator of the Grigorchuk group. The case of a general self-similar group G was studied later in [Nek04].

A natural question to ask is whether the isomorphism problem is solvable for the groups $\mathcal{V}_f$. Equivalently, is the topological conjugacy problem for expanding maps algorithmically solvable? Note that expanding maps can be given in different ways by a finite amount of information: using finite presentations in the sense of [Fri87], using combinatorial models in the sense of [IS10, Nek14], using iterated monodromy groups, e.t.c..

Another natural question is whether the groups $\mathcal{V}_f$ satisfy the finiteness condition F_∞, similarly to the Higman-Thompson groups (see [Bro87]). I.e., if they have classifying spaces with finite n-dimensional skeleta for all n. It would be also interesting to study homology of $\mathcal{V}_f$ in relation with homological properties of the dynamical system.

The structure of the paper is as follows. In "Definition of the groups $\mathcal{V}_f$" we give a review of terminology related to rooted trees, and define the groups $\mathcal{V}_f$. In the next section "Symbolic coding" we encode the vertices of the tree of preimages T_t by finite words over an alphabet X, and show that $\mathcal{V}_f$ contains a copy of the Higman-Thompson group, and that $\mathcal{V}_f$ is generated by the Higman-Thompson group and the iterated monodromy group. We also give a review of the basic notions of the theory of self-similar groups, and define the groups $\mathcal{V}_G$ associated with self-similar groups, following [Nek04].

In Section 4 we prove that the commutator subgroup $\mathcal{V}_G'$ of $\mathcal{V}_G$ is simple for any self-similar group G. In particular, $\mathcal{V}_f'$ is simple for any map f. Note that the fact that every proper quotient of $\mathcal{V}_G$ is abelian, i.e., that every non-trivial normal subgroup of $\mathcal{V}_G$ contains $\mathcal{V}_G'$ was already proved in [Nek04], and we use this fact in our proof. Later, in the next section we give an interpretation of $\mathcal{V}_f/\mathcal{V}_f'$ in topological terms. Namely, we prove the following (see Proposition 5.7).

Proposition 1.2. *Suppose that $f : \mathcal{M}_1 \longrightarrow \mathcal{M}$ is expanding, $\mathcal{M}$ is path-connected and semi-locally simply connected, and $\mathcal{M}_1 \subseteq \mathcal{M}$.*

If $\deg f$ is even, then $\mathcal{V}_f/\mathcal{V}_f'$ is isomorphic to the quotient of $H_1(\mathcal{M})$ by the range of the endomorphism $1 - \iota_* \circ f^!$.

If $\deg f$ is odd, then $\mathcal{V}_f/\mathcal{V}_f'$ is isomorphic to the quotient of $\mathbb{Z}/2\mathbb{Z} \oplus H_1(\mathcal{M})$ by the range of the endomorphism $1 - \sigma_1$, where $\sigma_1(t, c) = (t + \mathrm{sign}(c), \iota_* \circ f^!(c))$.

Here $\iota_* : H_1(\mathcal{M}_1) \longrightarrow H_1(\mathcal{M})$ is the homomorphism induced by the identical embedding $\iota : \mathcal{M}_1 \longrightarrow \mathcal{M}$, the homomorphism $f^! : H_1(\mathcal{M}) \longrightarrow H_1(\mathcal{M}_1)$ maps a cycle c to its full preimage $f^{-1}(c)$, and $\mathrm{sign} : H_1(\mathcal{M}) \longrightarrow \mathbb{Z}/2\mathbb{Z}$ maps a cycle c defined by a loop γ to 1 if γ acts as an odd permutation on the fiber of f, and to 0 otherwise.

The main result of Section 5 is existence of finite presentation of $\mathcal{V}_f$ when f is expanding. More generally, we show that $\mathcal{V}_G$ is finitely presented, if G is a *contracting self-similar group*, see Theorem 5.9. We also give (in Subsection 5.2) a general definition of the groups $\mathcal{V}_f$ for expanding maps $f : \mathcal{M} \longrightarrow \mathcal{M}$ (where $\mathcal{M}$ is not required to be path connected).

Section "Dynamical systems and groupoids" is an overview of the theory of limit dynamical systems of contracting self-similar groups and basic results on hyperbolic groupoids, following [Nek05] and [Nek15]. These results are needed for the proof the fact that $\mathcal{V}_f$ is a complete invariant of the dynamical system in the expanding case, which is proved (Theorem 7.1) in the last section of the paper.

The general scheme of the proof of Theorem 7.1 is as follows. First, we show, using a theorem of M. Rubin [Rub89], that two groups $\mathcal{V}_{f_1}$ and $\mathcal{V}_{f_2}$ are isomorphic if and only if their actions on the corresponding boundaries of trees are topologically conjugate. This implies, that the groupoid of germs $\mathfrak{G}$ of the action of the group $\mathcal{V}_f$ on the boundary of the tree is uniquely determined by the group $\mathcal{V}_f$.

The groupoid $\mathfrak{G}$ is hyperbolic, and hence it uniquely determines the equivalence class of its dual (see [Nek15]), which is the groupoid generated by the germs of $f : \mathcal{M} \longrightarrow \mathcal{M}$. It remains to show that the dynamical system $f : \mathcal{M} \longrightarrow \mathcal{M}$ is uniquely determined (up to topological conjugacy) by the equivalence class of the groupoid of germs generated by it. This is proved using the techniques of hyperbolic groupoids. Connectedness of $\mathcal{M}$ is used in the proof in an essential way.

It is an interesting question if Theorem 7.1 is true in general (without the condition that $\mathcal{M}$ is path connected).

As a corollary of the proof of Theorem 7.1, we clarify the relation between the groupoid-theoretic equivalence of groupoids of germs associated with expanding dynamical systems, and their topological conjugacy, see Theorem 7.7.

Acknowledgments

I am grateful to the organizers of the LMS Durham Symposium "Geometric and Cohomological Group Theory" for inviting me to give a talk, which inspired me to return to the topics of this paper.

The paper is based in part on work supported by NSF grant DMS1006280.

2 Definition of the groups $\mathcal{V}_f$

2.1 Rooted trees

Let T be a locally finite rooted tree, and let v, u be its vertices. We write $v \preceq u$ if the path connecting the root to u passes through v. This defines a partial order on the set of vertices of T, and T is its Hasse diagram (though we tend to draw rooted trees "upside down" with the root on top).

We denote by T_v the sub-tree with root v spanned by all vertices u such that $v \preceq u$. We have $v \preceq u$ if and only if $T_v \supseteq T_u$. If v and u are incomparable, then T_v and T_u are disjoint.

Boundary ∂T of the tree T is the set of all infinite simple paths starting at the root of T. The boundary ∂T_v is naturally identified with the set of

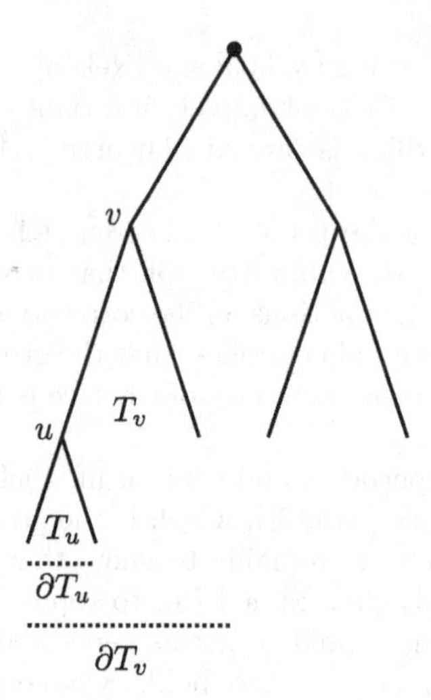

Figure 1: Rooted tree

paths $w \in \partial T$ passing through v. The sets ∂T_v form a basis of open sets of a natural topology on ∂T. The subsets ∂T_v are clopen (closed and open), and every clopen subset of ∂T_v is disjoint union of a finite number of sets of the form ∂T_v.

The *nth level* of T is the set of vertices that are on distance n from the root of the tree. An *antichain* is a set of pairwise incomparable vertices. A finite antichain A is said to be *complete* if it is maximal, i.e., if every set B of vertices of T properly containing A is not an antichain. For example, every level of T is a complete antichain.

A set A is an antichain if and only if the sets ∂T_v for $v \in A$ are disjoint. It follows that A is a complete antichain if and only if ∂T is disjoint union of the sets ∂T_v for $v \in A$.

Let X be a finite set. We denote by X^* the free monoid generated by X, i.e., the set of all finite words $x_1 x_2 \dots x_n$ for $x_i \in X$, together with the empty word $\varnothing$. *Length* of a word $v \in X^*$ is the number of its letters.

We introduce on the set X^* structure of a rooted tree coinciding with the left Cayley graph of the free monoid. Namely, two finite words are connected by an edge if and only if they are of the form vx and v for $v \in X^*$ and $x \in X$. The empty word is the root of the tree X^*. We have $v \preceq u$ for $v, u \in X^*$ if and only if v is a beginning of u. The subtree T_v of $T = X^*$ for $v \in X^*$ is the set vX^* of all words starting with v. The nth level of X^* is the set X^n of words of length n.

The boundary of the tree X^* is naturally identified with the space X^ω of right-infinite sequences $x_1 x_2 \dots$ of elements of X. The topology on the

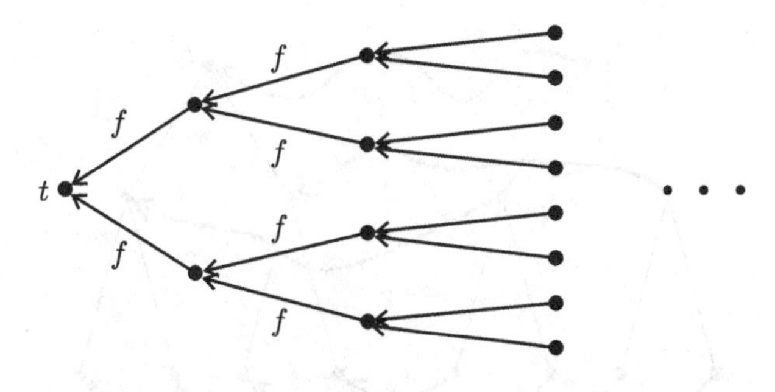

Figure 2: Tree T_t

boundary coincides with the direct product topology on X^ω.

2.2 Definition

Let $\mathcal{M}$ be a topological space. A *partial self-covering* is a finite degree covering map $f : \mathcal{M}_1 \longrightarrow \mathcal{M}$, where $\mathcal{M}_1 \subseteq \mathcal{M}$.

If $f : \mathcal{M}_1 \longrightarrow \mathcal{M}$ is a partial self-covering, then we can iterate it as a partial self-map of $\mathcal{M}$. Then the nth iteration $f^n : \mathcal{M}_n \longrightarrow \mathcal{M}$ is also a partial self-covering. Here $\mathcal{M}_n \subset \mathcal{M}_{n-1} \subset \ldots \subset \mathcal{M}_1$ are domains of the iterations of f, defined by the condition $\mathcal{M}_{n+1} = f^{-1}(\mathcal{M}_n)$.

For a point $t \in \mathcal{M}$ denote by T_t the *tree of preimages* of t under the iterations of f, i.e., the tree with the set of vertices equal to the formal disjoint union $\bigsqcup_{n\geq 0} f^{-n}(t)$ of the sets of preimages of t under the iterations f^n : $\mathcal{M}_n \longrightarrow \mathcal{M}$. Here $f^{-0}(t) = \{t\}$ consists of the root of the tree, and a vertex $v \in f^{-n}(t)$ is connected by an edge to the vertex $f(v) \in f^{-(n-1)}(t)$, see Figure 2. If v is a vertex of T_t, then the tree of preimages T_v is in a natural way a sub-tree of the tree T_t, and our notation agrees with the notation of the previous subsection.

Assume now that $\mathcal{M}$ is path connected. Let $t_1, t_2 \in \mathcal{M}$, and let γ be a path from t_1 to t_2 in $\mathcal{M}$. Then for every $n \geq 0$ and every $v \in f^{-n}(t_1)$ there is a unique lift by the covering $f^n : \mathcal{M}_n \longrightarrow \mathcal{M}$ of γ starting in v. Let $S_\gamma(v) \in f^{-n}(t_2)$ be its end. It is easy to see that the map $S_\gamma : T_{t_1} \longrightarrow T_{t_2}$ is an isomorphism of the rooted trees, see Figure 3. It defines a homeomorphism of their boundaries $S_\gamma : \partial T_{t_1} \longrightarrow \partial T_{t_2}$, which we will denote by the same letter.

Definition 2.1. Let $f : \mathcal{M}_1 \longrightarrow \mathcal{M}$ be a partial self-covering, and let $t \in \mathcal{M}$. Denote by $\mathcal{T}_f$ the semigroup of partial homeomorphisms of ∂T_t generated by the homeomorphisms of the form $S_\gamma : \partial T_{v_1} \longrightarrow \partial T_{v_2}$, where γ is a path connecting points $v_1, v_2 \in \bigsqcup_{n\geq 0} f^{-n}(t)$.

The semigroup $\mathcal{T}_f$ contains the *zero* map between empty subsets of T_t. A product $S_{\gamma_1} S_{\gamma_2}$ is zero if the range of S_{γ_2} is disjoint from the domain of S_{γ_1}.

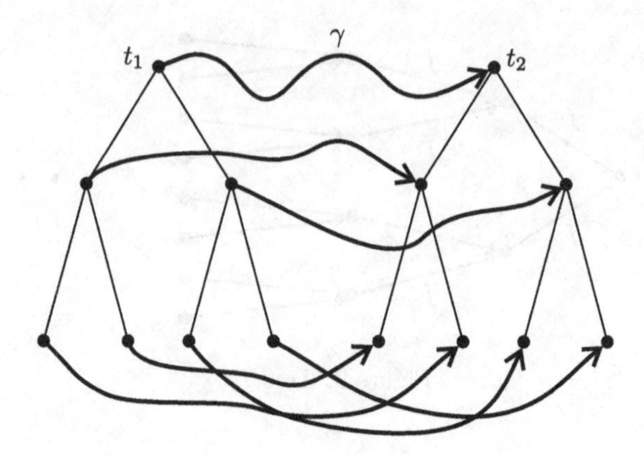

Figure 3: Isomorphism S_γ

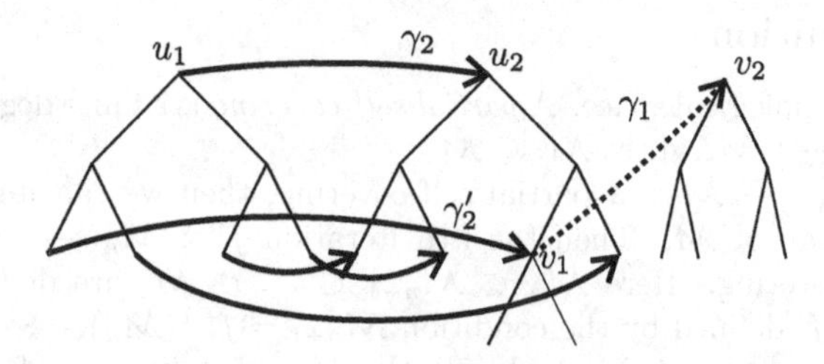

Figure 4: Composition $S_{\gamma_1} S_{\gamma_2}$

Let $S_{\gamma_1} : \partial T_{v_1} \longrightarrow \partial T_{v_2}$ and $S_{\gamma_2} : \partial T_{u_1} \longrightarrow \partial T_{u_2}$ be two generators of $\mathcal{T}_f$. The product $S_{\gamma_1} S_{\gamma_2}$ is non-zero if and only if T_{u_2} and T_{v_1} are not disjoint, i.e., if either $T_{u_2} \supseteq T_{v_1}$, or $T_{u_2} \subseteq T_{v_1}$. It the first case, v_1 is a preimage of u_2 under some iteration f^k of f. Let γ_2' be the unique lift of γ_2 by f^k that ends in v_1. Then it follows from the definition of the transformations S_γ that

$$S_{\gamma_1} S_{\gamma_2} = S_{\gamma_1 \gamma_2'}, \tag{1}$$

see Figure 4. Here and in the sequel, we multiply paths as we compose functions: in the product $\gamma_1 \gamma_2'$ the path γ_2' is passed before the path γ_1.

Similarly, if $T_{u_2} \subseteq T_{v_1}$, then u_2 is a f^k-preimage of v_1 for some $k \geq 0$, and

$$S_{\gamma_1} S_{\gamma_2} = S_{\gamma_1' \gamma_2}, \tag{2}$$

where γ_1' is the lift of γ_1 by f^k starting in u_2.

It follows that all non-zero elements of $\mathcal{T}_f$ are of the form S_γ for some path γ in $\mathcal{M}$ connecting two vertices of T_t. Note also that $\mathcal{T}_f$ is an inverse semigroup, where $S_\gamma^* = S_{\gamma^{-1}}$.

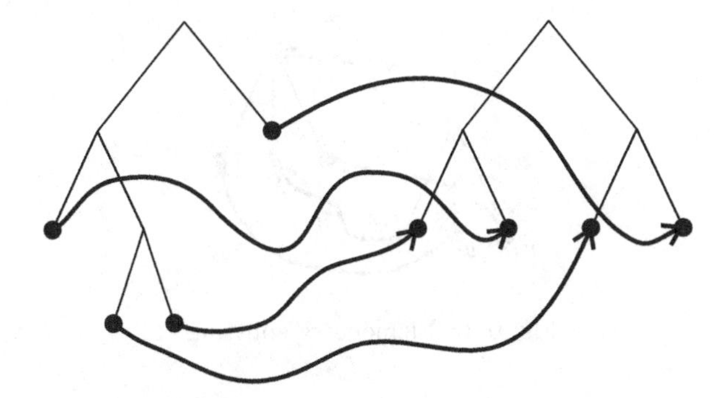

Figure 5: Elements of $\mathcal{V}_f$

Let A_1, A_2 be two complete antichains of T_t of equal cardinality. Choose a bijection $\alpha : A_1 \longrightarrow A_2$ and a collection of paths γ_a from $a \in A_1$ to the corresponding vertex $\alpha(a)$, see Figure 5. Let $g : \partial T_t \longrightarrow \partial T_t$ be the map given by the rule

$$g(w) = S_{\gamma_v}(w), \qquad \text{if } w \in \partial T_v.$$

It is easy to see that $g : \partial T_t \longrightarrow \partial T_t$ is a homeomorphism. Denote by $\mathcal{V}_f$ the set of all such homeomorphisms.

We will represent homeomorphisms $g \in \mathcal{V}_f$ by tables of the form

$$\begin{pmatrix} v_1 & v_2 & \dots & v_n \\ \gamma_{v_1} & \gamma_{v_2} & \dots & \gamma_{v_n} \\ \alpha(v_1) & \alpha(v_2) & \dots & \alpha(v_n) \end{pmatrix}, \tag{3}$$

where the first row is the list of vertices of a complete antichain, $g(w) = S_{\gamma_{v_i}}(w)$ for all $w \in \partial T_{v_i}$, and $\alpha(v_i)$ is the end of γ_{v_i}. Note that the vertices v_i and $\alpha(v_i)$ are uniquely determined by the paths γ_{v_i}, but it is convenient for us to keep information about them in the tables.

An *elementary splitting* of such a table is the operation of replacing a column

$$\begin{pmatrix} u \\ \gamma_u \\ \alpha(u) \end{pmatrix} \quad \text{by the array} \quad \begin{pmatrix} u_1 & u_2 & \dots & u_d \\ \gamma_{u_1} & \gamma_{u_2} & \dots & \gamma_{u_d} \\ w_1 & w_2 & \dots & w_d \end{pmatrix},$$

where $\{u_1, u_2, \dots, u_d\} = f^{-1}(u)$, γ_{u_i} is the lift of γ_u by f starting at u_i, and w_i is the end of γ_{u_i}, see Figure 6. A *splitting* of a table is the results of a finite sequence of elementary splittings.

It follows directly from the definition that splitting of a table does not change the homeomorphism $g \in \mathcal{V}_f$ that it defines. It is obvious that if g_1

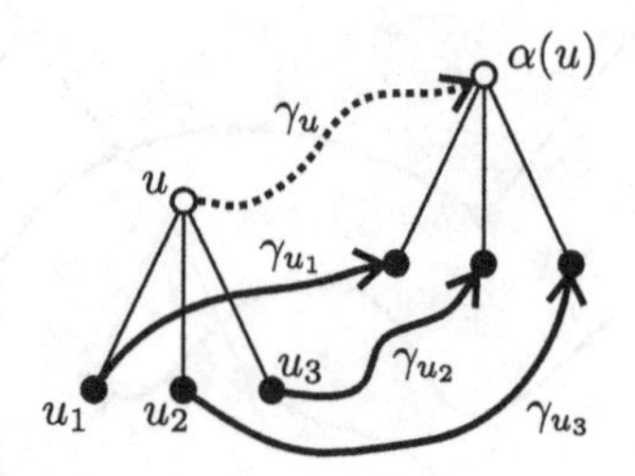

Figure 6: Elementary splitting

and g_2 are defined by tables of the form

$$\begin{pmatrix} v_1 & v_2 & \dots & v_n \\ \gamma_{v_1} & \gamma_{v_2} & \dots & \gamma_{v_n} \\ u_1 & u_2 & \dots & u_n \end{pmatrix}$$

and

$$\begin{pmatrix} w_1 & w_2 & \dots & w_n \\ \gamma_{w_1} & \gamma_{w_2} & \dots & \gamma_{w_n} \\ v_1 & v_2 & \dots & v_n \end{pmatrix},$$

then the composition $g_1 g_2$ is defined by the table

$$\begin{pmatrix} w_1 & w_2 & \dots & w_n \\ \gamma_{v_1}\gamma_{w_1} & \gamma_{v_2}\gamma_{w_2} & \dots & \gamma_{v_n}\gamma_{w_n} \\ u_1 & u_2 & \dots & u_n \end{pmatrix}.$$

Let

$$\begin{pmatrix} a_1 & a_2 & \dots & a_n \\ \gamma_{a_1} & \gamma_{a_2} & \dots & \gamma_{a_n} \\ b_1 & b_2 & \dots & b_n \end{pmatrix}, \quad \begin{pmatrix} c_1 & c_2 & \dots & c_m \\ \gamma_{c_1} & \gamma_{c_2} & \dots & \gamma_{c_m} \\ a'_1 & a'_2 & \dots & a'_m \end{pmatrix}$$

be tables defining elements of $\mathcal{V}_f$. We can find a complete antichain A such that for every $v \in A$ the subtree T_v is contained in a subtree T_{a_i} and a subtree $T_{a'_j}$ for some i and j. For example, we can take A to be equal to the kth level of the tree T_t for k big enough. Then there exists a splitting of the first table such that its first row is A, and there exists a splitting of the second table such that its last row is A.

It follows that if $g_1, g_2 \in \mathcal{V}_f$, then their composition $g_1 g_2$ also belongs to $\mathcal{V}_f$. As a corollary, we get the following proposition.

Proposition 2.1. *The set $\mathcal{V}_f$ is a group.*

We will use sometimes, instead of tables, the following notation. If F and G are two partial transformations with disjoint domains, then we denote by

$F + G$ their union, i.e., the map equal to F on the domain of F and equal to G on the domain of G. Then the transformation defined by a table

$$\begin{pmatrix} v_1 & v_2 & \dots & v_n \\ \gamma_1 & \gamma_2 & \dots & \gamma_n \\ u_1 & u_2 & \dots & u_n \end{pmatrix}$$

is written $S_{\gamma_1} + S_{\gamma_2} + \dots + S_{\gamma_n}$.

An elementary splitting of a table is equivalent then to application of the identity

$$S_\gamma = \sum_{\delta \in f^{-1}(\gamma)} S_\delta,$$

where $f^{-1}(\gamma)$ is the set of all lifts of γ by f.

3 Symbolic coding

3.1 Two trees

Let $\{t_1, t_2, \dots, t_d\} = f^{-1}(t)$, and choose paths ℓ_i in $\mathcal{M}$ from t to t_i. Then $S_{\ell_i} : T_t \longrightarrow T_{t_i}$ is an isomorphism. The elements $S_{\ell_i} \in \mathcal{T}_f$ satisfy the relations:

$$S_{\ell_i}^* S_{\ell_i} = 1, \qquad \sum_{i=1}^{d} S_{\ell_i} S_{\ell_i}^* = 1,$$

where S_γ^* denotes inversion in the inverse semigroup $\mathcal{T}_f$, so that $S_\gamma^* = S_{\gamma^{-1}}$.

The C^*-algebra defined by these relations is known as the *Cuntz algebra* [Cun77]. If we denote by S the row $(S_{\ell_1}, S_{\ell_2}, \dots, S_{\ell_d})$, and we denote by S^* the column $(S_{\ell_1}^*, S_{\ell_2}^*, \dots, S_{\ell_d}^*)^\top$, then the relations can be written as matrix equalities

$$S^* S = I_d, \qquad SS^* = I_1,$$

where I_n denotes the $n \times n$ identity matrix. Rings given by these and similar defining relations were studied by W. Leavitt [Lea56, Lea65].

Let Γ_t be the graph with the set of vertices equal to the set of vertices of T_t in which two vertices $v \in f^{-n}(t)$ and $v \in f^{-(n+1)}(t)$ are connected by an edge if and only if they are connected by a path equal to a lift of a path ℓ_i by the covering f^n. In other words, the graph Γ_t is obtained by taking preimages of the paths ℓ_i under all iterations of f, see Figure 7.

It is easy to see that Γ_t is a tree, and its nth level is equal to the nth level of the tree T_t. It follows that any two vertices of T_t are connected by a unique simple path in Γ_t.

Let $i_{n-1} \dots i_1 i_0 \in \{1, 2, \dots, d\}^n$, and consider the product

$$S_{\ell_{i_{n-1}}} \cdots S_{\ell_{i_1}} S_{\ell_{i_0}} \in \mathcal{T}_f.$$

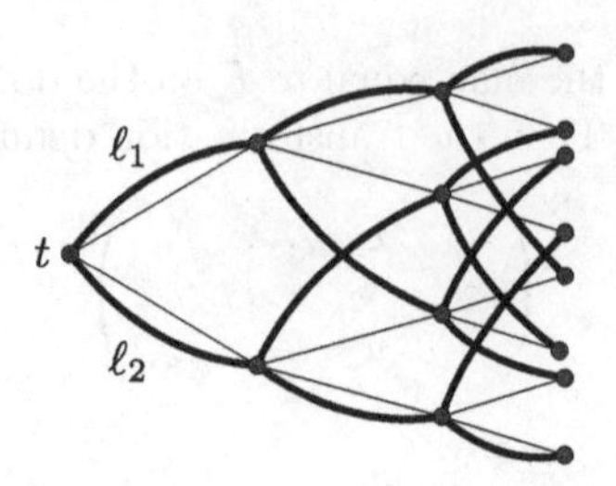

Figure 7: Trees Γ_t and T_t

According to the multiplication rule (2), it is equal to $S_{\lambda_{n-1}\ldots\lambda_1\lambda_0}$, where $\lambda_0 = \ell_{i_0}$, and λ_k is the lift of ℓ_k by f^k starting at the end of λ_{k-1}. Denote by $\Lambda(i_{n-1}\ldots i_1 i_0)$ the end of the last path λ_{n-1}. Then $S_{\ell_{i_{n-1}}}\cdots S_{\ell_{i_1}} S_{\ell_{i_0}} = S_{\ell_{i_{n-1}\ldots i_2 i_1}}$, where $\ell_{i_{n-1}\ldots i_2 i_1}$ is unique simple path in Γ_t starting at the root and ending in $\Lambda(i_{n-1}\ldots i_1 i_0)$. The path $\ell_{i_{n-1}\ldots i_2 i_1}$ and its end $\Lambda(i_{n-1}\ldots i_1 i_0)$ satisfy the recurrent rule:

$$\ell_{i_{n-1}\ldots i_2 i_1} = \lambda_{i_{n-1},\Lambda(i_{n-2}\ldots i_1 i_0)}\ell_{i_{n-2}\ldots i_2 i_1},$$

where $\lambda_{i_{n-1},\Lambda(i_{n-2}\ldots i_1 i_0)}$ is the lift of $\ell_{i_{n-1}}$ by f^{n-1} starting at $\Lambda(i_{n-2}\ldots i_1 i_0)$ (and hence ending in $\Lambda(i_{n-1}\ldots i_1 i_0)$).

The map Λ is a bijection between $\{1, 2, \ldots, d\}^n$ and the nth level $f^{-n}(t)$ of the trees T_t and Γ_t. It follows directly from the description of the path $\ell_{i_{n-1}\ldots i_1 i_0}$ that $\Lambda(i_{n-1}\ldots i_1 i_0)$ is adjacent to $\Lambda(i_{n-1}\ldots i_2 i_1)$ in T_t and to $\Lambda(i_{n-2}\ldots i_1 i_1)$ in Γ_t. In other words, T_t and Γ_t are identified by Λ with the right and the left Cayley graphs of the free monoid generated by $X = \{1, 2, \ldots, d\}$, respectively.

For any two sequences $i_1 i_2 \ldots i_n$ and $j_1 j_2 \ldots j_m \in X^*$, the product

$$S_{\ell_{i_1}} S_{\ell_{i_2}} \cdots S_{\ell_{i_n}} (S_{\ell_{j_1}} S_{\ell_{j_2}} \ldots S_{\ell_{j_m}})^*$$

is equal to S_γ, where γ is the path in Γ_t from $\Lambda(j_1 j_2 \ldots j_m)$ to $\Lambda(i_1 i_2 \ldots i_n)$. This follows directly from the definitions of Γ_t, Λ, and rules (1) and (2).

We will use notation $S_x = \Lambda^{-1} S_{\ell_x} \Lambda$ and $S_{x_1 x_2 \ldots x_n} = S_{x_1} S_{x_2} \cdots S_{x_n}$, for $x, x_i \in X$. Then, by the definition of Λ, the transformations $S_{x_1 x_2 \ldots x_n}$ of X^ω are given by the rule

$$S_{x_1 x_2 \ldots x_n}(v) = x_1 x_2 \ldots x_n v.$$

For $v, u \in X^*$, the transformation $S_v S_u^*$ is defined on uX^ω, and acts by the rule

$$S_v S_u^*(uw) = vw.$$

In particular, $\sum_{x\in X} S_x S_x^* = 1$, and we obviously have $S_x^* S_x = 1$.

3.2 The Higman-Thompson group

Let A_1 and A_2 be complete antichains in X^*, and let $\alpha : A_1 \longrightarrow A_2$ be a bijection. Then $g_\alpha = \sum_{v \in A_1} S_{\alpha(v)} S_v^*$ is a homeomorphism of X^ω defined by the rule

$$g_\alpha(vw) = \alpha(v)w,$$

for all $v \in A_1$ and $w \in X^\omega$. The set of all such homeomorphisms g_α is the *Higman-Thompson group* group $G_{|X|,1}$, see [Hig74], which we will denote by $\mathcal{V}_X$ or $\mathcal{V}_d$, where $d = |X|$.

Its copy $\Lambda \cdot \mathcal{V}_X \cdot \Lambda^{-1}$ in $\mathcal{V}_f$ is the group defined by the paths in the tree Γ_t. Namely, for any bijection $\alpha : A_1 \longrightarrow A_2$ between complete antichains of T_t there exist unique simple paths γ_v connecting $v \in A_1$ to $\alpha(v) \in A_2$ inside the tree Γ_t. Then the corresponding element of $\mathcal{V}_f$ is equal to $\sum_{v \in A_1} S_{\gamma_v}$.

The following simple lemma will be useful later (for a proof, see, for example [Nek04, Lemma 9.12]).

Lemma 3.1. *Let $A_1, A_2 \subset X^*$ be finite incomplete (i.e., non-maximal) antichains, and let $\alpha : A_1 \longrightarrow A_2$ be a bijection. Then there exists $g \in \mathcal{V}_X$ such that $g(vw) = \alpha(v)w$ for all $v \in A_1$ and $w \in X^\omega$.*

3.3 The iterated monodromy group

Every element γ of the fundamental group $\pi_1(\mathcal{M}, t)$ defines an element $S_\gamma : \partial T_t \longrightarrow \partial T_t$ of $\mathcal{V}_f$. We get in this way a natural homomorphism $\gamma \mapsto S_\gamma$ from $\pi_1(\mathcal{M}, t)$ to $\mathcal{V}_f$. Its image is called the *iterated monodromy group* of f and is denoted $\mathrm{IMG}\,(f)$. It acts on T_t by automorphisms, so that the action on the nth level coincides with the natural *monodromy action* associated with the covering $f^n : \mathcal{M}_n \longrightarrow \mathcal{M}$, see [Nek05, Chapter 5], [BGN03, Nek11].

Let us choose paths ℓ_i connecting the root t to the vertices of the first level $f^{-1}(t)$ of the tree T_t. Let Γ_t be the tree obtained by taking lifts of the paths ℓ_i by iterations of f, as in Subsection 3.1.

For a vertex v of T_t, denote by ℓ_v the unique simple path inside Γ_t from t to v. Then for an arbitrary path γ in $\mathcal{M}$ starting in a vertex v and ending in a vertex u of T_t, the path $\ell_u^{-1}\gamma\ell_v$ is a loop based at t. Let $g = S_{\ell_u^{-1}\gamma\ell_v}$ be the corresponding element of $\mathrm{IMG}\,(f)$. Then we have

$$S_\gamma = S_{\ell_u} S_{\ell_u^{-1}\gamma\ell_v} S_{\ell_v}^* = S_{\ell_u} g S_{\ell_v}^*.$$

Thus we obtain the following description of the elements of $\mathcal{V}_f$.

Lemma 3.2. *Let $g \in \mathcal{V}_f$ be defined by a table $\begin{pmatrix} v_1 & v_2 & \dots & v_n \\ \gamma_1 & \gamma_2 & \dots & \gamma_n \\ u_1 & u_2 & \dots & u_n \end{pmatrix}$. Denote $g_i = S_{\ell_{u_i}}^* S_{\gamma_i} S_{\ell_{v_i}}$. Then $g_i \in \mathrm{IMG}\,(f)$, and $g = \sum_{i=1}^n S_{\ell_{u_i}} g_i S_{\ell_{v_i}}^*$.*

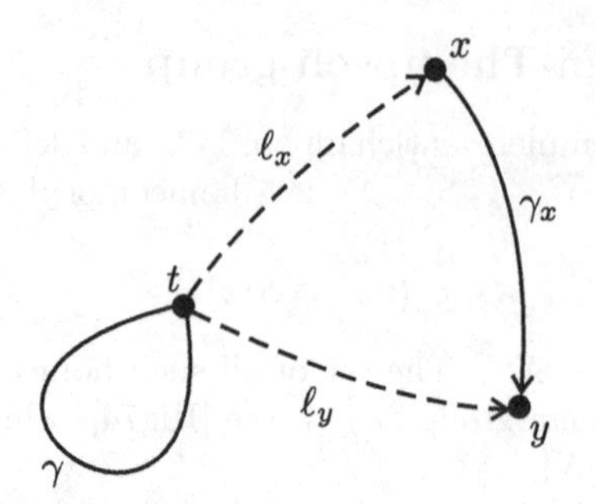

Figure 8: Iterated monodromy recursion

Let $g \in \mathrm{IMG}(f)$ be defined by a loop γ, and let $x \in f^{-1}(t)$ be a vertex of the first level. Let γ_x be the lift of γ by f starting at x, and let y be its end. Then we have

$$gS_{\ell_x} = S_\gamma S_{\ell_x} = S_{\gamma_x \ell_x} = S_{\ell_y \ell_y^{-1} \gamma_x \ell_x} = S_{\ell_y} S_{\ell_y^{-1} \gamma_x \ell_x}. \tag{4}$$

Note that $\ell_y^{-1}\gamma_x\ell_x$ is a loop based at t, i.e., an element of $\pi_1(\mathcal{M}, t)$, see Figure 8.

Let us conjugate the action of $\mathcal{V}_f$ on ∂T_t (and the action of $\mathrm{IMG}(f)$ on T_t) to an action on X^ω (and X^*) using the isomorphism $\Lambda : X^* \longrightarrow T_t$. We call the obtained actions of $\mathcal{V}_f$ and $\mathrm{IMG}(f)$ *standard*. Then formula (4) proves the following lemma, see also [Nek05, Proposition 5.2.2].

Lemma 3.3. *For every $g \in \mathrm{IMG}(f)$ and every $x \in f^{-1}(t)$ there exist $h \in \mathrm{IMG}(f)$ and $y \in f^{-1}(t)$ such that*

$$gS_x = S_y h.$$

Moreover, if g is defined by a loop γ, then h is defined by the loop $\ell_y^{-1}\gamma_x\ell_x$, where γ_x is the lift of γ by f starting at x.

If $gS_x = S_y h$ for $g, h \in \mathrm{IMG}(f)$ and $x, y \in X$, then we denote $h = g|_x$ and $y = g(x)$. Note that the last equality agrees with the definition of g as an automorphism S_γ of the tree T_t.

Since $1 = \sum_{x \in X} S_x S_x^*$, we have

$$g = \sum_{x \in X} gS_x S_x^* = \sum_{x \in X} S_{g(x)} g|_x S_x^*, \tag{5}$$

which gives us a splitting rule for the expressions for elements of $\mathcal{V}_f$ given in Lemma 3.2.

Namely, we get the following description of $\mathcal{V}_f$ in terms of $\mathrm{IMG}(f)$ and formula (5).

Proposition 3.4. *The group $\mathcal{V}_f$ is isomorphic to the group of homeomorphisms of X^ω of the form*

$$\sum_{v \in A_1} S_{\alpha(v)} g_v S_v^*, \tag{6}$$

where $g_v \in \mathrm{IMG}(f)$, A_1 is a complete antichain in X^*, and $\alpha : A_1 \longrightarrow A_2$ is a bijection of A_1 with a complete antichain A_2. Two elements of $\mathcal{V}_f$ given by expressions of the form (6) are equal if and only if they can be made equal after repeated applications of the splitting rules (5) to the elements g_v.

Equivalently, we can use the table notation, and represent the element (6) by the table

$$\begin{pmatrix} v_1 & v_2 & \dots & v_m \\ g_{v_1} & g_{v_2} & \dots & g_{v_m} \\ \alpha(v_1) & \alpha(v_2) & \dots & \alpha(v_m) \end{pmatrix}, \tag{7}$$

where the splitting rule is the operation of replacing a column by the array

$$\begin{pmatrix} v \\ g \\ u \end{pmatrix} \mapsto \begin{pmatrix} vx_1 & vx_2 & \dots & vx_d \\ g|_{x_1} & g|_{x_2} & \dots & g|_{x_d} \\ ug(x_1) & ug(x_2) & \dots & ug(x_d) \end{pmatrix}. \tag{8}$$

Example 3.1. Consider the self-covering $f : x \mapsto 2x$ of the circle $\mathbb{R}/\mathbb{Z}$. Take $t = 0$ as the basepoint. Its preimages are 0 and $1/2$. Let ℓ_0 be the trivial path at 0, and let ℓ_1 be the path from 0 to $1/2$ equal to the image of the segment $[0, 1/2] \subset \mathbb{R}$. Let γ be the generator of $\pi_1(\mathbb{R}/\mathbb{Z}, 0)$ equal to the image of the segment $[0, 1] \subset \mathbb{R}$ with the natural (increasing on $[0, 1]$) orientation. It has two lifts by the covering f:

$$\gamma_0 = [0, 1/2], \qquad \gamma_1 = [1/2, 1].$$

Note that $\gamma_0 = \ell_1$.

By (2),

$$S_\gamma S_{\ell_0} = S_{\gamma_0 \ell_0} = S_{\ell_1}, \tag{9}$$

and

$$S_\gamma S_{\ell_1} = S_{\gamma_1 \ell_1} = S_{\ell_0 \gamma} = S_{\ell_0} S_\gamma. \tag{10}$$

Let $X = \{0, 1\}$, and consider the corresponding standard actions on X^* and X^ω. Denote by a the generator of $\mathrm{IMG}(f)$ corresponding to S_γ. Then, by (9) and (10),

$$aS_0 = S_1, \qquad aS_1 = S_0 a.$$

In other words, the action of a on X^ω is given by the recurrent formulas

$$a(0v) = 1v, \qquad a(1v) = 0a(v).$$

We see that a acts as the *binary adding machine*, see [Nek05, Section 1.7.1].

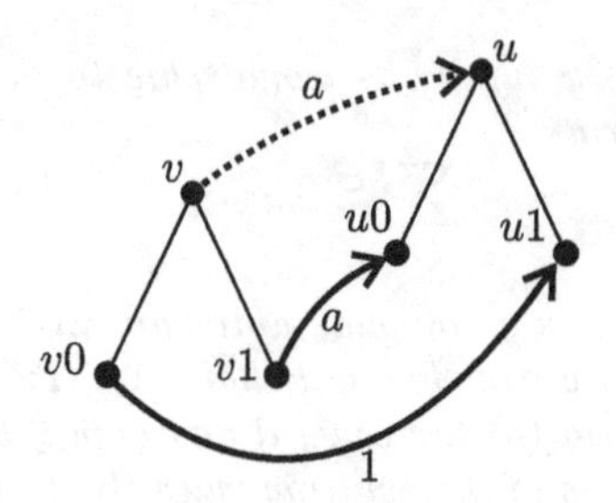

Figure 9: The adding machine

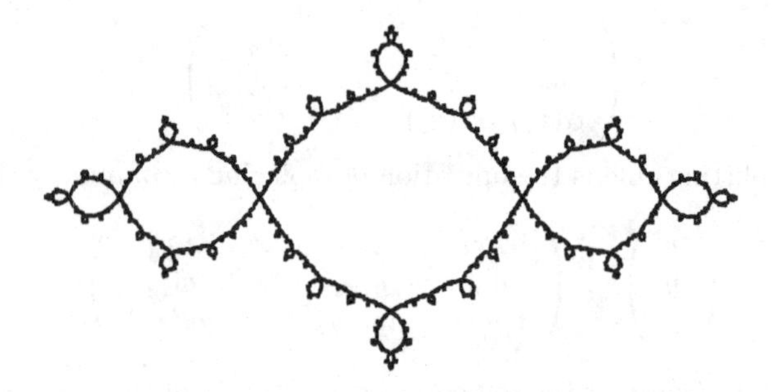

Figure 10: Julia set of $z^2 - 1$

The group $\mathcal{V}_f$ is generated by the Higman-Thompson group $\mathcal{V}_2$ (also coinciding with the Thompson group V, see [CFP96]) and an element a satisfying the splitting rule $a = S_1 S_0^* + S_0 a S_1^*$, i.e.,

$$\begin{pmatrix} v \\ a \\ u \end{pmatrix} = \begin{pmatrix} v0 & v1 \\ 1 & a \\ u1 & u0 \end{pmatrix}.$$

See Figure 9.

Example 3.2. Consider the complex polynomial $z^2 - 1$ as a partial self-covering $f : \mathcal{M}_1 \longrightarrow \mathcal{M}$, where $\mathcal{M} = \mathbb{C} \setminus \{0, -1\}$, and $\mathcal{M}_1 = \mathbb{C} \setminus \{0, \pm 1\}$. Alternatively, we can consider it as a self-covering of its Julia set, see Figure 10.

Then $\mathcal{V}_{z^2-1}$ is generated by the Thompson group $\mathcal{V}_2$ and two elements a, b satisfying

$$a = S_1 S_0^* + S_0 b S_1^*, \qquad b = S_0 S_0^* + S_1 a S_1^*.$$

For a detailed proof of the recurrent definitions of the elements a and b, see [Nek05, Subsection 5.2.2.].

3.4 Self-similar groups

The description of $\mathcal{V}_f$ in terms of the iterated monodromy group given in Proposition 3.4 can be generalized in the following way.

Definition 3.1. Let G be a group acting faithfully by automorphisms of the tree X^*. We say that G is a *self-similar group* if for all $g \in G$ and $x \in X$ there exist $h \in G$ and $y \in X$ such that

$$g(xw) = yh(w)$$

for all $w \in X^\omega$.

We will usually denote self-similar groups as pairs (G, X). Note that the equation in Definition 3.1 is equivalent to the equality

$$g \cdot S_x = S_y \cdot h,$$

of compositions of self-maps of X^ω, where S_x is, as before, the transformation $S_x(w) = xw$ of X^ω.

Definition 3.2. Let G be a self-similar group acting on X^*. The group $\mathcal{V}_G$ is the set of all homeomorphisms g of X^ω for which there exist complete antichains $A_1, A_2 \subset X^*$, a bijection $\alpha : A_1 \longrightarrow A_2$, and elements $g_v \in G$, for $v \in A_1$ such that

$$g = \sum_{v \in A_1} S_{\alpha(v)} g_v S_v^*,$$

i.e.,

$$g(vw) = \alpha(v) g_v(w)$$

for all $v \in A_1$ and $w \in X^\omega$.

If G is a self-similar group acting on X^*, then for every $v \in X^*$ there exists a unique element of G, denoted $g|_v$, such that

$$g(vw) = g(v) g|_v(w)$$

for all $w \in X^\omega$. We call $g|_v$ the *section* of g in v.

The elements of $\mathcal{V}_G$ are represented by tables of the form (7) with the splitting rule (8). The following proposition follows now directly from the described constructions.

Proposition 3.5. *Consider* IMG (f) *as a self-similar group with respect to a standard action on* X^*. *Then* $\mathcal{V}_f$ *(with the corresponding standard action on* X^ω*) is equal to* $\mathcal{V}_{\mathrm{IMG}(f)}$.

Let us describe more examples of groups $\mathcal{V}_f$ and $\mathcal{V}_G$.

Example 3.3. The Grigorchuk group G is generated by the transformations

$$a = S_1S_0^* + S_0S_1^*, \quad b = S_0aS_0^* + S_1cS_1^*, c = S_0aS_0^* + S_1dS_1^*, d = S_0S_0^* + S_1bS_1^*,$$

see [Gri80].

The corresponding group $\mathcal{V}_G$ was defined by C. Röver in [Röv99, Röv02]. This was the first example of a group $\mathcal{V}_G$. C. Röver proved that $\mathcal{V}_G$ is isomorphic to the abstract commensurator of G, that it is finitely presented, and simple. We will study the last two properties of the groups $\mathcal{V}_G$, generalizing the results of C. Röver for a wide class of self-similar groups.

Example 3.4. Let $f(z) = z^2 + c$ be a complex quadratic polynomial such that $f^n(0) = 0$ for some n (we assume that n is the smallest number with this property). Then f is a self-covering of its Julia set, which is path connected. The iterated monodromy groups $\mathrm{IMG}(f)$ associated with such polynomials were described in [BN08]. There exists a sequence $v = x_1x_2\ldots x_{n-1} \in \{0,1\}^{n-1}$ such that $\mathrm{IMG}(f)$ is isomorphic to the group $\mathfrak{K}_v$ generated by n elements $a_0, a_1, \ldots, a_{n-1}$ given by the recurrent relations

$$a_0 = S_1S_0^* + S_0a_{n-1}S_1^*,$$

and

$$a_i = \begin{cases} S_0a_{i-1}S_0^* + S_1S_1^* & \text{if } x_i = 0 \\ S_0S_0^* + S_1a_{i-1}S_1^* & \text{if } x_i = 1, \end{cases}$$

for $i = 1, 2, \ldots, n-1$. For example, $\mathrm{IMG}(z^2 - 1) = \mathfrak{K}_0$.

3.5 Wreath recursions

Let (G, X) be a self-similar group. Every element $g \in G$ defines a permutation σ_g of $X = X^1 \subset X^*$, and an element of G^X equal to the function $f_g : x \mapsto g|_x$. It is easy to check that the map $\psi : G \longrightarrow \mathrm{Symm}(X) \ltimes G^X$ mapping g to (σ_g, f_g) is a homomorphism of groups, which we call the *wreath recursion* associated with the self-similar group.

Let $X = \{1, 2, \ldots, d\}$. We will write elements of $\mathrm{Symm}(d) \ltimes G^d = \mathrm{Symm}(X) \ltimes G^X$ as products $\sigma(g_1, g_2, \ldots, g_d)$, where $\sigma \in \mathrm{Symm}(d)$ and $(g_1, g_2, \ldots, g_d) \in G^d$. Multiplication rule for elements of the wreath product $G \wr \mathrm{Symm}(d) = \mathrm{Symm}(d) \ltimes G^d$ is given by the formula

$$\sigma(g_1, g_2, \ldots, g_d)\pi(h_1, h_2, \ldots, h_d) = \sigma\pi(g_{\pi(1)}h_1, g_{\pi(2)}h_2, \ldots, g_{\pi(d)}h_d). \quad (11)$$

The wreath recursion completely describes the self-similar group G by giving recurrent formulas for the action of its elements on X^*.

Example 3.5. The adding machine, see Example 3.1, is given by the recursion $a = \sigma(1, a)$, where σ is the transposition $0 \leftrightarrow 1$. The generators of $\mathrm{IMG}(z^2 - 1)$ are given by

$$a = \sigma(1, b), \qquad b = (1, a),$$

see Example 3.2.

Any homomorphism $\psi : G \longrightarrow \mathrm{Symm}(d) \ltimes G^d$ defines an action of G on X^* (for $X = \{1, 2, \ldots, d\}$) by the recurrent rule:

$$g(xw) = \sigma(x) g_x(w),$$

where σ and g_x are defined by the condition $\psi(g) = \sigma(g_1, g_2, \ldots, g_d)$. This action is not faithful in general. The quotient of G by the kernel of its action on X^* is called the *faithful quotient* of G, and it is a self-similar group in the sense of Definition 3.1.

Let $\psi : G \longrightarrow \mathrm{Symm}(d) \ltimes G^d$ be an arbitrary homomorphism. Then we can define the group $\mathcal{V}_\psi$ associated with it in the same way as the groups $\mathcal{V}_G$ were defined for self-similar groups. Namely, elements of $\mathcal{V}_\psi$ are defined by tables of the form

$$\begin{pmatrix} v_1 & v_2 & \ldots & v_n \\ g_1 & g_2 & \ldots & g_n \\ u_1 & u_2 & \ldots & u_n \end{pmatrix},$$

where $g_i \in G$, and $\{v_1, v_2, \ldots, v_n\}$ and $\{u_1, u_2, \ldots, u_n\}$ are complete antichains of X^*. Two tables define the same element if they can be made equal (up to permutations of the columns) by iterated replacement of a column $\begin{pmatrix} v \\ g \\ u \end{pmatrix}$ by the columns

$$\begin{pmatrix} v1 & v2 & \ldots & vd \\ g_1 & g_2 & \ldots & g_d \\ u\sigma(1) & u\sigma(2) & \ldots & u\sigma(d) \end{pmatrix},$$

where $\psi(g) = \sigma(g_1, g_2, \ldots, g_d)$. Multiplication of the tables is defined by the rule

$$\begin{pmatrix} w_1 & \ldots & w_n \\ g_1 & \ldots & g_n \\ u_1 & \ldots & u_n \end{pmatrix} \begin{pmatrix} v_1 & \ldots & v_n \\ h_1 & \ldots & h_n \\ w_1 & \ldots & w_n \end{pmatrix} = \begin{pmatrix} v_1 & \ldots & v_n \\ g_1 h_1 & \ldots & g_n h_n \\ u_1 & \ldots & u_n \end{pmatrix}.$$

3.6 Bisets

A formalism equivalent to wreath recursions is provided by the notion of a *covering biset*. If (G, X) is a self-similar group, then the set of transformations $S_x g : w \mapsto xg(w)$ of X^ω is invariant under the left and right multiplications by elements of G:

$$S_x g \cdot h = S_x(gh), \qquad h \cdot S_x g = S_{h(x)}(h|_x g).$$

We get therefore commuting left and right actions of G on the set $\Phi = \{S_x g : x \in X, g \in G\}$. We will write elements $S_x \cdot g$ of Φ just as $x \cdot g$.

We adopt the following definition.

Definition 3.3. Let G be a group. A *G-biset* is a set Φ together with commuting left and right G-actions. It is called a *covering biset* if the right action is free (i.e., if $x \cdot g = x$ for $x \in \Phi$ and $g \in G$ implies $g = 1$) and has a finite number of orbits.

Let Φ_1, Φ_2 be G-bisets. Then their tensor product $\Phi_1 \otimes \Phi_2$ is defined as the quotient of the set $\Phi_1 \times \Phi_2$ by the identifications

$$(x \cdot g) \otimes y = x \otimes (g \cdot y), \qquad g \in G.$$

Let Φ be the biset $\{x \cdot g \,:\, x \in X, g \in G\}$ associated with a self-similar group. Then every element of $\Phi^{\otimes n}$ can be uniquely written in the form $x_1 \otimes x_2 \otimes \cdots \otimes x_n \cdot g$, where $x_i = x_i \cdot 1$ are elements of X. It follows that nth tensor power $\Phi^{\otimes n}$ is naturally identified with the set of pairs $v \cdot g$, for $v \in X^n$ and $g \in G$, with the actions

$$h \cdot (v \cdot g) = h(v) \cdot (h|_v g), \qquad (v \cdot g) \cdot h = v \cdot (gh).$$

Let Φ be an arbitrary covering G-biset. Choose a transversal $X \subset \Phi$ of the orbits of the right action. Then every element of Φ is uniquely written in the form $x \cdot g$ for $x \in X$ and $g \in G$. For every $g \in G$ and $x \in X$ there exist $h \in G$ and $y \in X$ such that

$$g \cdot x = y \cdot h,$$

and the elements y, h are uniquely determined by g and x. We have therefore a homomorphism $\psi : G \longrightarrow \mathrm{Symm}\,(X) \ltimes G^X$, called the *wreath recursion* associated with Φ and X. Namely, $\psi(g) = \sigma \cdot f$, where $\sigma \in \mathrm{Symm}\,(X)$ and $f \in G^X$ satisfy

$$g \cdot x = \sigma(x) \cdot f(x)$$

for all $x \in X$ (where G^X is seen as the set of functions $X \longrightarrow G$). If we change the orbit transversal X to an orbit transversal Y, then the homomorphism $\psi : G \longrightarrow \mathrm{Symm}\,(|X|) \ltimes G^{|X|}$ is composed with an inner automorphism of the wreath product (after we identify X with Y by a bijection). Note that the biset Φ is uniquely determined, up to an isomorphism of biset, by the homomorphism ψ.

Definition 3.4. Two self-similar actions (G, X_1) and (G, X_2) of a group G are called *equivalent* if their associated bisets $\Phi_i = X_i \cdot G$ are isomorphic, i.e., if there exists a bijection $F : \Phi_1 \longrightarrow \Phi_2$ such that $F(g_1 \cdot a \cdot g_2) = g_1 \cdot F(a) \cdot g_2$ for all $g_1, g_2 \in G$ and $a \in \Phi_1$. Two self-similar actions of groups G_1, G_2 are equivalent if they become equivalent after identification of the groups G_1, G_2 by an isomorphism $G_1 \longrightarrow G_2$.

The wreath recursion can be defined invariantly, without a choice of the orbit transversal. Namely, let $\mathrm{Aut}(\Phi_G)$ be the automorphism group of the right G-set Φ, i.e., the set of all bijections $\alpha : \Phi \longrightarrow \Phi$ such that $\alpha(x \cdot g) =$

$\alpha(x) \cdot g$. Then $\mathrm{Aut}(\Phi_G)$ is isomorphic to the wreath product $\mathrm{Symm}\,(d) \ltimes G^d$, where d is the number of the orbits of the right action on Φ, since the right G-set Φ is free and has d orbits, i.e., is isomorphic to the disjoint union of d copies of G. For every element $g \in G$ the map $\psi(g) : x \mapsto g \cdot x$ is an automorphism of the right G-set Φ. Then $\psi : G \longrightarrow \mathrm{Aut}(\Phi_G)$ is the wreath recursion. For more on wreath recursions and bisets, see [Nek05, Nek08, Nek14].

Example 3.6. Let $f : \mathcal{M}_1 \longrightarrow \mathcal{M}$ be a covering map, and let $\iota : \mathcal{M}_1 \longrightarrow \mathcal{M}$ be a continuous map (for example, f is a partial self-covering, and ι is the identical embedding).

Suppose that $\mathcal{M}$ is path-connected. Choose a basepoint $t \in \mathcal{M}$, and consider the set Φ of pairs (z, ℓ), where $z \in f^{-1}(t)$, and ℓ is a homotopy class of a path in $\mathcal{M}$ from t to $\iota(z)$. Then $\pi_1(\mathcal{M}, t)$ acts on $\mathcal{M}$ be appending loops to the beginning of the path ℓ:

$$(z, \ell) \cdot \gamma = (z, \ell\gamma).$$

It also acts by appending images of lifts of γ to the end of the path ℓ:

$$\gamma(z, \ell) = (z', \iota(\gamma_z)\ell),$$

where γ_z is the lift of γ by f starting at z, and z' is the end of γ_z. Here, as before, we multiply paths as functions (second path in a product is passed first).

Then Φ is a covering $\pi_1(\mathcal{M}, t)$-biset. The associated wreath recursion coincides with the wreath recursion associated with the standard action of $\mathrm{IMG}\,(f)$.

Let us show a more canonical definition of the groups $\mathcal{V}_\psi$ in terms of bisets. Let Φ be a covering G-biset. Consider the biset Φ^* equal to the disjoint union of the bisets $\Phi^{\otimes n}$ for all integers $n \geq 0$. Here $\Phi^{\otimes 0}$ is the group G with the natural G-biset structure. The set Φ^* is a semigroup with respect to the tensor product operation.

Let us order the semigroup Φ^* with respect to the left divisibility, i.e., $v \preceq u$ if and only if there exists w such that $u = v \otimes w$. It is easy to check that Φ^* is left-cancellative, i.e., that $v \otimes w_1 = v \otimes w_2$ implies $w_1 = w_2$.

The quotient of Φ^* by the right G-action is a rooted d-regular tree, and the image of $\preceq$ under the quotient map is the natural order on the rooted tree Φ^*/G. If X is a right orbit transversal of Φ, then $X^{\otimes n} = X^n$ is a right orbit transversal of $\Phi^{\otimes n}$, and the identical embedding of $X^* = \bigcup_{n \geq 0} X^{\otimes n}$ into Φ^* induces an isomorphism of the rooted tree X^* with Φ^*/G.

The left action of G on Φ^* permutes the orbits of the right action and preserves the relation $\preceq$, hence G acts on the tree Φ^*/G by automorphisms. The corresponding action on X^* is the self-similar action defined by the wreath recursion associated with X.

Let $A_1, A_2 \subset \Phi^*$ be finite maximal antichains with respect to the divisibility order $\preceq$. Note that a subset $A \subset \Phi^*$ is a finite maximal antichain if and only if its image in Φ^*/G is a maximal antichain. Choose a bijection $\alpha : A_1 \longrightarrow A_2$.

If $w \in \Phi^*$ is such that $v \preceq w$ for some $v \in A_1$, then there exists a unique $u \in \Phi^*$ such that $w = v \otimes u$. Consider then the transformation $h_{A_1,\alpha,A_2} : w \mapsto \alpha(v) \otimes u$. The map h_{A_1,α,A_2} is defined for all elements of Φ^* bigger than some element of A_1, hence for all elements of $\Phi^{\otimes n}$, where n is big enough. We will identify two transformation h_{A_1,α,A_2} and $h_{A_1',\alpha',A_2'}$ if their actions on the sets $\Phi^{\otimes n}$ agree for all n big enough. It is not hard to prove that the set of equivalence classes of such maps is a group, which we will denote $\mathcal{V}_\Phi$. It is also straightforward to show that $\mathcal{V}_\Phi$ coincides with $\mathcal{V}_\psi$, where ψ is the wreath recursion associated with Φ, and that if Φ is the usual biset associated with a self-similar group G, then $\mathcal{V}_\Phi$ coincides with $\mathcal{V}_G$.

3.7 Epimorphism onto the faithful quotient

Consider a covering G-biset Φ and the corresponding group $\mathcal{V}_\Phi$. The faithful quotient $\overline{G}$ is a self-similar group acting on X^*. We have, therefore two groups: $\mathcal{V}_\Phi$ and $\mathcal{V}_{\overline{G}}$, which are non-isomorphic in general (the group $\mathcal{V}_{\overline{G}}$ is a homomorphic image of $\mathcal{V}_\Phi$).

Proposition 3.6. *Let Φ be a covering biset. Denote by K_n the subgroup of elements of G acting trivially from the left on $\Phi^{\otimes n}$, i.e., the kernel of the wreath recursion associated with the biset $\Phi^{\otimes n}$. Then $K_n \supseteq K_{n-1}$. If $\bigcup_{n\geq 0} K_n$ is equal to the kernel of the epimorphism $G \longrightarrow \overline{G}$, then the natural epimorphism $\mathcal{V}_\Phi \longrightarrow \mathcal{V}_{\overline{G}}$ is an isomorphism.*

Proof. Suppose that an element g of the kernel of $\mathcal{V}_\Phi \longrightarrow \mathcal{V}_{\overline{G}}$ is defined by a table $\begin{pmatrix} v_1 & v_2 & \dots & v_n \\ g_1 & g_2 & \dots & g_n \\ u_1 & u_2 & \dots & u_n \end{pmatrix}$. Then the table $\begin{pmatrix} v_1 & v_2 & \dots & v_n \\ \overline{g_1} & \overline{g_2} & \dots & \overline{g_n} \\ u_1 & u_2 & \dots & u_n \end{pmatrix}$ represents the trivial element of $\mathcal{V}_{\overline{G}}$, where $g \mapsto \overline{g}$ is the epimorphism $G \longrightarrow \overline{G}$. But this means that $v_i = u_i$ and $\overline{g_i} = 1$ for all i. Consequently, there exists k such that $g_i \in K_k$ for all i. It follows that after applying elementary splittings k times to all column of the table defining g, we will get a table defining the trivial element of $\mathcal{V}_\Phi$, which means that $g = 1$. □

4 Simplicity of the commutator subgroup

4.1 Some general facts

Let G be a group acting faithfully by homeomorphisms on an infinite Hausdorff space $\mathcal{X}$. For an open subset $U \subset \mathcal{X}$, denote by $G_{(U)}$ the subgroup of elements of G acting trivially on $\mathcal{X} \setminus U$. Denote by R_U the normal closure in G of the derived subgroup $G_{(U)}' = [G_{(U)}, G_{(U)}]$.

The following simple lemma has appeared in many papers in different forms, see, for example [BGŠ03, Lemma 5.3], [Mat06, Theorem 4.9].

Lemma 4.1. *Let N be a non-trivial normal subgroup of G. Then there exists a non-empty open subset $U \subset \mathcal{X}$ such that $R_U \leq N$.*

Proof. It is sufficient to prove that there exists an open subset U such that $G_{(U)}' \leq N$.

Let $g \in N \setminus \{1\}$, and let $x \in \mathcal{X}$ be such that $g(x) \neq x$. Then there exists an open subset U such that $x \in U$ and $U \cap gU = \emptyset$. For example, find disjoint neighborhoods U_x and $U_{g(x)}$ of x and $g(x)$, and set $U = U_x \cap g^{-1}(U_{g(x)})$.

Let $h_1, h_2 \in G_{(U)}$. Then $gh_1^{-1}g^{-1}$ acts trivially outside gU. Consequently, $[g^{-1}, h_1] = gh_1^{-1}g^{-1}h_1$ acts as h_1 on U, as $gh_1^{-1}g^{-1}$ on gU, and trivially outside $U \cup gU$. It follows that $[[g^{-1}, h_1], h_2]$ acts as $[h_1, h_2]$ on U and trivially outside U, i.e., $[[g^{-1}, h_1], h_2] = [h_1, h_2]$. On the other hand, $[[g^{-1}, h_1], h_2] \in N$, since N is normal and $g \in N$. It follows that $[h_1, h_2] \in N$ for all $h_1, h_2 \in G_{(U)}$, i.e., that $G_{(U)}' \leq N$. □

Lemma 4.2. *Let U be an open subset of $\mathcal{X}$ such that its G-orbit is a basis of topology of $\mathcal{X}$. Then every non-trivial normal subgroup of G contains R_U.*

Proof. For any $g \in G$ and $U \subset \mathcal{X}$, we have $gG_{(U)}g^{-1} = G_{(gU)}$. Consequently, $gG_{(U)}'g^{-1} = G_{(gU)}'$, and R_U is equal to the group generated by $\bigcup_{g \in G} G_{(gU)}'$. In particular, $R_{gU} = R_U$ for all $g \in G$.

Since $\{gU : g \in G\}$ is a basis of topology, for every open subset $W \subset \mathcal{X}$ there exists $g \in G$ such that $gU \subseteq W$, hence $R_U = R_{gU} \leq R_W$. This implies, by Lemma 4.1, that R_U is contained in every non-trivial normal subgroup of G. □

Proposition 4.3. *Suppose that U is an open subset of $\mathcal{X}$ such that its R_U-orbit is equal to its G-orbit and is a basis of topology of $\mathcal{X}$. Then R_U is simple and is contained in every non-trivial normal subgroup of G.*

Proof. The group $G_{(U)}'$ is non-trivial, since otherwise its normal closure R_U is trivial, which contradicts the fact that the R_U-orbit of U is a basis of topology. It follows that there exists $g \in G_{(U)}'$ moving a point $x \in \mathcal{X}$. We have then $x \in U$ and $g(x) \in U$. There exists a neighborhood W of x such that $W \cap gW = \emptyset$. Then there exists an element $h \in R_U$ such that $hU \subset W$, and there exists a non-trivial element $g' \in G_{(hU)}' \leq G_{(U)}'$. We have $[g, g'] \neq 1$, since g' and $gg'g^{-1}$ have disjoint supports. This shows that $G_{(U)}'$ is non-abelian, i.e., that $G_{(U)}''$ is non-trivial.

The subgroup R_U is contained in every non-trivial normal subgroup of G, by Lemma 4.2. We also have that the normal closure in R_U of the group $(R_U)_{(U)}'$ is contained in every normal subgroup of R_U.

Note that $G_{(U)}'$ is contained in $G_{(U)} \cap R_U = (R_U)_{(U)}$, hence $G_{(U)}'' \leq (R_U)_{(U)}'$. Conjugating by an element $g \in G$ (and using that R_U is normal in G), we get that $G_{(gU)}'' \leq (R_U)_{(gU)}'$.

It follows that for any non-trivial subgroup $N \trianglelefteq R_U$ we have

$$N \supseteq \bigcup_{g \in R_U} (R_U)_{(gU)}' \supseteq \bigcup_{g \in R_U} G_{(gU)}'' = \bigcup_{g \in G} G_{(gU)}''.$$

Consequently, N contains the group generated by the set $\bigcup_{g \in G} G_{(gU)}''$, which is normal in G and non-trivial, hence contains R_U. We have proved that every non-trivial normal subgroup of R_U contains R_U, i.e., that R_U is simple. $\square$

4.2 Simplicity of $\mathcal{V}_G'$

Let (G, X) be a self-similar group, and let $\mathcal{V}_G$ be the corresponding group of homeomorphisms of the Cantor set X^ω.

Fix a linear ordering "<" of the elements of X. Extend it to the lexicographic ordering on X^*. Namely, if $x_1x_2 \ldots x_n$ and $y_1y_2 \ldots y_m$ are incomparable with respect to the order $\preceq$, then $x_1x_2 \ldots x_n < y_1y_2 \ldots y_m$ if and only if $x_i < y_i$, where i is the smallest index such that $x_i \neq y_i$. If a word v is a beginning of a word w, then $v \leq w$.

Suppose that $\begin{pmatrix} v_1 & v_2 & \ldots & v_n \\ u_1 & u_2 & \ldots & u_n \end{pmatrix}$ is a table defining an element $g \in \mathcal{V}_X$, i.e., that $g = \sum_{i=1}^n S_{u_i} S_{v_i}^*$. Assume that its first row is ordered in the increasing lexicographic order. If the permutation putting the second row into the increasing lexicographic order is even, then we say that the table is even.

Note that if $d = |X|$ is even, then every table has an even splitting. On the other hand, if d is odd, then the set of elements defined by even tables is a subgroup of index 2 in $\mathcal{V}_d$. The following is proved in [Hig74].

Theorem 4.4. *If d is even, then $\mathcal{V}_d$ is simple. If d is odd, then the commutator subgroup $\mathcal{V}_d'$ is the group of elements defined by even tables, and is a simple subgroup of index 2 in $\mathcal{V}_d$.*

The following theorem is proved in [Nek04, Theorem 9.11].

Theorem 4.5. *All proper quotients of $\mathcal{V}_G$ are abelian.*

Let us describe the $\mathcal{V}_d$-orbits of clopen subsets of X^ω.

Proposition 4.6. *Let $d = |X|$, and let U be a clopen subset of X^ω. Let us decompose U into a disjoint union $\bigsqcup_{i=1}^n v_i X^\omega$ for $v_i \in X^*$. Then the residue of n modulo $d-1$ does not depend on the decomposition. Let us denote it $m(U) \in \mathbb{Z}/(d-1)\mathbb{Z}$. Let $U_1, U_2 \subset X^\omega$ be non-empty clopen proper subsets. Then the following conditions are equivalent.*

1. *U_1 and U_2 belong to one $\mathcal{V}_X$-orbit.*

2. *U_1 and U_2 belong to one $\mathcal{V}'_X$-orbit.*

3. $m(U_1) = m(U_2)$.

Proof. Let $U = \bigsqcup_{i=1}^{n} v_i X^\omega$ be a decomposition of U. We can *split* it by replacing one set $v_i X^\omega$ by the collection of d sets $v_i x X^\omega$, for $x \in X$. This way we increase the number of sets in the decomposition by $d - 1$. It is easy to see that for any two decompositions of U into cylindrical sets there exist sequences of successive splittings of each of the decompositions leading to the same decomposition. This implies that $m(U)$ is well defined.

It is obvious that $m(U)$ is preserved under the action of $\mathcal{V}_X$. Suppose that $m(U_1) = m(U_2)$ for non-empty proper clopen subsets of X^ω. Then there exist decompositions $U_1 = \bigsqcup_{i=1}^{n} v_i X^\omega$ and $U_2 = \bigsqcup_{i=1}^{n} u_i X^\omega$ of the sets U_i into equal number of cylindrical subsets. The sets $\{v_i\}_{i=1}^{n}$ and $\{u_i\}_{i=1}^{n}$ are incomplete antichains in X^*, hence (see Lemma 3.1) there exists $g \in \mathcal{V}_X$ such that $g(v_i X^\omega) = u_i X^\omega$ for all i. If $g \notin \mathcal{V}'_X$, then we can compose it with an element $h \in \mathcal{V}_X \setminus \mathcal{V}'_X$ acting trivially on $\bigsqcup_{i=1}^{n} v_i X^\omega$, and get an element $gh \in \mathcal{V}'_X$ such that $gh(v_i X^\omega) = u_i X^\omega$ for all i. It follows that U_1 and U_2 belong to one $\mathcal{V}'_X$-orbit and to one $\mathcal{V}_X$-orbit. □

It easily follows from the definitions that $\mathcal{V}_X$-orbits of clopen subsets of X^ω coincide with their $\mathcal{V}_G$-orbits for any self-similar group (G, X).

Theorem 4.7. *The commutator subgroup of $\mathcal{V}_G$ is simple.*

Proof. Let $U = xX^\omega$ for some $x \in X$. Then the $\mathcal{V}_G$-orbit of U coincides with its $\mathcal{V}_X$-orbit and its $\mathcal{V}'_X$-orbit, and is a basis of the topology on X^ω, see Proposition 4.6. The group $\mathcal{V}'_X$ is contained in R_U, since the normal closure of $(\mathcal{V}_X)_{(U)}'$ in $\mathcal{V}_X$ is equal to $\mathcal{V}'_X$, by Theorem 4.4. It follows that the R_U-orbit of U is equal to its G-orbit, and is a basis of the topology.

Consequently, by Proposition 4.3, R_U is simple and is contained in every non-trivial normal subgroup of $\mathcal{V}_G$. On the other hand, by Theorem 4.5, every non-trivial normal subgroup of $\mathcal{V}_G$ contains the commutator subgroup $\mathcal{V}_G{}'$. It follows that $\mathcal{V}'_G = R_U$, and $\mathcal{V}'_G$ is simple. □

Abelianization of $\mathcal{V}_G$ is described in [Nek04, Theorem 9.14].

Theorem 4.8. *Let (G, X) be a self-similar group. Let $\pi : G \longrightarrow G/G'$ be the abelianization epimorphism.*

Suppose that d is even. Then $\mathcal{V}_G/\mathcal{V}'_G$ is isomorphic to the quotient of G/G' by the relations $\pi(g) = \sum_{x \in X} \pi(g|_x)$ for all $g \in G$.

Suppose that d is odd. Then $\mathcal{V}_G/\mathcal{V}'_G$ is isomorphic to $\mathbb{Z}/2\mathbb{Z} \oplus G/G'$ modulo the relations $\pi(g) = \mathrm{sign}(g) \oplus \sum_{x \in X} \pi(g|_x)$ for all $g \in G$; where $\mathrm{sign}(g) = 0$ *if g defines an even permutation on the first level X of X^*, and* $\mathrm{sign}(g) = 1$ *otherwise.*

The proof of the following lemma follows directly from the multiplication rule (11).

Lemma 4.9. *Let (G, X) be a self-similar group, and let $\pi : G \longrightarrow G/G'$ be the canonical homomorphism. Then $\sigma : \pi(g) \mapsto \sum_{x\in X} \pi(g|_x)$ is a well defined endomorphism of G/G'.*

For odd d denote by σ_1 the endomorphism $(t, g) \mapsto (t + \mathrm{sign}(g), \sigma(g))$ of $\mathbb{Z}/2\mathbb{Z} \oplus G/G'$, where $\mathrm{sign} : G/G' \longrightarrow \mathbb{Z}/2\mathbb{Z}$ gives the parity of the action of any preimage of g in G on the first level X of the tree X^*.

Then Theorem 4.8 can be reformulated as follows. (We denote here the identity automorphism of a group by 1.)

Theorem 4.10. *If d is even, then $\mathcal{V}_G/\mathcal{V}_G'$ is isomorphic to the quotient of G/G' by the range of the endomorphism $1 - \sigma$.*

If d is odd, then $\mathcal{V}_G/\mathcal{V}_G'$ is isomorphic to the quotient of $\mathbb{Z}/2\mathbb{Z} \oplus G/G'$ by the range of the endomorphism $1 - \sigma_1$.

Example 4.1. Consider the double self-covering $f(x) = 2x$ of the circle $\mathbb{R}/\mathbb{Z}$. Then $\mathrm{IMG}(f)$ is generated by the adding machine $a = S_1S_0^* + S_0aS_1^*$. It follows that $\mathcal{V}_f/\mathcal{V}_f'$ is the quotient of $\mathbb{Z} = \mathrm{IMG}(f) / \mathrm{IMG}(f)'$ by the range of $1-\sigma$, where $\sigma(a) = 0+a = a$ is the identity isomorphism. Hence, $1-\sigma = 0$, and $\mathcal{V}_f/\mathcal{V}_f' \cong \mathbb{Z}$.

Example 4.2. Let $G = \mathfrak{K}_{x_1x_2\ldots x_{n-1}}$ be the iterated monodromy group of a quadratic polynomial, as in Example 3.4. It is proved in [BN08, Proposition 3.3] that G/G' is the free abelian group $\mathbb{Z}^n$ freely generated by the images of the generators a_i of G. The recursive relations defining $\mathfrak{K}_{x_1x_2\ldots x_{n-1}}$ show that $\mathcal{V}_G/\mathcal{V}_G'$ is $\mathbb{Z}^n = \langle \pi(a_0), \ldots, \pi(a_{n-1})\rangle$ modulo the relations $\pi(a_0) = \pi(a_1)$, and $\pi(a_i) = \pi(a_{i-1})$ for all $i = 1, 2, \ldots, n-1$. Consequently, $\mathcal{V}_G/\mathcal{V}_G'$ is isomorphic to $\mathbb{Z}$.

Example 4.3. Let G be the Grigorchuk group, see Example 3.3. It is generated by

$$a = S_1S_0^* + S_0S_1^*, b = S_0aS_0^* + S_1cS_1^*, c = S_0aS_0^* + S_1dS_1^*, d = S_0S_0^* + S_1bS_1^*.$$

It follows that $\mathcal{V}_G/\mathcal{V}_G'$ is the group G/G' modulo the relations

$$\pi(a) = 0, \pi(b) = \pi(a) + \pi(c), \pi(c) = \pi(a) + \pi(d), \pi(d) = \pi(b),$$

which are equivalent to $\pi(a) = 0, \pi(b) = \pi(c) = \pi(d)$. It is easy to check that $b = cd$ and $b^2 = 1$ in G, which implies

$$\pi(c) = \pi(d) = \pi(b) = \pi(c) + \pi(d) = \pi(b) + \pi(b) = \pi(b^2) = 0.$$

Consequently, $\mathcal{V}_G/\mathcal{V}_G'$ is trivial, hence $\mathcal{V}_G$ is simple. Simplicity of $\mathcal{V}_G$ was proved in [Röv99].

5 Expanding maps and finite presentation

5.1 Contracting self-similar groups and expanding maps

Definition 5.1. Let (G, X) be a self-similar group. We say that it is *contracting* if there exists a finite set $\mathcal{N} \subset G$ such that for every $g \in G$ there exists n such that $g|_v \in \mathcal{N}$ for all $v \in X^*$, $|v| \geq n$.

More generally, let Φ be a covering G-biset, and let X be a right orbit transversal. Define then $g|_v$ for $g \in G$ and $v \in X^n \subset \Phi^{\otimes n}$ as the unique element of G such that $g \cdot v = u \cdot g|_v$ for some (also unique) $u \in X^n$. We say that Φ is contracting (or *hyperbolic*) if there exists a finite set $\mathcal{N}$ such that for every $g \in G$ there exists n such that $g|_v \in \mathcal{N}$ for all $v \in X^*$ such that $|v| \geq n$. One can show (see [Nek05, Corollary 2.11.7]) that this property depends only on Φ (but the set $\mathcal{N}$ will depend on X).

The smallest set $\mathcal{N}$ satisfying the conditions of Definition 5.1 is called the *nucleus* of the self-similar action.

Definition 5.2. Let $f : \mathcal{M}_1 \longrightarrow \mathcal{M}$ be a partial self-covering such that $\mathcal{M}$ is compact. The covering is called *expanding* if there exist $L > 1$, $\epsilon > 0$, a positive integer n, and a metric $|x - y|$ on $\mathcal{M}$ such that

$$|f^n(x) - f^n(y)| \geq L|x - y|$$

for all $x, y \in \mathcal{M}_n$ such that $|x - y| < \epsilon$.

For example, if $\mathcal{M}$ is a connected Riemann manifold, and $\|Df^n(\xi)\| \geq CL^n\|\xi\|$ for all $n \geq 1$ and every tangent vector ξ, then f is expanding.

The following proposition is proved in [Nek05, Theorem 5.5.3] for the case when $\mathcal{M}$ is a complete length metric space with finitely generated fundamental group. We will repeat here the proof in a more general situation (but avoiding orbispaces, which is one of more technical subjects of [Nek05]).

Proposition 5.1. *Let $f : \mathcal{M}_1 \longrightarrow \mathcal{M}$ be an expanding partial self-covering of a compact path connected space. Then* IMG (f) *is a contracting self-similar group (with respect to any standard action).*

Proof. Let $\{\ell_i\}_{i=1,\ldots,d}$ be paths connecting the basepoint $t \in \mathcal{M}$ to the preimages $z_i \in f^{-1}(t)$. We consider the standard action of IMG (f) on X^* defined by these connecting paths, where $X = \{1, \ldots, d\}$.

It follows from Definition 5.2 that there exist $\epsilon > 0, L > 1, C > 1$ such that for any subset $A \subset \mathcal{M}$ of diameter less than ϵ and every $n \geq 1$, the set $f^{-n}(A)$ is a disjoint union of d^n sets A_i such that $f^n : A_i \longrightarrow A$ are homeomorphisms, and diameters of A_i are less than CL^{-n}.

Let $g \in$ IMG (f) be defined by a loop γ. Since we can represent γ as a union of sub-paths of diameter less than ϵ, there exists $C_\gamma > 1$ such that diameter of any lift of γ by f^n is less than $C_\gamma L^{-n}$.

By Lemma 3.3, the section $g|_{i_1i_2\ldots i_n}$ is defined by the loop

$$\ell_{j_1j_2\ldots j_n}^{-1}\gamma_{i_1i_2\ldots i_n}\ell_{i_1i_2\ldots i_n},$$

where $\gamma_{i_1i_2\ldots i_n}$ is a lift of γ by f^n, and $\ell_{i_1i_2\ldots i_n}$, $\ell_{j_1j_2\ldots j_n}$ are paths inside the tree Γ_t formed by lifts of the connecting paths ℓ_i. Since diameters of lifts of paths by f^n exponentially decrease with n, there exists n_0 (depending only on the connecting paths ℓ_i) such that if $n_0 \leq n < m$, then for all sequences $i_1i_2\ldots i_m \in X^*$, the path $\ell_{i_1i_2\ldots i_m}$ is a continuation of the path $\ell_{i_1i_2\ldots i_n}$ by a path of diameter less than $\epsilon/6$.

There exists $n_1 \geq n_0$ (depending on γ) such that if $n \geq n_1$, then all lifts of γ by f^n have diameters less than $\epsilon/6$. Then for every $n \geq n_1$, the section $g|_{i_1i_2\ldots i_n}$ is defined by a loop of the form $\beta = \ell_{j_1j_2\ldots j_{n_0}}^{-1}\alpha\ell_{i_1i_2\ldots i_{n_0}}$, where α is a path of diameter less than $\epsilon/2$. If $\beta' = \ell_{j_1j_2\ldots j_{n_0}}^{-1}\alpha'\ell_{i_1i_2\ldots i_{n_0}}$ is another path such that α' has diameter less than $\epsilon/2$, then

$$\beta'\beta^{-1} = \ell_{j_1j_2\ldots j_{n_0}}^{-1}\alpha'\alpha^{-1}\ell_{j_1j_2\ldots j_{n_0}},$$

where $\alpha'\alpha^{-1}$ is a loop of diameter less than ϵ. By the choice of ϵ, all lifts of $\alpha'\alpha^{-1}$ by iterations of f are loops, which implies that β and β' define equal elements of IMG(f). Consequently, $g|_{i_1i_2\ldots i_n}$ is uniquely determined by the pair $(i_1i_2\ldots i_{n_0}, j_1j_2\ldots j_{n_0})$. Since n_0 does not depend on g, it follows that IMG(f) is contracting. □

5.2 General definition of $\mathcal{V}_f$ for expanding maps

The group $\mathcal{V}_f$ can be defined for an expanding self-covering $f : \mathcal{M} \longrightarrow \mathcal{M}$, even if $\mathcal{M}$ is not path connected.

Let $f : \mathcal{M} \longrightarrow \mathcal{M}$ be an expanding self-covering of a compact metric space $\mathcal{M}$, such that for every $t \in \mathcal{M}$ the set $\bigcup_{n\geq 0} f^{-n}(t)$ is dense in $\mathcal{M}$. This condition is always satisfied for a path-connected space $\mathcal{M}$.

Let $\delta > 0$ be such that for any two points $t_1, t_2 \in \mathcal{M}$ such that $t_1 \neq t_2$ and $f(t_1) = f(t_2)$ we have $|t_1 - t_2| > \delta$. It is easy to prove that such δ exists for any self-covering $f : \mathcal{M} \longrightarrow \mathcal{M}$ of a compact metric space. It follows from the definition of an expanding self-covering, that there exists $\epsilon > 0$ such that for any two points $z_1, z_2 \in \mathcal{M}$ and for any $n \geq 1$ there exists an isomorphism $S_{z_1,z_2} : T_{z_1} \longrightarrow T_{z_2}$ of the trees of preimages such that $|S_{z_1,z_2}(v) - v| < \delta/2$ for all $v \in T_{z_1}$. Moreover, it is easy to prove (by induction on the level number) that the isomorphism S_{z_1,z_2} is unique.

Fix a basepoint $t \in \mathcal{M}$, and define $\mathcal{V}_f$ as the group of homeomorphisms of ∂T_t piecewise equal to the isomorphisms $S_{z_1,z_2} : \partial T_{z_1} \longrightarrow \partial T_{z_2}$ for $z_1, z_2 \in T_t$.

Note that if $\gamma : [0, 1] \longrightarrow \mathcal{M}$ is a path in $\mathcal{M}$, then there exists n such that all lifts of γ by f^m for $m \geq n$ have diameter less than $\delta/2$. This implies that if $\mathcal{M}$ is path connected, then our original definition of $\mathcal{V}_f$ agrees with the given definition for expanding maps.

Example 5.1. Consider the one-sided shift $\mathsf{s} : X^\omega \longrightarrow X^\omega$

$$\mathsf{s}(x_1x_2\ldots) = x_2x_3\ldots.$$

Consider the metric $|w_1 - w_2| = 2^{-n}$, where n is the smallest index for which $x_n \neq y_n$, where $w_1 = x_1x_2\ldots$ and $w_2 = y_1y_2\ldots$.

Then $|\mathsf{s}(w_1)-\mathsf{s}(w_2)| = 2|w_1-w_2|$ whenever $|w_1-w_2| \le 1/2$. Consequently, s is expanding.

For any $w \in X^\omega$ the set $\mathsf{s}^{-n}(w)$ is equal to the set of sequences of the form vw, where $v \in X^n$. Hence, we can identify the nth level of the tree T_w with the set X^n by the map $vw \mapsto v$. Note that the tree T_w after this identification becomes the left Cayley graph of the monoid X^*: two vertices are connected by an edge if and only if they are of the form v, xv for $v \in X^*$, $x \in X$. In particular, the boundary ∂T_w is naturally identified with the space $X^{-\omega}$ of left-infinite sequences $\ldots x_2x_1$.

It follows directly from the definitions that the maps $S_{wv_1,wv_2} : T_{wv_1} \longrightarrow T_{wv_2}$ act by the rule

$$S_{wv_1,wv_2}(uv_1) = uv_2.$$

It follows that the homeomorphism $\ldots x_2x_1 \mapsto x_1x_2\ldots$ of $X^{-\omega} = \partial T_w$ with X^ω conjugates the action of $\mathcal{V}_\mathsf{s}$ with the Higman-Thompson group $\mathcal{V}_X$.

Example 5.2. If $\mathcal{M}$ is not connected, then a covering $f : \mathcal{M} \longrightarrow \mathcal{M}$ needs not to be of constant degree. For example, $f : \mathcal{M} \longrightarrow \mathcal{M}$ can be a one-sided shift of finite type. The corresponding group $\mathcal{V}_f$ is the topological full groups of a shift of finite type (the dual of f). These groups were studied in [Mat15].

Other examples of full groups of groupoids with their connection to semigroup theory were studied in [Law14, Law15].

5.3 Abelianization of $\mathcal{V}_f$ in expanding case

Self-similar contracting groups acting faithfully on X^* are typically infinitely presented. On the other hand, for every contracting group G there exists a finitely presented group $\tilde{G}$ and a hyperbolic covering $\tilde{G}$-biset Φ such that the faithful quotient of $\tilde{G}$ is G. More precisely, we have the following description of $\tilde{G}$, given in [Nek05, Section 2.13.2.].

Proposition 5.2. *Let (G, X) be a contracting group. Suppose that the nucleus $\mathcal{N}$ generates G.*

Let $\tilde{G}$ be the group given by the presentation $\langle \mathcal{N} \mid R \rangle$, where R is the set of all relations $g_1g_2g_3 = 1$ of length at most 3 that hold for elements of $\mathcal{N}$ in G. Let Φ be the $\tilde{G}$-biset of pairs $x \cdot g$, for $x \in X$ and $g \in \tilde{G}$ with the actions given by the usual rules:

$$(x \cdot g) \cdot h = x \cdot (gh), \qquad h \cdot (x \cdot g) = h(x) \cdot (h|_x g),$$

where $g \in \tilde{G}$, $h \in \mathcal{N}$, $x \in X$; and $h(x) \in X$, $h|_x \in \mathcal{N}$ are defined as in G. Then Φ is contracting.

Note that for any contracting group G, the group generated by the nucleus $\mathcal{N}$ is self-similar contracting, and $\mathcal{V}_{\langle\mathcal{N}\rangle} = \mathcal{V}_G$.

The following is proved in [Nek05, Proposition 2.13.2].

Proposition 5.3. *Let Φ be a contracting G-biset. Let $\rho : G \longrightarrow \overline{G}$ be the canonical epimorphism onto the faithful quotient of G. If $\rho(g) \neq 1$ for every non-trivial element of the nucleus of G (defined using some right orbit transversal X), then the kernel of ρ is equal to the union of the kernels K_n of the left actions of G on $\Phi^{\otimes n}$.*

Let Φ be a G-biset, and let d be the number of orbits of the right action of G on Φ. Choose a right orbit transversal $X \subset \Phi$, and define, for $g \in G$ and $x \in X$, the section $g|_x$ as the unique element of G such that $g \cdot x = y \cdot g|_x$ for $y \in X$. Let $\pi : G \longrightarrow G/G'$ be the abelianization epimorphism.

By the same arguments as in Lemma 4.9, the map $\sigma : \pi(g) \mapsto \sum_{x\in X} \pi(g|_x)$ is a well defined endomorphism of G/G'. It is also checked directly that it does not depend on the choice of the right orbit transversal X. If d is odd, then define homomorphism $\mathrm{sign} : G/G' \longrightarrow \mathbb{Z}/2\mathbb{Z}$ as in Theorem 4.8.

The following is a direct corollary of Propositions 5.3, 3.6, and Theorem 4.10.

Corollary 5.4. *Let Φ be a G-biset satisfying the conditions of Proposition 5.3.*

If d is even, then $\mathcal{V}_\Phi/\mathcal{V}'_\Phi$ is isomorphic to the quotient of G/G' by the range of the homomorphism $1 - \sigma$. If d is odd, then $\mathcal{V}_\Phi/\mathcal{V}'_\Phi$ is isomorphic to the quotient of $\mathbb{Z}/2\mathbb{Z} \oplus G/G'$ by the range of the endomorphism $1 - \sigma_1$, where $\sigma_1(t, g) = (t + \mathrm{sign}(g), \sigma(g))$.

Proposition 5.5. *Suppose that $f : \mathcal{M}_1 \longrightarrow \mathcal{M}$ is expanding, $\mathcal{M}$ is path-connected and semi-locally simply connected. Then the $\pi_1(\mathcal{M})$-biset associated with f is contracting.*

If $g \in \pi_1(\mathcal{M})$ has trivial image in $\mathrm{IMG}(f)$, then there exists n such that g acts trivially from the left on $\Phi_f^{\otimes n}$.

The first paragraph of the proposition is proved in the same way as Proposition 5.1. The second paragraph follows directly from exponential decreasing of diameters of lifts of paths by iterations of f and the condition that $\mathcal{M}$ is semi-locally simply connected.

Corollary 5.6. *Suppose that $f : \mathcal{M}_1 \longrightarrow \mathcal{M}$ satisfies the conditions of Proposition 5.5, and let Φ be the $\pi_1(\mathcal{M})$-biset associated with f. Then $\mathcal{V}_\Phi$ is isomorphic to $\mathcal{V}_f = \mathcal{V}_{\mathrm{IMG}(f)}$.*

Let $f : \mathcal{M}_1 \longrightarrow \mathcal{M}$ be a partial self-covering satisfying the conditions of Proposition 5.5. Let $\iota : \mathcal{M}_1 \longrightarrow \mathcal{M}$ be the identical embedding.

The group $\pi_1(\mathcal{M})/\pi_1(\mathcal{M})'$ is naturally isomorphic to the first homology group $H_1(\mathcal{M})$. The map $\sigma : H_1(\mathcal{M}) \longrightarrow H_1(\mathcal{M})$ from Corollary 5.4 is equal to $\iota_* \circ f^!$, where $f^! : H_1(\mathcal{M}) \longrightarrow H_1(\mathcal{M}_1)$ is the map (called the *transfer map*) given by the condition that image of a chain c is equal to its full preimage $f^{-1}(c)$.

Suppose that $c \in H_1(\mathcal{M})$ is defined by a loop γ. Then parity of the monodromy action of γ on fibers of f is well defined and generates a homomorphism from $H_1(\mathcal{M})$ to $\mathbb{Z}/2\mathbb{Z}$. Let us denote it by $\mathrm{sign} : H_1(\mathcal{M}) \longrightarrow \mathbb{Z}/2\mathbb{Z}$.

Then the following description of $\mathcal{V}_f/\mathcal{V}_f'$ follows directly from Corollary 5.4.

Proposition 5.7. *Suppose that $f : \mathcal{M}_1 \longrightarrow \mathcal{M}$ is expanding, $\mathcal{M}$ is path-connected and semi-locally simply connected.*

If $\deg f$ is even, then $\mathcal{V}_f/\mathcal{V}_f'$ is isomorphic to the quotient of $H_1(\mathcal{M})$ by the range of the endomorphism $1 - \iota_* \circ f^!$.

If $\deg f$ is odd, then $\mathcal{V}_f/\mathcal{V}_f'$ is isomorphic to the quotient of $\mathbb{Z}/2\mathbb{Z} \oplus H_1(\mathcal{M})$ by the range of the endomorphism $1 - \sigma_1$, where $\sigma_1(t, c) = (t + \mathrm{sign}(c), \iota_* \circ f^!(c))$.

5.4 Example: Hyperbolic rational functions

Let f be a complex rational function, and let C_f be the set of critical points of $f : \widehat{\mathbb{C}} \longrightarrow \widehat{\mathbb{C}}$. The *post-critical set* of f is the union $P_f = \bigcup_{n \geq 1} f^n(C_f)$ of forward orbits of critical values. Suppose that P_f is finite (we say then that f is *post-critically finite*).

Let us additionally suppose that every cycle of $f : P_f \longrightarrow P_f$ contains a critical point. Then f is *hyperbolic*, i.e., is expanding on a neighborhood of its Julia set, see [Mil06, Section 19].

One can find disjoint open topological discs around points of P_f such that if $\mathcal{M}$ is the complement of the union of these discs in the Riemann sphere, then $\mathcal{M}$ contains the Julia set of f, $\mathcal{M}_1 = f^{-1}(\mathcal{M}) \subset \mathcal{M}$, and there exists a metric on $\mathcal{M}$ such that $f : \mathcal{M}_1 \longrightarrow \mathcal{M}$ is strictly expanding. For instance, one can take discs bounded by the equipotential lines of the basins of attraction, see [Mil06, Section 9].

Then $H_1(\mathcal{M})$ is isomorphic to the quotient of the free group $\mathbb{Z}^{|P_f|}$ generated by elements a_z, $z \in P_f$, corresponding to the boundaries of the discs, modulo the relation $\sum_{z \in P_f} a_z = 0$. It is easy to see that the map $\sigma = \iota_* \circ f^!$ acts by the rule

$$\sigma(a_z) = \sum_{y \in f^{-1}(z) \cap P_f} a_y.$$

Suppose now that $\deg f$ is odd. We say that z is a *critical value mod 2* if $|f^{-1}(z)|$ is even. Note that if z is a critical value mod 2, then it is a critical value, since then $|f^{-1}(z)| \neq \deg f$. In particular, all critical values mod 2 belong to P_f. It is also easy to see that z is a critical value mod 2 if and only if the monodromy action of a small simple loop around z is an odd permutation. Namely, lengths of cycles of the monodromy action are equal to the local degrees of f at the preimages of z. The sum of local degrees is equal to $\deg f$, i.e., is odd, hence the number of odd local degrees is odd. Parity of the monodromy action is equal to parity of the number of cycles of even length, which is equal to parity of $|f^{-1}(z)|$ minus the number of odd local degrees, which is equal to parity of $|f^{-1}(z)| + 1$.

Proposition 5.8. *Let f be a hyperbolic post-critically finite rational function. Let k be the number of attracting cycles of f. Let l be the greatest common divisor of their lengths.*

If $\deg f$ is even, then $\mathcal{V}_f/\mathcal{V}'_f$ is isomorphic to $\mathbb{Z}^{k-1} \oplus \mathbb{Z}/l\mathbb{Z}$.

If $\deg f$ is odd, and there exists an attracting cycle C such that the number of critical values mod 2 whose forward f-orbits are attracted to C is odd, then $\mathcal{V}_f/\mathcal{V}'_f$ is also isomorphic to $\mathbb{Z}^{k-1} \oplus \mathbb{Z}/l\mathbb{Z}$. Otherwise, $\mathcal{V}_f/\mathcal{V}'_f \cong \mathbb{Z}/2\mathbb{Z} \oplus \mathbb{Z}^{k-1} \oplus \mathbb{Z}/l\mathbb{Z}$.

Note that $\mathbb{Z}^{k-1} \oplus \mathbb{Z}/l\mathbb{Z}$ coincides with the K_1-group of the C^*-algebraic analog of $\mathcal{V}_f$, see [Nek09]. Its K_0-group $\mathbb{Z}/(d-1)\mathbb{Z}$ has also appeared in our paper, see Proposition 4.6.

Proof. Every attracting cycle of f is superattracting (i.e., contains critical points of f), hence it belongs to the post-critical set P_f.

Suppose at first that $\deg f$ is even. If $z \in P_f$ does not belong to a cycle of $f : P_f \longrightarrow P_f$, then there exists n such that $f^{-n}(z) \cap P_f = \emptyset$ and hence $\sigma^n(a_z) = 0$. It follows that the images of such elements a_z under the epimorphism $\pi : H_1(\mathcal{M}) \longrightarrow \mathcal{V}_f/\mathcal{V}'_f$ are equal to zero.

If C is a cycle of $f : P_f \longrightarrow P_f$, then the images of a_z in $\mathcal{V}_f/\mathcal{V}'_f$, for $z \in C$, satisfy the relations $\pi(a_z) = \pi(a_{f(z)})$, since we have $\sigma(a_{f(z)}) = a_z$. It follows that $\mathcal{V}_f/\mathcal{V}'_f$ is the quotient of $H_1(\mathcal{M})$ by the relations making elements corresponding to the points of each cycle of $f : P_f \longrightarrow P_f$ equal, and making equal to zero all elements corresponding to elements of P_f not belonging to cycles. It follows that $\mathcal{V}_f/\mathcal{V}'_f$ is the quotient of the free abelian group $\mathbb{Z}^k = \langle e_1, e_2, \ldots, e_k \rangle$ by the relation $l_1 e_1 + l_2 e_2 + \cdots + l_k e_k = 0$, where l_i are the lengths of the corresponding cycles of $f : P_f \longrightarrow P_f$. Consequently, $\mathcal{V}_f/\mathcal{V}'_f \cong \mathbb{Z}^{k-1} \oplus \mathbb{Z}/l\mathbb{Z}$, where l is the g.c.d. of $l_1, l_2, \ldots, l_k$.

Suppose now that $\deg f$ is odd. Then $\mathcal{V}_f/\mathcal{V}'_f$ is isomorphic to the quotient of $\mathbb{Z}/2\mathbb{Z} \oplus H_1(\mathcal{M})$ by the relations $\sigma_1(a) = a$, where $\sigma_1(t, g) = (t + \mathrm{sign}(g), \sigma(g))$, where $\mathrm{sign}(g)$ is the parity of the monodromy action of g for the covering map $f : \mathcal{M}_1 \longrightarrow \mathcal{M}$.

It follows that σ_1 acts on the elements of the form $(0, a_z)$, for $z \in P_f$, by the rule

$$\sigma_1(0, a_z) = \begin{cases} \left(1, \sum_{y \in f^{-1}(z) \cap P_f} a_y\right) & \text{if } z \text{ is a critical value mod 2,} \\ \left(0, \sum_{y \in f^{-1}(z) \cap P_f} a_y\right) & \text{otherwise.} \end{cases}$$

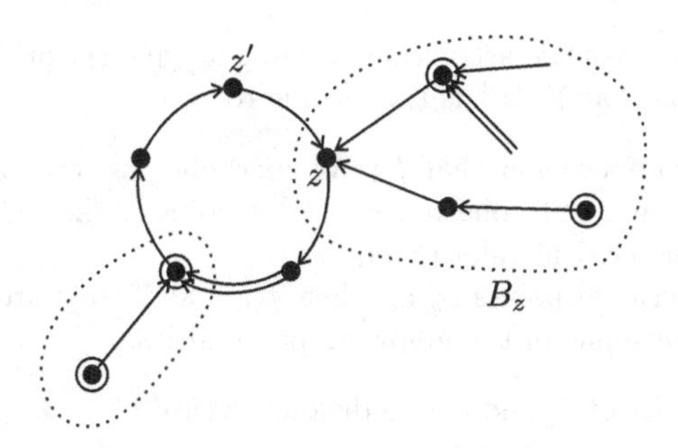

Figure 11: A post-critical cycle

Suppose that $z \in P_f$ is such that no point of $\bigcup_{n\geq 0} f^{-n}(z)$ is a critical value mod 2, and z does not belong to a cycle. Then there exists n such that $\sigma_1^n(0, a_z) = 0$, hence image of $(0, a_z)$ in $\mathcal{V}_f/\mathcal{V}'_f$ is zero.

If $z \in P_f$ is a critical value mod 2, but no point of $\bigcup_{n\geq 1} f^{-n}(z)$ is a critical value mod 2, then $\sigma_1(0, a_z) = (1, \sigma(a_z))$, and hence the image of $(0, a_z)$ in $\mathcal{V}_f/\mathcal{V}'_f$ is equal to the image of $(1, 0) \in \mathbb{Z}/2\mathbb{Z} \oplus H_1(\mathcal{M})$.

It follows by induction that if $z \in P_f$ does not belong to a cycle, then the image of $(0, a_z)$ in $V_f/\mathcal{V}'_f$ is equal to the image of $(m, 0) \in \mathbb{Z}/2\mathbb{Z} \oplus H_1(\mathcal{M})$, where m is the parity of the number of critical values mod 2 in the set $\bigcup_{n\geq 0} f^{-n}(z)$. In particular, $\mathcal{V}_f/\mathcal{V}'_f$ is a quotient of $\mathbb{Z}/2\mathbb{Z} \oplus H$, where $H \leq H_1(\mathcal{M})$ is the subgroup generated by a_z for z belonging to cycles of $f : P_f \longrightarrow P_f$.

Suppose now that C is a cycle of length r of the map $f : P_f \longrightarrow P_f$. For $z \in C$, denote by z' the unique element of C such that $f(z') = z$, and by t_z the parity of the number of critical values mod 2 in the set $B_z = \{z\} \cup \bigcup_{y\in f^{-1}(z)\setminus z'} \bigcup_{n\geq 0} f^{-1}(y)$, see Figure 11. Then $\mathcal{V}_f/\mathcal{V}'_f$ is isomorphic to the quotient of $\mathbb{Z}/2\mathbb{Z} \oplus H$ by the relations $(0, a_z) = (t_z, a_{z'})$.

Note that $t_C = \sum_{z\in C} t_z$ is the number of points y that are critical values mod 2 and $f^n(y) \in C$ for all n big enough. If t_C is odd, then we have a relation $(0, a_z) = (1, a_z)$, which implies that $(1, 0)$ belongs to the kernel of the epimorphism $\mathbb{Z}/2\mathbb{Z} \oplus H_1(\mathcal{M}) \longrightarrow \mathcal{V}_f/\mathcal{V}'_f$, and the arguments for even $\deg f$ show that $\mathcal{V}_f/\mathcal{V}'_f \cong \mathbb{Z}^{k-1} \oplus \mathbb{Z}/l\mathbb{Z}$.

Suppose that t_C is even for every cycle C of P_f. Order elements of every cycle $C \subset P_f$ into a sequence $z_0, z_1, \ldots, z_{r-1}$ so that $f(z_i) = z_{i-1}$ for all $i = 1, 2, \ldots, r-1$, and $f(z_0) = z_{r-1}$. Denote then $b_{z_0} = (0, a_{z_0}), b_{z_1} = (t_{z_0}, a_{z_1}), b_{z_2} = (t_{z_0} + t_{z_1}, a_{z_2}), \ldots, b_{z_{r-1}} = (t_{z_0} + t_{z_1} + \cdots + t_{z_{r-2}}, a_{z_0})$. Then $\mathcal{V}_f/\mathcal{V}'_f$ is the quotient of $\mathbb{Z}/2\mathbb{Z} \oplus H$ by the relations $b_{z_i} = b_{z_{i+1}}$ and $b_{z_{r-1}} = b_{z_0}$. It follows that $\mathcal{V}_f/\mathcal{V}'_f$ is isomorphic to $\mathbb{Z}/2\mathbb{Z} \oplus \mathbb{Z}^{k-1} \oplus \mathbb{Z}/l\mathbb{Z}$. □

Example 5.3. If $f(z) = z^2 + c$ is a hyperbolic post-critically finite quadratic polynomial, then it has two attracting cycles: $\{\infty\}$ and the orbit of the critical point 0. It follows that $\mathcal{V}_f/\mathcal{V}_f'$ is isomorphic to $\mathbb{Z}$.

Example 5.4. Suppose now that f is a hyperbolic post-critically finite cubic polynomial. If it has only one finite critical point c, then $|f^{-1}(f(c))| = 1$, hence there are no critical values mod 2.

If f has two critical points c_1, c_2, then $f(c_1)$ and $f(c_2)$ are critical values mod 2, and we have one of the following possibilities:

(a) forward orbits of c_1 and c_2 are disjoint cycles;

(b) both points c_1 and c_2 belong to a common cycle;

(c) one of the critical points belongs to a cycle C, and the forward orbit of the other critical point eventually belongs to C.

It follows now from Proposition 5.8 that $\mathcal{V}_f/\mathcal{V}_f' \cong \mathbb{Z}/2\mathbb{Z} \oplus \mathbb{Z}$ if the number of finite attracting cycles of f is 1, and $\mathcal{V}_f/\mathcal{V}_f' \cong \mathbb{Z}^2$ if it is 2.

5.5 Finite presentation

Theorem 5.9. *If G is a contracting self-similar group, then the group $\mathcal{V}_G$ is finitely presented.*

In particular, if $f : \mathcal{M} \longrightarrow \mathcal{M}$ is an expanding self-covering of a compact path connected metric space, then the group $\mathcal{V}_f$ is finitely presented.

Proof. Let $\mathcal{N}$ be the nucleus of G. We may assume that $\mathcal{N}$ is a generating set of G, since otherwise we can replace G by $\langle\mathcal{N}\rangle$ without changing $\mathcal{V}_G$.

For $v \in X^*$, and $g \in \mathcal{V}_G$, denote by $L_v(g)$ the element of $\mathcal{V}_G$ defined by the rule

$$L_v(g)(w) = \begin{cases} vg(u) & \text{if } w = vu \text{ for some } u \in X^\omega \\ w & \text{if } w \text{ does not begin with } v. \end{cases}$$

The following is straightforward.

Proposition 5.10. *For every $v \in X^*$ the map $L_v : \mathcal{V}_G \longrightarrow \mathcal{V}_G$ is a group monomorphism. If $v, u \in X^*$ are not comparable, then the subgroups $L_v(\mathcal{V}_G)$ and $L_u(\mathcal{V}_G)$ of $\mathcal{V}_G$ commute. If $v, u \in X^*$ are non-empty, and $h \in \mathcal{V}_G$ is such that $h(vw) = uw$ for all $w \in X^\omega$, then $h \cdot L_v(g) \cdot h^{-1} = L_u(g)$ for all $g \in \mathcal{V}_G$.*

Fix a letter $x_1 \in X$. We will denote $L(g) = L_{x_1}(g)$. For every pair $x, y \in X$ choose elements $A_{x,y}$ and B_x of $\mathcal{V}_X$ such that

$$A_{x,y}(yw) = xyw, \quad B_x(x_1w) = xw,$$

for all $w \in X^\omega$. We assume that $B_{x_1} = 1$.

Let $\langle S \mid R\rangle$ be a finite presentation of the Higman-Thompson group $\mathcal{V}_X$, see [Hig74]. Let S_1 be the set of elements of $\mathcal{V}_G$ of the form $L(g)$ for $g \in \mathcal{N}$.

Lemma 5.11. *The set $S \cup S_1$ generates $\mathcal{V}_G$.*

Proof. For every non-empty $v \in X^*$ we can find an element $h_v \in \mathcal{V}_X$ such that $h_v(vw) = x_1 w$ for all $w \in X^\omega$, see Lemma 3.1. Then $h_v^{-1} L(g) h_v = L_v(g)$ for all $g \in \mathcal{V}_G$. It follows that $L_v(g) \in \langle S \cup S_1 \rangle$ for all $g \in \mathcal{N}$ and $v \in X^* \setminus \{\varnothing\}$. Every element $g \in \mathcal{V}_G$ can be represented by a table

$$\begin{pmatrix} v_1 & v_2 & \dots & v_n \\ g_1 & g_2 & \dots & g_n \\ u_1 & u_2 & \dots & u_n \end{pmatrix},$$

where v_i, u_i are non-empty, and $g_i \in \mathcal{N}$. But then

$$g = \begin{pmatrix} v_1 & v_2 & \dots & v_n \\ 1 & 1 & \dots & 1 \\ u_1 & u_2 & \dots & u_n \end{pmatrix} L_{u_1}(g_1) L_{u_2}(g_2) \cdots L_{u_n}(g_n) \in \langle S \cup S_1 \rangle.$$

□

Represent each $A_{x,y}$, B_x as group words $\overline{A}_{x,y}$, $\overline{B}_x$ in S, and denote, for $y_1 y_2 \dots y_n \in X^n$ and $y \in X$,

$$\overline{A}_{y_1 y_2 \dots y_n, y} = \overline{A}_{y_1, y_2} \cdots \overline{A}_{y_{n-1}, y_n} \overline{A}_{y_n, y}.$$

Let $A_{v,y}$ be the image of $\overline{A}_{v,y}$ in $\mathcal{V}_X$. Then $A_{v,y}$ satisfies

$$A_{v,y}(yw) = vyw$$

for all $w \in X^\omega$.

For every word $v = y_1 y_2 \dots y_n$ of length at least 2 and every $g \in \mathcal{V}_G$ we have

$$L_v(g) = A_{y_1 y_2 \dots y_{n-1}, y_n} B_{y_n} L(g) B_{y_n}^{-1} A_{y_1 y_2 \dots y_{n-1}, y_n}^{-1}.$$

For every $v = y_1 y_2 \dots y_n \in X^*$ and $g \in \mathcal{N}$ denote by $\overline{L}_v(g)$ the word

$$\overline{A}_{y_1 y_2 \dots y_{n-1}, y_n} \overline{B}_{y_n} L(g) \overline{B}_{y_n}^{-1} \overline{A}_{y_1 y_2 \dots y_{n-1}, y_n}^{-1}$$

in generators $S \cup S_1$. Also denote by $\overline{L}_x(g)$ the word $\overline{B}_x L(g) \overline{B}_x^{-1}$.

Denote by $\mathrm{Symm}(d^n)$ the subgroup of $\mathcal{V}_X$ consisting of all elements of the form $\sum_{i=1}^{d^n} S_{u_i} S_{v_i}^*$, where $\{v_1, v_2, \dots, v_{d^n}\} = \{u_1, u_2, \dots, u_{d^n}\} = X^n$. It is isomorphic to the symmetric group of degree d^n. Here $d = |X|$.

For every $x \in X$ choose a finite generating set W_x (as a set of group words in S) of the group $(\mathcal{V}_X)_{(X^\omega \setminus x X^\omega)}$ of elements of $\mathcal{V}_X$ acting trivially on xX^ω. This group is isomorphic to the Higman-Thompson group $G_{d,d-1}$, hence is finitely generated (see [Hig74]).

Let R_1 be the union of the following sets of relations.

(C) **Commutation.** Relations of the form

$$[\overline{L}_x(g_1), \overline{L}_y(g_2)] = [\overline{L}_{v_1}(g_1), \overline{L}_{v_2}(g_2)] = [L(g), h] = 1$$

for all $g, g_1, g_2 \in \mathcal{N}, x, y \in X, v_1, v_2 \in X^2, h \in W_{x_1}$, where $x \neq y$ and $v_1 \neq v_2$.

(N) **Nucleus.** Relations of the form

$$L(g_1)L(g_2)L(g_3) = 1$$

for all $g_1, g_2, g_3 \in \mathcal{N}$ such that $g_1 g_2 g_3 = 1$ in G.

(S) **Splitting.** Relations of the form

$$L(g) = \overline{h} \cdot \overline{L}_{x_1 y_1}(g|_{y_1}) \overline{L}_{x_1 y_2}(g|_{y_2}) \cdots \overline{L}_{x_1 y_d}(g|_{y_d}),$$

for all $g \in \mathcal{N}$, where $\overline{h}$ is a word in the generators S representing an element $h \in \mathrm{Symm}\,(d^2)$ such that

$$L(g) = h L_{x_1 y_1}(g|_{y_1}) L_{x_1 y_2}(g|_{y_2}) \cdots L_{x_1 y_d}(g|_{y_d}).$$

Let us show that the set $R \cup R_1$ is a set of defining relations for the group $\mathcal{V}_G$. Denote by $\hat{\mathcal{V}}_G$ the group defined by the presentation $\langle S \cup S_1 \mid R \cup R_1 \rangle$. All relations $R \cup R_1$ hold in $\mathcal{V}_G$, hence $\mathcal{V}_G$ is a quotient of $\hat{\mathcal{V}}_G$, and it is enough to show that all relations in $\mathcal{V}_G$ also hold in $\hat{\mathcal{V}}_G$.

Note that since R is a set of defining relations of $\mathcal{V}_X$, a group word in S is trivial in $\hat{\mathcal{V}}_G$ if and only if it is trivial in $\mathcal{V}_X$. We will identify, therefore, the elements of the subgroup $\langle S \rangle \leq \hat{\mathcal{V}}_G$ with the corresponding elements of $\mathcal{V}_X$.

Lemma 5.12. *Suppose that $h \in \mathcal{V}_X$ and $u, v \in X^*$ are such that $h(uw) = vw$ for all $w \in X^\omega$. Then $h\overline{L}_u(g)h^{-1} = \overline{L}_v(g)$ holds in $\hat{\mathcal{V}}_G$.*

Proof. Let $u = a_1 a_2 \ldots a_n$ and $v = b_1 b_2 \ldots b_m$ for $a_i, b_i \in X$. Then

$$\overline{L}_u(g) = A_{a_1 a_2 \ldots a_{n-1}, a_n} B_{a_n} L(g) B_{a_n}^{-1} A_{a_1 a_2 \ldots a_{n-1}, a_n}^{-1}$$

and

$$\overline{L}_v(g) = A_{b_1 b_2 \ldots b_{m-1}, b_m} B_{b_m} L(g) B_{b_m}^{-1} A_{b_1 b_2 \ldots b_{m-1}, b_m}^{-1},$$

by definition. We have dropped the lines above the letters A and B, because the corresponding elements belong to $\mathcal{V}_X$.

We have then

$$\begin{aligned} h\overline{L}_u(g)h^{-1} = hA_{a_1 a_2 \ldots a_{n-1}, a_n} B_{a_n} L(g) B_{a_n}^{-1} A_{a_1 a_2 \ldots a_{n-1}, a_n}^{-1} h^{-1} = \\ A_{b_1 b_2 \ldots b_{m-1}, b_m} B_{b_m} \cdot B_{b_m}^{-1} A_{b_1 b_2 \ldots b_{m-1}, b_m}^{-1} h A_{a_1 a_2 \ldots a_{n-1}, a_n} B_{a_n} \cdot \\ L(g) \cdot \\ B_{a_n}^{-1} A_{a_1 a_2 \ldots a_{n-1}, a_n}^{-1} h^{-1} A_{b_1 b_2 \ldots b_{m-1}, b_m} B_{b_m} \cdot B_{b_m}^{-1} A_{b_1 b_2 \ldots b_{m-1}, b_m}^{-1}. \end{aligned}$$

The element

$$f = B_{b_m}^{-1} A_{b_1 b_2 \dots b_{m-1}, b_m}^{-1} h A_{a_1 a_2 \dots a_{n-1}, a_n} B_{a_n}$$

satisfies

$$f(x_1 w) = A_{v,y_m}^{-1} h A_{u,x_n} a_{y_m,x_n}(y_m w) = x_1 w,$$

for all $w \in X^\omega$. Hence, by relations (**C**), $L(g)$ commutes with f, i.e., $f \cdot L(g) \cdot f^{-1} = L(g)$ in $\hat{\mathcal{V}}_G$, which finishes the proof. □

Lemma 5.13. *If $v, u \in X^*$ are incomparable, then $\overline{L}_v(g_1)$ and $\overline{L}_u(g_2)$ commute in $\hat{\mathcal{V}}_G$ for all $g_1, g_2 \in \mathcal{N}$.*

Proof. Let $x, y \in X$ be a pair of different letters. Then $[\overline{L}_{xx}(g_1), \overline{L}_{xy}(g_2)] = 1$ in $\hat{\mathcal{V}}_G$, by (**C**). Since v, u are incomparable, either they both have length 1, or they form an incomplete antichain. In the first case commutation of $\overline{L}_v(g_1)$ and $\overline{L}_u(g_2)$ is a part of relations (**C**). In the second case, there exists $a \in \mathcal{V}_X$ such that $a(uw) = xxw$ and $a(vw) = xyw$ for all w (see Lemma 3.1). Then, by Lemma 5.12,

$$[\overline{L}_v(g_1), \overline{L}_u(g_2)] = a[\overline{L}_{xx}(g_1), \overline{L}_{xy}(g_2)]a^{-1} = 1$$

in $\hat{\mathcal{V}}_G$. □

Let us prove now that any group word in $S \cup S_1$ that is trivial in $\mathcal{V}_G$ is trivial in $\hat{\mathcal{V}}_G$. Note that relations (**S**) and Lemma 5.12 imply relations

(**S'**)

$$\overline{L}_v(g) = h \cdot \overline{L}_{vy_1}(g|_{y_1}) \overline{L}_{vy_2}(g|_{y_2}) \cdots \overline{L}_{vy_d}(g|_{y_d})$$

for all $g \in \mathcal{N}$ and non-empty $v \in X^*$, where h is an element of $\mathcal{V}_d$ such that $L_v(g) = h L_{vy_1}(g|_{y_1}) L_{vy_2}(g|_{y_2}) \cdots L_{vy_d}(g|_{y_d})$.

Each element of $\hat{\mathcal{V}}_G$ can be written in the form $hL(g_1)^{h_1} L(g_2)^{h_2} \cdots L(g_n)^{h_n}$ for $h, h_i \in \mathcal{V}_X$, and $g_i \in \mathcal{N}$.

Let n_1 be such that the element h_n can be written as $\sum_{i=1}^{d^{n_1}} S_{u_i} S_{v_i}^*$, where $\{u_1, u_2, \dots, u_{d^{n_1}}\} = X^{n_1}$. Using relations (**S'**) and Lemma 5.12, we can rewrite the element $L(g_n)$ in the form $\alpha \prod_{v \in X^{n_1 - 1}} \overline{L}_{x_1 v}(g_n|_v)$ for some $\alpha \in \mathrm{Symm}(d^{n_1})$. (Note that the factors $\overline{L}_{x_1 v}(g_n|_v)$ commute with each other, by Lemma 5.13.) Then for every $v \in X^{n_1 - 1}$ there exists i such that $x_1 v = u_i$, and then by Lemma 5.12, we have

$$\overline{L}_{x_1 v}(g_n|_v)^{h_n} = \overline{L}_{v_i}(g_n|_v),$$

so that $L(g_n)^{h_n}$ can be rewritten as a product of α^{h_n} followed by a product of elements of the form $\overline{L}_v(g_{v,n})$ for some $v \in X^{n_1}$ and $g|_{v,n} \in \mathcal{N}$.

It follows by induction that every element of $\hat{\mathcal{V}}_G$ can be written in the form

$$g = h\overline{L}_{v_1}(g_1)\overline{L}_{v_2}(g_2)\cdots\overline{L}_{v_m}(g_m) \tag{12}$$

for some $v_i \in X^*$, $g_i \in \mathcal{N}$, and $h \in \mathcal{V}_X$.

Suppose that not all words v_i are of the same length. Let v_i be the shortest, and let $k > |v_i|$ be the shortest length of words v_j strictly longer than v_i. Using relations (**S'**), we can rewrite $\overline{L}_{v_i}(g_i)$ as $\alpha_i \prod_{u\in X^{k-|v_i|}} L_{v_i u}(g_i|_u)$, and then, using Lemma 5.12, move $\alpha_i \in \mathrm{Symm}\left(d^k\right)$ to the beginning of the product (12). This procedure will increase the length of the shortest word v_i in (12) without changing the length of the longest one. Repeating this procedure, we will change (12) to a product of the same form, but in which all words v_i are of the same length.

Therefore, we may assume that in (12) all words v_i are of the same length k. Note that then $\overline{L}_{v_1}(g_1)\overline{L}_{v_2}(g_2)\cdots\overline{L}_{v_m}(g_m)$ does not change the beginning of length k in any word $w \in X^\omega$. Since g is trivial in $\mathcal{V}_G$, this implies that h does not change the beginning of length k in any word. It follows that we can write h as a product $\prod_{v\in X^k} L_v(h_v)$ for some $h_v \in \mathcal{V}_X$. Using Lemmas 5.12 and 5.13 we can now rearrange the factors of (12) in such a way that $g = \prod_{v\in X^k} f_v$, where $f_v = L_v(h_v)\overline{L}_v(g_{v,1})\overline{L}_v(g_{v,2})\cdots\overline{L}_v(g_{v,m_v})$ for $g_{v,i} \in \mathcal{N}$ and $h_v \in \mathcal{V}_X$. Note that f_v are trivial in $\mathcal{V}_G$. The latter implies that $h_v \in \mathrm{Symm}\left(d^l\right)$ for some l, and that the action of $h_v g_{v,1} g_{v,2}\cdots g_{v,m_v}$ on X^l is trivial. Consequently, using relations (**S'**), we can rewrite f_v as a product of elements of the form $\overline{L}_{vu}(g)$ for $u \in X^l$. Therefore, we may assume that h_v are trivial. Then $g_{v,1}g_{v,2}\cdots g_{v,m_v}$ is trivial in G. Relations (**N**), (**S**), and Propositions 5.2, 5.3 finish the proof. □

6 Dynamical systems and groupoids

This section is an overview of relations between expanding dynamical systems and self-similar groups, basic definitions of the theory of étale groupoids, and properties of hyperbolic groupoids. For more details and proofs, see [Nek05, Nek08, Nek14] and [Ren80, Pat99, Hae01, Nek15].

6.1 Limit dynamical system of a contracting group

Let (G, X) be a contracting self-similar group. Denote by $X^{-\omega}$ the space of all left-infinite sequences $\ldots x_2x_1$ of elements of X with the direct product topology.

Definition 6.1. Sequences $\ldots x_2x_1, \ldots y_2y_1 \in X^{-\omega}$ are *asymptotically equivalent* if there exists a finite set $N \subset G$ and a sequence $g_n \in N$ such that

$$g_n(x_n \ldots x_2x_1) = y_n \ldots y_2y_1,$$

for all $n \in \mathbb{N}$.

Denote by $\mathcal{J}_G$ the quotient of the space $X^{-\omega}$ by the asymptotic equivalence relation.

The asymptotic equivalence relation on $X^{-\omega}$ is invariant with respect to the shift $\ldots x_2x_1 \mapsto \ldots x_3x_2$. Therefore, the shift induces a continuous map $f : \mathcal{J}_G \longrightarrow \mathcal{J}_G$. The dynamical system $(f, \mathcal{J}_G)$ is called the *limit dynamical system* of G.

The map $f : \mathcal{J}_G \longrightarrow \mathcal{J}_G$ is expanding in the sense of Definition 5.2, if it is a covering. Namely, we can represent $\mathcal{J}_G$ as the boundary of a naturally defined Gromov hyperbolic graph (see [Nek03] and [Nek05, Section 3.8]), and some iteration of f will be locally uniformly expanding with respect to the visual metric on the boundary.

Definition 6.2. A self-similar group (G, X) is said to be *regular* if for every $g \in G$ there exists a positive integer n such that for every $v \in X^n$ either $g(v) \neq v$, or $g|_v = 1$.

Note that it is enough to check the conditions of Definition 6.2 for elements g of the nucleus of G.

The following proposition is proved in [Nek09, Proposition 6.1].

Proposition 6.1. *The shift map $f : \mathcal{J}_G \longrightarrow \mathcal{J}_G$ is a covering if and only if G is regular.*

Definition 6.3. A self-similar action of G on X^* is said to be *self-replicating* (*recurrent* in [Nek05]) if the left action of G on the associated biset is transitive, i.e., if for every $x, y \in X$ there exists $g \in G$ such that $g \cdot x = y \cdot 1$.

An automorphism group G of the rooted tree X^* is said to be *level-transitive* if it is transitive on X^n for every n.

Note that every self-replicating group is level transitive.

Theorem 6.2. *Let G be a contracting group. The space $\mathcal{J}_G$ is connected if and only if G is level-transitive. It is path connected if and only if G is self-replicating. If G is self-replicating, then $\mathcal{J}_G$ is also locally path connected.*

Proof. Proof of connectedness, local connectedness, and path connectedness of $\mathcal{J}_G$ under the appropriate conditions is given in [Nek05, Theorem 3.6.3.].

If $\ldots x_2x_1$ and $\ldots y_2y_1$ are asymptotically equivalent elements of $X^{-\omega}$, then for any n the words $x_n \ldots x_2x_1$ and $y_n \ldots y_2y_1$ belong to the same G-orbit. If there exists n such that the action of G on X^n is not transitive, then partition of X^n into G-orbits defines a partition of $X^{-\omega}$ into clopen sets such that the asymptotic equivalence relation identifies only points inside the sets of the partition. This implies that $\mathcal{J}_G$ is disconnected.

The same arguments shows that if G is not self-replicating, then $\mathcal{X}_G$ is disconnected (see Subsection 6.2). Moreover, if $v_1, v_2 \in X^n \cdot G$ belong to different orbits of the left action, then for any $k, m \geq 0$, and any $u_1, u_2 \in X^k \cdot G$

and $w_1, w_2 \in X^m \cdot G$ the elements $u_1 \otimes v_1 \otimes w_1$ and $u_2 \otimes v_2 \otimes w_2$ belong to different orbits of the left action. It follows that the set of connected components of $\mathcal{X}_G$ is then uncountable. Consequently, the set of path connected components of $\mathcal{X}_G$ is uncountable, and since G is countable, the set of path connected components of $\mathcal{J}_G = \mathcal{X}_G/G$ is also uncountable. □

The following theorem is proved in [Nek05, Sections 5.3, 5.5] (in the context of length metric spaces, but all the arguments remain to be valid in the general case, if we use diameters of paths instead of their lengths, as in the proof of Proposition 5.1).

Theorem 6.3. *Suppose that $f : \mathcal{J} \longrightarrow \mathcal{J}$ is an expanding self-covering of a path connected space. Then* IMG (f) *is contracting, regular, self-replicating, and the limit dynamical system of* IMG (f) *is topologically conjugate to $(f, \mathcal{J})$.*

Let G be a contracting regular self-replicating group. Then it is equivalent, as a self-similar group (see Definition 3.4), to the iterated monodromy group of its limit dynamical system.

Corollary 6.4. *Let $f_i : \mathcal{J}_i \longrightarrow \mathcal{J}_i$, for $i = 1, 2$, be expanding self-coverings of path connected compact spaces. Then $(f_1, \mathcal{J}_1)$ and $(f_2, \mathcal{J}_2)$ are topologically conjugate if and only if* IMG (f_1) *and* IMG (f_2) *are equivalent as self-similar groups.*

6.2 Limit solenoid and the limit G-space

Let $X^{\mathbb{Z}}$ be the space of all bi-infinite sequences $\ldots x_{-2}x_{-1}.x_0x_1\ldots$, where the dot denotes the place between the coordinates number 0 and -1. Sequences $(x_n)_{n\in\mathbb{Z}}, (y_n)_{n\in\mathbb{Z}} \in X^{\mathbb{Z}}$ are *asymptotically equivalent* if there exists a finite set $N \subset G$ and a sequence $g_n \in N$ such that

$$g_n(x_nx_{n+1}\ldots) = y_ny_{n+1}\ldots$$

for all $n \in \mathbb{Z}$.

The quotient $\mathcal{S}_G$ of $X^{\mathbb{Z}}$ by the asymptotic equivalence relation is called the *limit solenoid* of the group G. The shift map

$$\ldots x_{-2}x_{-1}.x_0x_1\ldots \mapsto \ldots x_{-3}x_{-2}.x_{-1}x_0\ldots$$

induces a homeomorphism of $\mathcal{S}_G$, which we will denote by $\hat{f}$.

It is shown in [Nek05, Proposition 5.7.8.] that the space $\mathcal{S}_G$ is naturally homeomorphic to the inverse limit of the backward iterations of the limit dynamical system $f : \mathcal{J}_G \longrightarrow \mathcal{J}_G$:

$$\mathcal{J}_G \longleftarrow \mathcal{J}_G \longleftarrow \mathcal{J}_G \longleftarrow \cdots,$$

and the map $\hat{f}$ is conjugate to the map induced by f on the inverse limit. In other words, $(\hat{f}, \mathcal{S}_G)$ is the *natural extension* of the limit dynamical system

$(f, \mathcal{J}_G)$. The point of $\mathcal{S}_G$ represented by a sequence $\ldots x_{-2}x_{-1}.x_0x_1\ldots \in X^{\mathbb{Z}}$ corresponds to the point of the inverse limit represented by the sequence

$$\cdots \mapsto \ldots x_{-2}x_{-1}x_0x_1x_2 \mapsto \ldots x_{-2}x_{-1}x_0x_1 \mapsto \ldots x_{-2}x_{-1}x_0 \mapsto \ldots x_{-2}x_{-1}.$$

Another natural dynamical system associated with a contracting group G is the limit G-space $\mathcal{X}_G$. Consider the topological space $X^{-\omega} \times G$, where G is discrete. Two pairs $(\ldots x_2x_1, g), (\ldots y_2y_1, h) \in X^{-\omega} \times G$ are *asymptotically equivalent* if there exists a sequence $g_n \in G$ taking a finite set of values such that for all $n \geq 1$

$$g_n \cdot x_n \ldots x_2x_1 \cdot g = y_n \ldots y_2y_1 \cdot h$$

in the nth tensor power $\Phi^{\otimes n}$ of the associated G-biset, i.e., if

$$g_n(x_n \ldots x_2x_1) = y_n \ldots y_2y_1, \qquad g_n|_{x_n \ldots x_2x_1} g = h.$$

The quotient of $X^{-\omega} \times G$ by the asymptotic equivalence relation is called the *limit G-space* $\mathcal{X}_G$. We represent the points of the space $\mathcal{X}_G$ by the sequences $\ldots x_2x_1 \cdot g$, where $x_i \in X$ and $g \in G$.

The asymptotic equivalence relation on $X^{-\omega} \times G$ is invariant with respect to the right action of G on the second factor of the direct product. It follows that this action induces a right action of G on $\mathcal{X}_G$ by homeomorphisms. The action of G on $\mathcal{X}_G$ is proper, and the space of orbits $\mathcal{X}_G/G$ is naturally homeomorphic to $\mathcal{J}_G$.

The spaces $\mathcal{J}_G, \mathcal{S}_G, \mathcal{X}_G$ and the corresponding dynamical systems depend only on the biset Φ associated with the self-similar group. For example, $\mathcal{X}_G$ can be constructed in the following way.

Let Ω be the direct limit of the spaces $A^{-\omega}$, where A runs through all finite subsets of Φ. We write a sequence $(\ldots, a_2, a_1) \in A^{-\omega}$ as $\ldots \otimes a_2 \otimes a_1$. Two sequences $\ldots \otimes a_2 \otimes a_1, \ldots \otimes b_2 \otimes b_1 \in \Omega$ are said to be equivalent if there exist a sequence $g_n \in G$ taking values in a finite set, such that

$$g_n \cdot a_n \otimes \cdots \otimes a_2 \otimes a_1 = b_n \otimes \cdots \otimes b_2 \otimes b_1$$

in $\Phi^{\otimes n}$ for all n.

The quotient of Ω by this equivalence relation is naturally homeomorphic to $\mathcal{X}_G$. Moreover, the homeomorphism conjugates the natural action on $\mathcal{X}_G$ with the action induced by the natural right action of G on Ω.

For every $v \cdot g \in X^n \cdot G = \Phi^{\otimes |v|}$ we have the map $w \mapsto w \otimes (v \cdot g)$ on Ω, given in terms of $X^{-\omega} \times G$ by the rule

$$\ldots x_2x_1 \cdot h \mapsto \ldots x_2x_1h(v) \cdot h|_v g.$$

It induces a continuous map $F_{v \cdot g} : \mathcal{X}_G \longrightarrow \mathcal{X}_G$. If G is regular, then $F_{v \cdot g}$ is a covering map.

Since the limit dynamical system $f : \mathcal{J}_G \longrightarrow \mathcal{J}_G$ is induced by the shift on $X^{-\omega}$, the maps $F_{v \cdot g} : \mathcal{X}_G \longrightarrow \mathcal{X}_G$ are lifts of $f^{-|v|}$ by the quotient map $P : \mathcal{X}_G \longrightarrow \mathcal{X}_G/G = \mathcal{J}_G$, i.e., we have equality $f^{|v|} \circ P \circ F_{v \cdot g} = P$.

6.3 Groupoids of germs

Definition 6.4. Let $\mathcal{X}$ be a topological space. A *pseudogroup* acting on $\mathcal{X}$ is a set $\widetilde{\mathfrak{G}}$ of homeomorphisms between open subsets of $\mathcal{X}$ that is closed under

1. *compositions*: if $F_1 : U_1 \longrightarrow V_1$ and $F_2 : U_2 \longrightarrow V_2$ are elements of $\widetilde{\mathfrak{G}}$, then $F_1 \circ F_2 : F_2^{-1}(V_2 \cap U_1) \longrightarrow F_1(V_2 \cap U_1)$ is an element of $\widetilde{\mathfrak{G}}$;

2. *taking inverses*: if $F : U \longrightarrow V$ is an element of $\widetilde{\mathfrak{G}}$, then $F^{-1} : V \longrightarrow U$ is an element of $\widetilde{\mathfrak{G}}$;

3. *taking restrictions* to an open subset: if $F : V \longrightarrow U$ is an element of $\widetilde{\mathfrak{G}}$ and V' is an open subset of V, then $F|_{V'} \in \widetilde{\mathfrak{G}}$;

4. *taking unions*: if for a homeomorphism $F : U \longrightarrow V$ between open subsets of $\mathcal{X}$ there exists a cover U_i of U by open subsets, such that $F|_{U_i} \in \widetilde{\mathfrak{G}}$ for all i, then $F \in \widetilde{\mathfrak{G}}$.

We always assume that the identity homeomorphism $\mathcal{X} \longrightarrow \mathcal{X}$ belongs to the pseudogroup.

Let $\widetilde{\mathfrak{G}}$ be a pseudogroup acting on $\mathcal{X}$. A *germ* of $\widetilde{\mathfrak{G}}$ is an equivalence class of a pair (F, x), where $F \in \widetilde{\mathfrak{G}}$, and x is a point of the domain of F. Two pairs (F_1, x_1) and (F_2, x_2) represent the same germ (are equivalent) if and only if $x_1 = x_2$ and there exists a neighborhood U of x_1 such that $F_1|_U = F_2|_U$.

The set of all germs of $\widetilde{\mathfrak{G}}$ has a natural topology whose basis consists of sets of the form $\{(F, x) : x \in \mathrm{dom}(F)\}$, where $F \in \widetilde{\mathfrak{G}}$.

If (F_1, x_1) and (F_2, x_2) are germs such that $F_2(x_2) = x_1$, then we can compose them:

$$(F_1, x_1)(F_2, x_2) = (F_1 \circ F_2, x_2).$$

The *inverse* of a germ (F, x) is the germ $(F^{-1}, F(x))$. The set $\mathfrak{G}$ of all germs of $\widetilde{\mathfrak{G}}$ is a groupoid with respect to these operations (i.e., a small category of isomorphisms).

The groupoid $\mathfrak{G}$ is *topological*, i.e., the operations of composition and taking inverse are continuous.

Example 6.1. If $f : \mathcal{X} \longrightarrow \mathcal{X}$ is a covering map, then restrictions of f to open subsets $U \subset \mathcal{X}$ such that $f : U \longrightarrow f(U)$ is a homeomorphism generate a pseudogroup. Its groupoid of germs $\mathfrak{F}$ will be called the *groupoid of germs generated by* f. Every element of $\mathfrak{F}$ can be represented as a product $(f^n, x)^{-1}(f^m, y)$ for some $x, y \in \mathcal{X}$ such that $f^m(y) = f^n(x)$.

Example 6.2. If G is a group acting on a topological space $\mathcal{X}$, then every germ of the pseudogroup generated by G is a germ of an element of G. Therefore, the *groupoid of germs of* G is the set of equivalence classes of pairs $(g, x) \in G \times \mathcal{X}$, where (g_1, x_1) and (g_2, x_2) are equivalent if and only if $x_1 = x_2$, and $g_1^{-1} g_2$ fixes all points of a neighborhood of x_1. This groupoid is in general different from the *groupoid of the action*, which is equal as a set to $G \times \mathcal{X}$.

If $\mathfrak{G}$ is a groupoid of germs of a pseudogroup acting on a space $\mathcal{X}$, then we identify the germ $(1, x)$ of the identity homeomorphism $1 : \mathcal{X} \longrightarrow \mathcal{X}$ with the point x of $\mathcal{X}$, and call the elements of the form $(1, x)$ the *units* of the groupoid. We will sometimes denote $\mathcal{X}$ by $\mathfrak{G}^{(0)}$, as the space of units of $\mathfrak{G}$.

For $(F, x) \in \mathfrak{G}$, we denote by $\mathsf{o}(F, x) = x$ and $\mathsf{t}(F, x) = F(x)$ the *origin* and *target* of the germ. Two germs $g_1, g_2 \in \mathfrak{G}$ are *composable* (i.e., $g_1 g_2$ is defined) if and only if $\mathsf{t}(g_2) = \mathsf{o}(g_1)$.

We say that points $x, y \in \mathfrak{G}^{(0)}$ *belong to one orbit* if there exists $g \in \mathfrak{G}$ such that $x = \mathsf{o}(g)$ and $y = \mathsf{t}(g)$. This is an equivalence relation on $\mathfrak{G}^{(0)} = \mathcal{X}$, and this notion of orbits coincides with the natural notion of orbits of pseudogroups. A set $A \subset \mathfrak{G}^{(0)}$ is a *$\mathfrak{G}$-transversal* if it intersects every $\mathfrak{G}$-orbit.

If A is a subset of $\mathfrak{G}^{(0)}$, then *restriction* $\mathfrak{G}|_A$ of $\mathfrak{G}$ to A is the groupoid of all elements $g \in \mathfrak{G}$ such that $\mathsf{o}(g), \mathsf{t}(g) \in A$. The *isotropy* group $\mathfrak{G}_x$, for $x \in \mathfrak{G}^{(0)}$, is the group of elements $g \in \mathfrak{G}$ such that $\mathsf{o}(g) = \mathsf{t}(g) = x$.

Note that the pseudogroup $\widetilde{\mathfrak{G}}$ can be reconstructed from the groupoid of its germs $\mathfrak{G}$. Namely, a *bisection* is a subset $F \subseteq \mathfrak{G}$ of the groupoid, such that $\mathsf{o} : F \longrightarrow \mathsf{o}(F)$ and $\mathsf{t} : F \longrightarrow \mathsf{t}(F)$ are homeomorphisms. Every open bisection F defines a homeomorphism $\mathsf{o}(F) \longrightarrow \mathsf{t}(F)$ by the rule $\mathsf{o}(g) \mapsto \mathsf{t}(g)$ for $g \in F$. The set $\widetilde{\mathfrak{G}}$ of all open bisections is a pseudogroup, and if $\mathfrak{G}$ is the groupoid of germs of a pseudogroup $\widetilde{\mathfrak{G}}$, then the pseudogroup of bisections coincides with $\widetilde{\mathfrak{G}}$. We say that $\widetilde{\mathfrak{G}}$ is the *associated pseudogroup* of the groupoid.

Definition 6.5. Let $\mathfrak{G}_1, \mathfrak{G}_2$ be groupoids of germs. We say that they are *equivalent* if there exists a groupoid $\mathfrak{G}$ such that $\mathfrak{G}^{(0)}$ is the disjoint union $\mathfrak{G}_1^{(0)} \sqcup \mathfrak{G}_2^{(0)}$, the restriction of $\mathfrak{G}$ to $\mathfrak{G}_i^{(0)}$ is equal to $\mathfrak{G}_i$ for $i = 1, 2$, and the sets $\mathfrak{G}_i^{(0)}$ are $\mathfrak{G}$-transversals.

The following procedure is a standard way of constructing a groupoid equivalent to a given one. Namely, let $p : \mathcal{Y} \longrightarrow \mathfrak{G}^{(0)}$ be a local homeomorphism, i.e., for every $y \in \mathcal{Y}$ there exists a neighborhood U of y such that $p : U \longrightarrow p(U)$ is a homeomorphism. Suppose that $p(\mathcal{Y})$ is a $\mathfrak{G}$-transversal. Then *lift* of $\mathfrak{G}$ by p is the groupoid of germs of the pseudogroup generated by all local homeomorphisms of the form $p' \circ F \circ p : U \longrightarrow W$, where

- U is such that $p : U \longrightarrow p(U)$ is a homeomorphism,
- $p(U)$ is contained in the domain of F,
- W is such that $p : W \longrightarrow F(p(U))$ is a homeomorphism,
- p' is the inverse of $p : W \longrightarrow F(p(U))$.

Then the map p induces a morphism from the lift of $\mathfrak{G}$ to $\mathfrak{G}$, mapping the germ of $p' \circ F \circ p$ at x to the germ of F at $p(x)$.

Example 6.3. Consider the *trivial groupoid* on a manifold $\mathcal{M}$, i.e., the groupoid consisting of units only. Let $\pi : \widetilde{\mathcal{M}} \longrightarrow \mathcal{M}$ be the universal covering. Then lift of the trivial groupoid by π is the groupoid of germs of the action of the fundamental group on $\widetilde{\mathcal{M}}$. (In this case it coincides with the groupoid of the action.)

Definition 6.6. Let $\mathfrak{G}$ be a groupoid of germs. It is said to be *proper* if the map $(\mathsf{o}, \mathsf{t}) : \mathfrak{G} \longrightarrow \mathfrak{G}^{(0)} \times \mathfrak{G}^{(0)}$ is proper, i.e., if preimages of compact subsets of $\mathfrak{G}^{(0)} \times \mathfrak{G}^{(0)}$ under this map are compact.

The groupoid $\mathfrak{G}$ is proper if and only if for every compact subset C of $\mathfrak{G}^{(0)}$ the set of elements $g \in \mathfrak{G}$ such that $\mathsf{o}(g), \mathsf{t}(g) \in C$ is compact.

If $\mathfrak{G}$ is proper, then for every $x \in \mathfrak{G}^{(0)}$ the isotropy group $\mathfrak{G}_x$ is finite.

Every groupoid equivalent to a proper groupoid is proper. If $\mathfrak{G}$ is proper, then the space of orbits of $\mathfrak{G}$ is Hausdorff.

Let $\mathfrak{G}$ be a groupoid of germs. Its *topological full group* $[[\mathfrak{G}]]$ is the set of all bisections F such that $\mathsf{o}(F) = \mathsf{t}(F) = \mathfrak{G}^{(0)}$, i.e., the set of homeomorphisms $F : \mathfrak{G}^{(0)} \longrightarrow \mathfrak{G}^{(0)}$ such that all germs of F belong to $\mathfrak{G}$. See [GPS99], where the notion of a topological full group (for a groupoid of germs generated by one homeomorphism) was introduced.

Example 6.4. Let $f : \mathcal{M}_1 \longrightarrow \mathcal{M}$ be a partial self-covering. Then $\mathcal{V}_f$ is the full topological group of the groupoid of germs of the local homeomorphisms S_γ of the boundary of the tree T_t.

Example 6.5. Let G be a self-similar group acting on X^ω. Let $\mathfrak{G}$ be the groupoid of germs of the pseudogroup generated by the action of G and the germs of the homeomorphisms $S_x(x_1x_2\ldots) = xx_1x_2\ldots$ for $x \in X$. It is easy to see that the topological full group of $\mathfrak{G}$ is the group $\mathcal{V}_G$.

6.4 Hyperbolic groupoids

Here we present a very short overview of the basic definitions and results of the paper [Nek15].

Let $\mathfrak{G}$ be a groupoid of germs. A *compact generating pair* of $\mathfrak{G}$ is a pair $(S, \mathcal{X}_1)$, where $S \subset \mathfrak{G}$ and $\mathcal{X}_1 \subset \mathfrak{G}^{(0)}$ are compact, $\mathcal{X}_1$ contains an open $\mathfrak{G}$-transversal, $S \subset \mathfrak{G}|_{\mathcal{X}_1}$, and for every $g \in \mathfrak{G}|_{\mathcal{X}_1}$ there exists n such that $(S \cup S^{-1})^n$ is a neighborhood of g in $\mathfrak{G}|_{\mathcal{X}_1}$.

A groupoid is *compactly generated* if it has a compact generating pair. See a variant of this definition in [Hae02]. A groupoid equivalent to a compactly generated groupoid is also compactly generated.

Let $(S, \mathcal{X}_1)$ be a compact generating pair of $\mathfrak{G}$. Let $x \in \mathcal{X}_1$. Then the *Cayley graph* $\Gamma(x, S)$ is the oriented graph with the set of vertices

$$\{g \in \mathfrak{G} \ : \ \mathsf{o}(g) = x, \mathsf{t}(g) \in \mathcal{X}_1\},$$

in which there is an arrow from g_1 to g_2 if and only if $g_2 g_1^{-1} \in S$.

A path $v_1, v_2, \dots$ in a graph Γ (i.e., a sequence of vertices such that v_i is adjacent to v_{i+1}) is said to be a C-quasi-geodesic (where $C > 1$ is a constant) if $|v_i - v_j| \geq C^{-1}|i-j| + C$ for all i, j.

Definition 6.7. A Hausdorff groupoid of germs $\mathfrak{G}$ is *hyperbolic* if there is a compact generating pair $(S, \mathcal{X}_1)$ of $\mathfrak{G}$, a metric $|x-y|$ defined on a neighborhood of $\mathcal{X}_1$, and constants $L, C > 1, \Delta > 0$ such that

1. Every element $g \in S$ is a germ of a homeomorphism $F \in \widetilde{\mathfrak{G}}$ such that $|F(x) - F(y)| \leq L^{-1}|x-y|$ for all $x, y \in \operatorname{Dom} F$.
2. For every $x \in \mathcal{X}_1$ the Cayley graph $\Gamma(x, S)$ is Gromov Δ-hyperbolic.
3. For every $x \in \mathcal{X}_1$ there exists a point ω_x of the boundary of $\Gamma(x, S)$ such that every oriented path in the Cayley graph $\Gamma(x, S^{-1})$ is a C-quasi-geodesic converging to ω_x.
4. $\mathsf{o}(S) = \mathsf{t}(S) = \mathcal{X}_1$.
5. All elements of the pseudogroup $\widetilde{\mathfrak{G}}$ are locally bi-Lipschitz.

Example 6.6. Let $f : \mathcal{J} \longrightarrow \mathcal{J}$ be an expanding self-covering of a compact metric space. Then the groupoid of germs generated by f is hyperbolic. The corresponding generating pair is $(S, \mathcal{J})$, where S is the set of germs of f^{-1}. The corresponding Cayley graphs $\Gamma(x, S)$ are trees. The special point ω_x of the boundary is the limit of the forward germs (f^n, x) for $n \to +\infty$. See more in [Nek15, Section 5.2]

Let $\mathfrak{G}$ be a hyperbolic groupoid. For every $x \in \mathfrak{G}^{(0)}$ there exists a generating pair $(S, \mathcal{X}_1)$ satisfying the conditions of Definition 6.7 and such that $x \in \mathcal{X}_1$. Denote by $\partial\mathfrak{G}_x$ the boundary of the Cayley graph $\Gamma(x, S)$ minus the point ω_x. The space $\partial\mathfrak{G}_x$ does not depend on the generating pair.

Let $(S, \mathcal{X}_1)$ be a generating pair satisfying the conditions of Definition 6.7. Find a finite set of contractions $\mathcal{F} \subset \widetilde{\mathfrak{G}}$ such that $S \subset \bigcup_{F \in \mathcal{F}} F$, i.e., every element $s \in S$ is a germ of a contraction $F \in \mathcal{F}$. Every point $\xi \in \partial\mathfrak{G}_x$ can be represented as the limit of a sequence of vertices of the Cayley graph of the form $g, s_1 g, s_2 s_1 g, \dots, s_n \cdots s_2 s_1 g, \dots$, where $g \in \mathfrak{G}$, and $s_i \in S$. There exists $\epsilon > 0$ (not depending on ξ) and a sequence $F_i \in \mathcal{F}$ such that s_i is equal to a germ (F_i, x_i), and the ϵ-neighborhood of $x_i = \mathsf{o}(s_i)$ belongs to the domain of F_i. Then there exists δ (also depending only on S and $\mathcal{F}$) such that the δ-neighborhood of $\mathsf{t}(g)$ belongs to the domain of $F_n \circ \cdots \circ F_2 \circ F_1$ for all n.

Let $F \in \widetilde{\mathfrak{G}}$ be such that $g \in F$ and $\mathsf{t}(F)$ belongs to the δ-neighborhood of $\mathsf{t}(g)$. Then $\mathsf{o}(F_n \circ \cdots \circ F_2 \circ F_1 \circ F) = \mathsf{o}(F)$ for every n.

It is shown in [Nek15] that there exists a topology on the disjoint union $\partial\mathfrak{G}$ of the spaces $\partial\mathfrak{G}_x$, $x \in \mathfrak{G}^{(0)}$, which agrees with the topology on its subsets

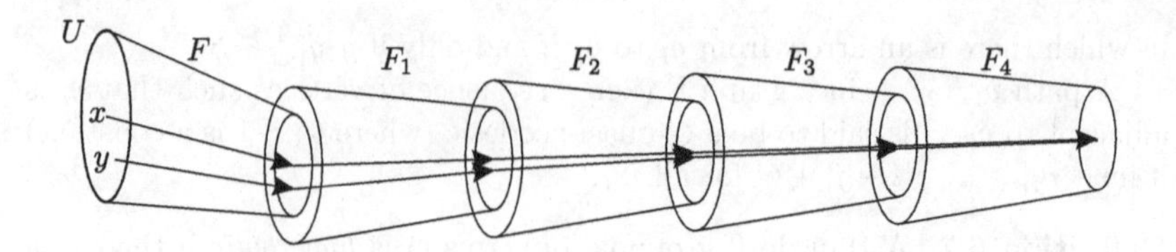

Figure 12: Composition of contractions

$\partial\mathfrak{G}_x$ and such that the map

$$(y, \xi) \mapsto \lim_{n\to\infty}(F_n \circ F_{n-1} \circ \cdots \circ F_1 \circ F, y) \in \partial\mathfrak{G}_y \tag{13}$$

is a well defined homeomorphism (if $U = \mathsf{o}(F)$ is small enough) from the direct product of $U = \mathsf{o}(F)$ and a subset of $\partial\mathfrak{G}_x$ to a subset of $\partial\mathfrak{G}$, see Figure 12.

Moreover, these homeomorphisms agree with a natural *local product structure* of $\partial\mathfrak{G}$. Namely, a basis of the topology on $\partial\mathfrak{G}$ consists of *rectangles*, i.e., sets with a decomposition into a direct product of topological spaces, such that the decompositions agree where they overlap, and locally coincide with the maps given by (13).

The groupoid $\mathfrak{G}$ acts on the space $\partial\mathfrak{G}$ from the right. Namely, every $g \in \mathfrak{G}$ defines a natural homeomorphism $\partial\mathfrak{G}_{\mathsf{t}(g)} \longrightarrow \partial\mathfrak{G}_{\mathsf{o}(g)}$ mapping the limit of a sequence $g_n \in \Gamma(\mathsf{t}(g), S)$ to the limit of the sequence $g_n g \in \Gamma(\mathsf{o}(g), S)$. This action is an action of the topological groupoid $\mathfrak{G}$ on the topological space $\partial\mathfrak{G}$ over the projection map $P : \partial\mathfrak{G} \longrightarrow \mathfrak{G}^{(0)}$ mapping all points of $\partial\mathfrak{G}_x$ to x.

The action of $\mathfrak{G}$ on $\partial\mathfrak{G}$ (i.e., the associated action of $\tilde{\mathfrak{G}}$ on $\partial\mathfrak{G}$ by local homeomorphisms) preserves the local product structure of $\partial\mathfrak{G}$. Naturally defined projection of the action of $\mathfrak{G}$ onto the first coordinate of the local product decomposition is equivalent to $\mathfrak{G}$, while the projection onto the second coordinate is the *dual groupoid* of $\mathfrak{G}$.

Let us give an equivalent, and maybe more intuitive, definition of the dual groupoid.

Definition 6.8. Let $\overline{\Gamma(x, S)}$ and $\overline{\Gamma(y, S)}$ be the Cayley graphs of $\mathfrak{G}$ with adjoined boundaries $\partial\mathfrak{G}_x$ and $\partial\mathfrak{G}_y$. A homeomorphism $F : U \longrightarrow V$ between open neighborhoods $U \subset \overline{\Gamma(x, S)}$ and $V \subset \overline{\Gamma(y, S)}$ of points of $\partial\mathfrak{G}_x$ and $\partial\mathfrak{G}_y$ is an *asymptotic morphism* if for every sequence of pairwise different edges $(g_1, h_1), (g_2, h_2), \ldots$ in U the distance between $g_i h_i^{-1}$ and $F(g_i)F(h_i)^{-1}$ goes to zero.

Note that $g_i h_i^{-1}$ and $F(g_i)F(h_i)^{-1}$ belong to a compact subset of $\mathfrak{G}$, hence the notion of convergence of their distance to zero does not depend on the choice of a metric on $\mathfrak{G}$.

Definition 6.9. The groupoid $\partial\mathfrak{G}$ of germs of restrictions of the asymptotic morphisms to the spaces $\partial\mathfrak{G}_x$, $x \in \mathfrak{G}^{(0)}$, is the *dual groupoid* of $\mathfrak{G}$.

The space of units of $\partial\mathfrak{G}$ is the topologically disjoint union of the spaces $\partial\mathfrak{G}_x$. (In particular, it is not separable.) If $\mathfrak{G}$ is minimal (i.e., if all orbits are dense), then $\partial\mathfrak{G}_x$ is an open transversal of the dual groupoid for any $x \in \mathfrak{G}^{(0)}$, hence the dual groupoid can be defined as the groupoid of germs on $\partial\mathfrak{G}_x$ of the asymptotic morphisms. We will denote it $\partial\mathfrak{G}_x$.

We will denote by $\mathfrak{G}^\top$ any groupoid equivalent to $\partial\mathfrak{G}$. The following theorem is proved in [Nek15].

Theorem 6.5. *Let $\mathfrak{G}$ be a minimal Hausdorff hyperbolic groupoid. Then the dual groupoid $\mathfrak{G}^\top$ is minimal, Hausdorff, and hyperbolic, and $(\mathfrak{G}^\top)^\top$ is equivalent to $\mathfrak{G}$.*

6.5 Groupoid of germs generated by an expanding self-covering

Let $f : \mathcal{J} \longrightarrow \mathcal{J}$ be an expanding self-covering of a path connected compact metric space. Denote by $\mathfrak{F}$ the groupoid of germs generated by f. Every element of $\mathfrak{F}$ can be written as a product $(f^n, x)^{-1}(f^m, y)$, for $n, m \in \mathbb{N}$, and $x, y \in \mathcal{J}$ such that $f^n(x) = f^m(y)$.

A *natural extension* of f is the inverse limit $\hat{\mathcal{J}}$ of the maps f together with the homeomorphism $\hat{f}$ of $\hat{\mathcal{J}}$ induced by f, see 6.2. Let $P_S : \hat{\mathcal{J}} \longrightarrow \mathcal{J}$ be the natural projection.

For every point $x \in \mathcal{J}$ there exists a neighborhood U that is evenly covered by each map $f^n : \mathcal{J} \longrightarrow \mathcal{J}$, since f is expanding. It follows that the set $P^{-1}(U)$ is naturally decomposed into the direct product of U with the boundary ∂T_x of the tree of preimages of a point $x \in U$.

The groupoid $\mathfrak{F}$ is hyperbolic, and we can consider the space $\partial\mathfrak{F}$ together with the projection $P : \partial\mathfrak{F} \longrightarrow \mathfrak{F}^{(0)} = \mathcal{J}$. Let us use the generating set S of $\mathfrak{F}$ equal to the set of germs of the inverse map f^{-1}. Then the Cayley graphs $\Gamma(x, S)$ are regular trees such that every vertex has one incoming and $d = \deg f$ outgoing arrows. The fiber $\partial\mathfrak{F}_x$ is equal to the boundary of this tree minus the limit of the path (f^n, x), $n \geq 0$. In other words, it is the natural inductive limit of the boundaries of the preimage trees $T_{f^n(x)}$ for $n \geq 0$.

Let $\nu : \mathfrak{F} \longrightarrow \mathbb{Z}$ be the homomorphism (*cocycle*) given by the rule $\nu(f, x) = -1$, so that $\nu((f^n, x)^{-1}(f^m, y)) = n - m$. See Figure 13, where the Cayley graph $\Gamma(x, S)$ together with the levels of the cocycle ν are shown.

Every point of $\partial\mathfrak{F}_x$ can be uniquely represented as the limit of a sequence $s_n \cdots s_2 s_1 \cdot g$, where $s_i \in S$, and $g \in \mathfrak{F}$ is such that $\mathsf{o}(g) = x$ and $\nu(g) = 0$. Note that the set of limits of the sequences $s_n \cdots s_2 s_1 \cdot g$ for a fixed g and all possible choices of $s_i \in S$ is naturally identified with the fiber $P_S^{-1}(\mathsf{t}(g))$ of the solenoid $\hat{\mathcal{J}}$ (i.e., with $\partial T_{\mathsf{t}(g)}$). It follows then directly from the definitions that the space $\partial\mathfrak{F}$ is homeomorphic to the subset $\{(\zeta, g) : P_S(\zeta) = \mathsf{t}(g)\}$ of the direct product $\hat{\mathcal{J}} \times \mathfrak{F}_0$, where $\mathfrak{F}_0$ is the subgroupoid $\nu^{-1}(0) \subset \mathfrak{F}$. The

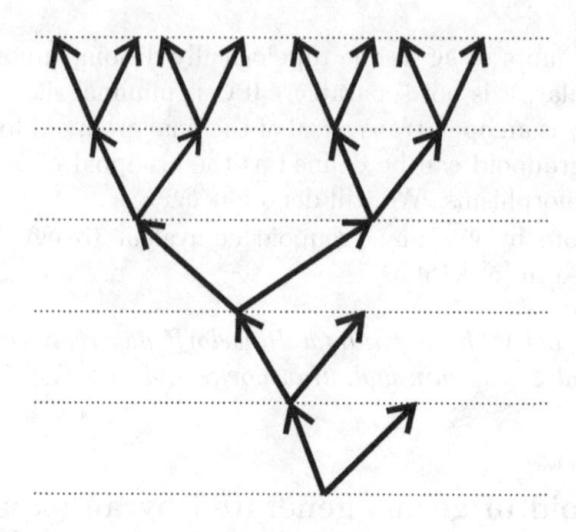

Figure 13: Cayley graph of $\mathfrak{F}$

action of $\mathfrak{F}$ on $\partial\mathfrak{F}$ is given in these terms by the rules

$$(\zeta, g) \cdot h = \begin{cases} (\hat{f}^n(\zeta), f^n \circ gh), & \text{if } \nu(h) = n > 0, \\ (\hat{f}^{-n}(\zeta), s_n \cdots s_2 s_1 gh), & \text{if } \nu(h) = -n < 0, \\ (\zeta, gh), & \text{if } \nu(h) = 0, \end{cases}$$

where $s_i \in S$ are such that $\zeta = \lim_{m \to \infty} s_m \cdots s_2 s_1$.

Therefore, we have the following description of the natural extension $\hat{f} : \hat{\mathcal{J}} \longrightarrow \hat{\mathcal{J}}$ in terms of the $\mathfrak{F}$-space $\partial\mathfrak{F}$.

Proposition 6.6. *The quotient of the space $\partial\mathfrak{F}$ by the action of $\mathfrak{F}_0 = \nu^{-1}(0)$ is homeomorphic to $\hat{\mathcal{J}}$. If $F \in \tilde{\mathfrak{F}}$ is such that $\nu(F) = \{n\}$, then the germs of the map induced by F on the quotient space $\hat{\mathcal{J}} = \partial\mathfrak{F}/\mathfrak{F}_0$ are germs of the map $\hat{f}^{-n}$.*

We say that points $\xi, \zeta \in \hat{\mathcal{J}}$ are *unstably equivalent* if the distance between $\hat{f}^{-n}(\xi)$ and $\hat{f}^{-n}(\zeta)$ goes to zero as $n \to +\infty$. They are said to be *stably equivalent* if the distance between $\hat{f}^n(\xi)$ and $\hat{f}^n(\zeta)$ goes to zero as $n \to +\infty$.

A *leaf* of $\hat{\mathcal{J}}$ is its path connected component. (Recall that we assume that $\mathcal{J}$ is path connected.) Every leaf is an equivalence class of the unstable equivalence relation on $\hat{\mathcal{J}}$.

Each leaf is dense in $\hat{\mathcal{J}}$, and it is more natural to consider it with the *inductive limit topology*. Namely, a subset U of a leaf $\mathcal{L}$ is open if and only if its intersection with every compact subset $C \subset \mathcal{L}$ is open in C. Note that for every compact set $C \subset \mathcal{L}$ the map $P : C \longrightarrow \mathcal{J}$ is finite-to-one.

The restriction of the map $P : \hat{\mathcal{J}} \longrightarrow \mathcal{J}$ to any leaf $\mathcal{L}$ of $\hat{\mathcal{J}}$ is a covering map. Let $\mathfrak{F}_{\mathcal{L}}$ be the lift of the groupoid $\mathfrak{F}$ to the leaf $\mathcal{L}$ by this covering. Then the groupoid $\mathfrak{F}_{\mathcal{L}}$ is equivalent to $\mathfrak{F}$. The cocycle ν lifts to the cocycle $\nu \circ P$ on $\mathfrak{F}_{\mathcal{L}}$, which we will denote by $\nu_{\mathcal{L}}$ or just ν.

It follows then from the definition of $\partial\mathfrak{G}$ for a hyperbolic groupoid $\mathfrak{G}$, that the space $\partial\mathfrak{F}_{\mathcal{L}}$ is the fiber product of the maps $P_S : \mathcal{L} \longrightarrow \mathcal{J}$ and $P : \partial\mathfrak{F} \longrightarrow \mathcal{J}$, i.e., the subset $\{(x, y) : P_S(x) = P(y)\}$ of $\mathcal{L} \times \partial\mathfrak{F}$. We get the following corollary of Proposition 6.6.

Corollary 6.7. *Let $\mathcal{L}$ be a leaf of $\hat{\mathcal{J}}$, and let $\mathfrak{F}_{\mathcal{L}}$ be the lift of $\mathfrak{F}$ by the covering $P_S : \mathcal{L} \longrightarrow \mathcal{J}$. Then the quotient of the space $\partial\mathfrak{F}_{\mathcal{L}}$ by the action of $\mathfrak{F}_{\mathcal{L},0} = \nu_{\mathcal{L}}^{-1}(0)$ is homeomorphic to $\hat{\mathcal{J}}$. If $F \in \tilde{\mathfrak{F}}_{\mathcal{L}}$ is such that $\nu_{\mathcal{L}}(F) = \{n\}$, then the germs of the map induced by F on the quotient space $\hat{\mathcal{J}} = \partial\mathfrak{F}_{\mathcal{L}}/\mathfrak{F}_{\mathcal{L},0}$ are germs of the map $\hat{f}^{-n}$.*

Let G be a contracting regular self-replicating group. Let $\mathfrak{G}$ be the groupoid of germs of the action of the group $\mathcal{V}_G$ on X^ω. It is generated by the groupoid of germs of G and the germs of the maps $S_x : x_1x_2\ldots \mapsto xx_1x_2\ldots$. It is shown in [Nek15, Subsection 5.3.] that $\mathfrak{G}$ is hyperbolic, and its dual is the groupoid generated by the limit dynamical system $f : \mathcal{J}_G \longrightarrow \mathcal{J}_G$.

More explicitly, let $w \in X^\omega$. Then the boundary $\partial\mathfrak{G}_w$ is the leaf of the limit solenoid $\mathcal{S}_G$ consisting of points representable by sequences $\ldots x_{-2}x_{-1} \cdot x_0x_1\ldots$, where $x_0x_1\ldots$ belongs to the G-orbit of w. The groupoid $\mathfrak{d}\mathfrak{G}_w$ is equal to the lift of the groupoid generated by the limit dynamical system $f : \mathcal{J}_G \longrightarrow \mathcal{J}_G$ to the leaf $\partial\mathfrak{G}_w$ (by the covering induced by the projection $\mathcal{S}_G \longrightarrow \mathcal{J}_G$ of the natural extension onto $\mathcal{J}_G$).

7 Reconstruction of the dynamical system from $\mathcal{V}_f$

The main result of this section is the following classification of the groups $\mathcal{V}_f$.

Theorem 7.1. *Let $f_i : \mathcal{J}_i \longrightarrow \mathcal{J}_i$, for $i = 1, 2$, be expanding self-coverings of path connected compact metric spaces. Then $\mathcal{V}_{f_1}$ and $\mathcal{V}_{f_2}$ are isomorphic as abstract groups if and only if the dynamical systems $(f_1, \mathcal{J}_1)$ and $(f_2, \mathcal{J}_2)$ are topologically conjugate.*

Example 7.1. One can show that the limit dynamical systems of two groups $\mathfrak{K}_{v_i}$, $i = 1, 2$, are topologically conjugate if and only if either $v_1 = v_2$, or v_1 can be obtained from v_2 by replacing each 0 by 1 and each 1 by 0. Namely, the sequence of letters of v_i can be interpreted as a *kneading sequence* of the dynamical system, which in turn can be defined in purely topological terms. This gives a complete classification of the groups $\mathcal{V}_{\mathfrak{K}_v}$ up to isomorphism.

7.1 M. Rubin's theorem

Recall that if G is a group acting on a topological space $\mathcal{X}$, and $U \subseteq \mathcal{X}$ is an open subset, then we denote by $G_{(U)}$ the group of elements $g \in G$ acting trivially outside U, see Subsection 4.1.

Resuts on M. Rubin [Rub89] make it possible to reconstruct the topological space $\mathcal{X}$ from a group G of homeomorphisms of $\mathcal{X}$. Namely, the following theorem is a corollary of Theorems 0.2 and Theorem 3.3 of [Rub89].

We say that a group G acting on a topological space $\mathcal{X}$ is *locally minimal* if there exists a basis of open sets $\mathcal{U}$ such that for every $U \in \mathcal{U}$ every orbit of the group $G_{(U)}$ acting on U is dense in U.

Theorem 7.2. *If G_i are locally minimal groups of homeomorphisms of topological spaces $\mathcal{X}_i$, then every isomorphism $\phi : G_1 \longrightarrow G_2$ is induced by a homeomorphism $F : \mathcal{X}_1 \longrightarrow \mathcal{X}_2$.*

Similar results (with simpler proofs), which can be applied to many groups $\mathcal{V}_G$, are proved in [GPS99, Med11, Mat15].

It is easy to see that the Higman-Thompson group $\mathcal{V}_{|X|}$ acting on the space X^ω is locally minimal. It follows that every group of homeomorphisms of X^ω containing the Higman-Thompson group is locally minimal, which implies the following fact.

Theorem 7.3. *Let G_i be groups acting on the Cantor sets X_i^ω and containing the Higman-Thompson groups $\mathcal{V}_{|X_i|}$. Then every isomorphism $\phi : G_1 \longrightarrow G_2$ is induced by a homeomorphism $F : X_1^\omega \longrightarrow X_2^\omega$.*

7.2 Proof of Theorem 7.1

If $f_1 : \mathcal{J}_1 \longrightarrow \mathcal{J}_1$ and $f_2 : \mathcal{J}_2 \longrightarrow \mathcal{J}_2$ are topologically conjugate self-coverings of path-connected spaces, then the groups $\mathcal{V}_{f_1}$ and $\mathcal{V}_{f_2}$ are obviously isomorphic, since they were defined in purely topological terms.

Let us prove the converse implication for expanding maps. By Theorem 7.3, if groups $\mathcal{V}_{f_1}$ and $\mathcal{V}_{f_2}$ are isomorphic, then their action on the corresponding spaces X_i^ω are topologically conjugate, hence the groupoid of germs of the action of $\mathcal{V}_{f_i}$ on X_i^ω are isomorphic.

Therefore, it is enough to show that if $f : \mathcal{J} \longrightarrow \mathcal{J}$ is an expanding self-covering of a compact path-connected metric space, then the dynamical system $(f, \mathcal{J})$ can be reconstructed from the topological groupoid $\mathfrak{G}$ of germs of the action of $\mathcal{V}_f$ on X^ω.

Denote by $\mathfrak{F}$ the groupoid of germs generated by $f : \mathcal{J} \longrightarrow \mathcal{J}$. We identify $\mathcal{J}$ with the limit space $\mathcal{J}_G$ of the self-similar group $G = \mathrm{IMG}(f)$, and hence encode points of $\mathcal{J}$ by sequences $\ldots x_2x_1 \in X^{-\omega}$. Recall that f acts then by the shift $\ldots x_2x_1 \mapsto \ldots x_3x_2$. Let $\nu : \mathfrak{F} \longrightarrow \mathbb{Z}$ be the cocycle (groupoid homomorphism) defined by the condition that $\nu(f, x) = -1$ for all $x \in \mathcal{J}$.

The groupoids $\mathfrak{F}$ and $\mathfrak{G}$ are hyperbolic and mutually dual. Let $w \in X^\omega$ be an arbitrary point, and denote $\mathfrak{H} = \mathfrak{d}\mathfrak{G}_w$ and $\mathcal{H} = \partial\mathfrak{G}_w = \mathfrak{H}^{(0)}$. It is enough to show that $(f, \mathcal{J})$ is uniquely determined (up to a topological conjugacy) by the groupoid $\mathfrak{H}$.

Denote by Ω_w the set of bi-infinite sequences $\ldots x_{-2}x_{-1}.x_0x_1\ldots$ such that $x_0x_1\ldots$ belongs to the G-orbit of w. Note that $\mathcal{J}$ is path connected, G is self-replicating, hence G-orbits coincide with the $\mathcal{V}_f = \mathcal{V}_G$-orbits. We consider Ω_w with the topology of the disjoint union of the set of the form $X^{-\omega}.x_0x_1\ldots$.

Then the space $\mathcal{H} = \partial\mathfrak{G}_w$ is naturally identified with the quotient of the space Ω_w by the asymptotic equivalence relation (defined in the same way as on $X^{\mathbb{Z}}$, see Subsection 6.2). Let $P_S : \mathcal{H} \longrightarrow \mathcal{J}$ be the natural projection induced by $\ldots x_{-2}x_{-1}.x_0x_1\ldots \mapsto \ldots x_{-2}x_{-1}$. It is a covering map, and $\mathfrak{H}$ is the lift of $\mathfrak{F}$ by P_S. We will also denote by P_S the corresponding functor (homomorphism of groupoids) $P_S : \mathfrak{H} \longrightarrow \mathfrak{F}$.

Let us show at first that the cocycle $\nu : \mathfrak{H} \longrightarrow \mathbb{Z}$ (equal to the lift of the cocycle $\nu : \mathfrak{F} \longrightarrow \mathbb{Z}$) is uniquely determined by the structure of the topological groupoid $\mathfrak{H}$.

Proposition 7.4. *Let $\mathcal{C}$ be a connected component of $\mathfrak{H}$. Then $\mathsf{o} : \mathcal{C} \longrightarrow \mathcal{H}$, $\mathsf{t} : \mathcal{C} \longrightarrow \mathcal{H}$ are coverings.*

If $\nu(\mathcal{C}) \neq 0$, then $\mathcal{C}$ contains a non-trivial element of infinite order in the isotropy group $\mathfrak{H}_x$ of a point.

If $\nu(\mathcal{C}) = 0$, then the groupoid generated by $\mathcal{C}$ is proper.

Proof. Let $\mathcal{X}_G$ be the limit G-space. The action of G on $\mathcal{X}_G$ is free, and the maps $F_v : \xi \mapsto \xi \otimes v$ are coverings for all $v \in X^n \cdot G$.

For $w \in X^\omega$, the leaf $\partial\mathfrak{G}_w = \mathcal{H}$ is the image of $\mathcal{X}_G$ under the map P_w : $\xi \mapsto \xi \cdot w$. This map coincides with the quotient of $\mathcal{X}_G$ by the action of the stabilizer G_w.

Let $\mathfrak{X}$ be the groupoid of germs with the space of units $\mathcal{X}_G$ generated by the germs of the action of G and the germs of the maps $F_v(\xi) = \xi \otimes v$ for $v \in X^* \cdot G$. Then $\mathfrak{X}$ is the lift of $\mathfrak{H}$ by the quotient map $P_w : \mathcal{X}_G \longrightarrow \mathcal{H}$.

Every element of $\mathfrak{H}$ is a germ of the transformation

$$F_{v\cdot g, u\cdot h} : \xi \otimes v.g(w) \mapsto \xi \otimes u.h(w),$$

for some $g, h \in G$ and $u, v \in X^*$.

The germ $(F_{v\cdot g,u\cdot h}, \zeta \otimes v.g(w))$ can be lifted to the germ $(\tilde{F}_{v\cdot g,u\cdot h}, \zeta \otimes v \cdot g)$ of the local homeomorphism

$$\tilde{F}_{v\cdot g,u\cdot h} : \xi \otimes v \cdot g \mapsto \xi \otimes u \cdot h$$

of $\mathfrak{X}$. It follows that every element of $\mathfrak{H}$ is a germ of $P_w F_{u\cdot h} F_{v\cdot g}^{-1} P_w^{-1}$. The space $\mathcal{X}_G$ is connected, the maps $F_{u\cdot h}, F_{v\cdot g}, P_w$ are coverings, hence if $\mathcal{C}$ is the connected component of the germ $(F_{v\cdot g,u\cdot h}, \zeta \otimes v.g(w))$, then $\mathsf{o} : \mathcal{C} \longrightarrow \mathcal{H}$ and $\mathsf{t} : \mathcal{C} \longrightarrow \mathcal{H}$ are covering maps.

Suppose that $\nu(\mathcal{C}) \neq 0$. It means that every germ $(F_{v\cdot g,u\cdot h}, \zeta \otimes v.g(w)) \in \mathcal{C}$ is such that $|v| \neq |u|$. Without loss of generality, we may assume that $|u| > |v|$. Let $u = u_1v_1$, where $|v_1| = |v|$. Since G is self-replicating, there exists $g_1 \in G$ such that $g_1 \cdot v \cdot g = v_1 \cdot h$ in the biset. Then a lift of the germ $(F_{v\cdot g, u\cdot h}, \zeta \otimes v.g(w))$ to $\mathfrak{X}$ is a germ of the transformation

$$\xi \otimes v \cdot g \mapsto \xi \otimes u_1 \otimes v_1 \cdot h = \xi \otimes u_1 \cdot g_1 \otimes v \cdot g.$$

The point $\zeta = \ldots u_1 \cdot g_1 \otimes u_1 \cdot g_1 \otimes u_1 \cdot g_1 \in \mathcal{X}_G$ is well defined (as it is the image of a point of Ω, see Subsection 6.2), and it satisfies $\zeta = \zeta \otimes u_1 \cdot g_1$. Then the germ of the transformation

$$\xi \otimes v \cdot g \mapsto \xi \otimes u_1 \cdot g \otimes v \cdot g$$

at $\zeta \otimes v \cdot g$ is a non-trivial contracting element of the isotropy group of $\zeta \otimes v \cdot g$. It is contained in the connected component of the germs of the transformation $\xi \otimes v \cdot g \mapsto \xi \otimes u \cdot h$. Mapping everything to $\mathfrak{H}$ by P_w, we find a non-trivial contracting (hence infinite order) element of an isotropy group.

If $\nu(\mathcal{C}) = 0$, then elements of $\mathcal{C}$ are germs of transformations of the form $\xi \otimes v.g(w) \mapsto \xi \otimes u.h(w)$, where $g, h \in G$ and $v, u \in X^*$ are such that $|v| = |u|$. There exists $g_1 \in G$ such that $u \cdot h = g_1 \cdot v \cdot g$ in $X^n \cdot G$. It follows that elements of $\mathcal{C}$ are lifted by $P_w F_{v\cdot g} : \mathfrak{X} \longrightarrow \mathfrak{H}$ to the action of g_1 on $\mathcal{X}_G$. It follows that the groupoid generated by $\mathcal{C}$ lifts by $P_w F_{v\cdot g}$ to a subgroupoid of the action of G on $\mathcal{X}_G$, and hence is proper. □

Proposition 7.5. *The cocycle $\nu : \mathfrak{H} \longrightarrow \mathbb{Z}$ is uniquely determined by the topological groupoid $\mathfrak{H}$.*

Proof. It follows from 7.4 that the value of ν on a connected component $\mathcal{C}$ of $\mathfrak{H}$ is zero if and only if $\mathcal{C}$ generates a proper groupoid. (Since isotropy groups of a proper groupoid are finite.)

Let g, h be arbitrary elements of $\mathfrak{H}$. By the first claim of Proposition 7.4, there exist g' and h' in the components of g and h, respectively, such that the products $g'h$ and gh' are defined.

It follows that $\nu(g) = \nu(h)$ if and only if there exists an element h' in the connected component of h such that $\nu(g^{-1}h') = 0$. Consequently, the set of homomorphisms $\{\nu, -\nu\}$ is uniquely determined by the structure of the topological groupoid $\mathfrak{H}$. But we can distinguish between ν and $-\nu$ using [Nek15, Proposition 3.4.1.] (informally, since the germs are contracting for positive values and expanding for negative values of ν). □

The next statement follows now directly from Proposition 7.5 and Corollary 6.7.

Proposition 7.6. *The natural extension $\hat{f} : \hat{\mathcal{J}} \longrightarrow \hat{\mathcal{J}}$ is uniquely determined, up to topological conjugacy, by the groupoid $\mathfrak{G}$.*

Suppose that $f_1 : \mathcal{J}_1 \longrightarrow \mathcal{J}_1$ and $f_2 : \mathcal{J}_2 \longrightarrow \mathcal{J}_2$ are two expanding homeomorphisms with the same natural extension $\hat{f} : \mathcal{S} \longrightarrow \mathcal{S}$. It remains to prove that $(f_1, \mathcal{J}_1)$ and $(f_2, \mathcal{J}_2)$ are topologically conjugate.

Denote by $P_i : \mathcal{S} \longrightarrow \mathcal{J}_i$ the corresponding projections. Let $\tilde{\mathcal{J}}$ be the image of $\mathcal{S}$ in $\mathcal{J}_1 \times \mathcal{J}_2$ under the map (P_1, P_2). It is compact and connected, since such is $\mathcal{S}$. We will denote by $\tilde{P}_i : \tilde{\mathcal{J}} \longrightarrow \mathcal{J}_i$ the restrictions of the projections $\mathcal{J}_1 \times \mathcal{J}_2 \longrightarrow \mathcal{J}_i$.

Since P_i locally are projections on the unstable coordinate of the local product decomposition of $\mathcal{S}$ (which depends only on the conjugacy class of $(\hat{f}, \mathcal{S})$), for every $\xi \in \mathcal{S}$ there exists a rectangular neighborhood $U \ni \xi$ such that $P_i : U \longrightarrow P_i(U)$ is decomposed into the composition of projection of U onto its unstable direction and a homeomorphism of this direction with $P_i(U)$. Moreover, since $\mathcal{S}$ is compact, we can cover $\mathcal{S}$ by a finite number of such rectangles U.

The map $\hat{f} : \mathcal{S} \longrightarrow \mathcal{S}$ induces a map $\tilde{f} : \tilde{\mathcal{J}} \longrightarrow \tilde{\mathcal{J}}$ by the rule $\tilde{f}(\xi_1, \xi_2) = (f_1(\xi_1), f_2(\xi_2))$. The projections $\tilde{P}_i$ are semi-conjugacies of $\tilde{f}$ with f_i.

Let $(\xi_1, \xi_2) \in \tilde{\mathcal{J}}$, i.e., there exists $\xi \in \mathcal{S}$ such that $\xi_i = P_i(\xi)$. There exists a rectangular neighborhood U of ξ such that the unstable direction of U is connected, and P_i are projections onto the unstable direction composed with homeomorphisms. If U is small enough, then $(\hat{f})^{-1}(U)$ is decomposed into a union of a finite set $\mathcal{R}$ of rectangles on which each of P_i is a homeomorphism with projection onto the unstable direction. Consider the sets $(P_1, P_2)(R)$ for $R \in \mathcal{R}$. We get a finite number of components of $(\tilde{f})^{-1}((P_1, P_2)(U))$ such that $\tilde{f}$ is a homeomorphism on each of them. It follows that $\tilde{f}$ is a finite degree covering map.

For every $\xi \in \mathcal{J}_1$ the set $P_1^{-1}(\xi)$ is a compact subset of $\mathcal{S}$ contained in one stable equivalence class. Consequently, there exists n_0 such that $P_2(\hat{f}^n(P_1^{-1}(\xi)))$ is a single point for all $n \geq n_0$. It follows that there exists a small neighborhood U of ξ such that the map $P_2 \circ \hat{f}^{n_0} \circ P_1^{-1} = f_2^{n_0} \circ P_2 \circ P_1^{-1}$ is a homeomorphism on U. By compactness, there exists n_1 such that $P_2 \circ f^{n_1} \circ P_1^{-1} = f_2^{n_1} \circ P_2 \circ P_1^{-1}$ is a well defined covering map from $\mathcal{J}_1$ to $\mathcal{J}_2$.

It follows that the projections $\tilde{P}_i : \tilde{\mathcal{J}} \longrightarrow \mathcal{J}_i$ are finite degree covering maps. For every point $t_i^{(j)} \in \tilde{P}_i^{-1}(t_i)$ we have the corresponding tree $T_{t_i^{(j)}}$ of preimages under iterations of $\tilde{f}$. They are disjoint (more precisely, for every n the sets $\tilde{f}^{-n}(t^{(j)})$ are disjoint for different $t^{(j)}$).

By the arguments above, there exists n_1 such that $\tilde{P}_i(z_1) = \tilde{P}_i(z_2)$ implies $\tilde{f}^{n_1}(z_1) = \tilde{f}^{n_1}(z_2)$. But this contradicts the fact that the trees T_j are disjoint. It follows that $\tilde{P}_i$ have degree 1, i.e., are homeomorphisms conjugating f with f_i.

7.3 Equivalence of groupoids

Theorem 7.7. *Let $f_i : \mathcal{J}_i \longrightarrow \mathcal{J}_i$, for $i = 1, 2$, be expanding self-coverings of connected and locally connected compact metric spaces. Then the following conditions are equivalent.*

1. *The dynamical systems* $(f_1, \mathcal{J}_1)$ *and* $(f_2, \mathcal{J}_2)$ *are topologically conjugate.*
2. *The groupoids generated by germs of* f_1 *and* f_2 *are equivalent.*
3. *The natural extensions of* f_1 *and* f_2 *are topologically conjugate.*
4. *The natural extensions of* f_1 *and* f_2 *generate equivalent groupoids of germs.*
5. *The actions of* $\mathcal{V}_{f_1}$ *and* $\mathcal{V}_{f_2}$ *on the corresponding Cantor sets are topologically conjugate.*
6. *The groupoids of germs generated by the actions of* $\mathcal{V}_{f_1}$ *and* $\mathcal{V}_{f_2}$ *on the corresponding Cantor sets are equivalent.*
7. *The self-similar groups* $\mathrm{IMG}(f_1)$ *and* $\mathrm{IMG}(f_2)$ *are equivalent.*
8. *The groups* $\mathcal{V}_{f_1}$ *and* $\mathcal{V}_{f_2}$ *are isomorphic as abstract groups.*

Proof. The groupoid of germs $\mathfrak{F}_i$ generated by f_i, the groupoid of germs $\mathfrak{G}_i$ generated by $\mathcal{V}_{f_i}$, and the groupoid of germs generated by the natural extension uniquely determine each other, up to equivalence of groupoids, since the first two are mutually dual hyperbolic groupoids, and the third one is their geodesic flow, see [Nek15]. Equivalence of (7) and (1) is proved in [Nek05].

It remains, therefore, to prove that the equivalence class of $\mathfrak{F}_i$ uniquely determines $(f_i, \mathcal{J}_i)$.

Suppose that w_1 and w_2 belong to one orbit of the groupoid $\mathfrak{G}$ from Definition 6.5. Let $g \in \mathfrak{G}$ be such that $\mathsf{o}(g) = w_2$ and $\mathsf{t}(g) = w_1$. Then the map $h \mapsto hg$ is a quasi-isometry between the Cayley graphs of $\mathfrak{G}_1$ and $\mathfrak{G}_2$ based at w_1 and w_2 respectively, inducing an isomorphism $\partial\mathfrak{G}_{w_1} \longrightarrow \partial\mathfrak{G}_{w_2}$. We have shown during the proof of Theorem 7.1 that the dynamical systems $(f_i, \mathcal{J}_i)$ can be uniquely reconstructed from the topological groupoids $\partial\mathfrak{G}_{w_i}$, which implies that $(f_1, \mathcal{J}_1)$ and $(f_2, \mathcal{J}_2)$ are topologically conjugate. □

References

[BGN03] Laurent Bartholdi, Rostislav Grigorchuk, and Volodymyr Nekrashevych, *From fractal groups to fractal sets*, Fractals in Graz 2001. Analysis – Dynamics – Geometry – Stochastics (Peter Grabner and Wolfgang Woess, eds.), Birkhäuser Verlag, Basel, Boston, Berlin, 2003, pp. 25–118.

[BGŠ03] Laurent Bartholdi, Rostislav I. Grigorchuk, and Zoran Šunić, *Branch groups*, Handbook of Algebra, Vol. 3, North-Holland, Amsterdam, 2003, pp. 989–1112.

[BN08] Laurent Bartholdi and Volodymyr V. Nekrashevych, *Iterated monodromy groups of quadratic polynomials I*, Groups, Geometry, and Dynamics **2** (2008), no. 3, 309–336.

[Bro87] Kenneth S. Brown, *Finiteness properties of groups*, Proceedings of the Northwestern conference on cohomology of groups (Evanston, Ill., 1985), vol. 44, 1987, pp. 45–75.

[CFP96] John W. Cannon, William I. Floyd, and Walter R. Parry, *Introductory notes on Richard Thompson groups*, L'Enseignement Mathematique **42** (1996), no. 2, 215–256.

[CP93] Michel Coornaert and Athanase Papadopoulos, *Symbolic dynamics and hyperbolic groups*, Lecture Notes in Mathematics, vol. 1539, Springer Verlag, 1993.

[Cun77] Joachim Cuntz, *Simple C^*-algebras generated by isometries*, Comm. Math. Phys. **57** (1977), 173–185.

[Fri87] David Fried, *Finitely presented dynamical systems*, Ergod. Th. Dynam. Sys. **7** (1987), 489–507.

[GPS99] Thierry Giordano, Ian F. Putnam, and Christian F. Skau, *Full groups of Cantor minimal systems*, Israel J. Math. **111** (1999), 285–320.

[Gri80] Rostislav I. Grigorchuk, *On Burnside's problem on periodic groups*, Functional Anal. Appl. **14** (1980), no. 1, 41–43.

[Gro87] Mikhael Gromov, *Hyperbolic groups*, Essays in Group Theory (S. M. Gersten, ed.), M.S.R.I. Pub., no. 8, Springer, 1987, pp. 75–263.

[Hae01] André Haefliger, *Groupoids and foliations*, Groupoids in Analysis, Geometry, and Physics. AMS-IMS-SIAM joint summer research conference, University of Colorado, Boulder, CO, USA, June 20-24, 1999, Contemp. Math, vol. 282, Providence, RI: A.M.S., 2001, pp. 83–100.

[Hae02] André Haefliger, *Foliations and compactly generated pseudogroups*, Foliations: geometry and dynamics (Warsaw, 2000), World Sci. Publ., River Edge, NJ, 2002, pp. 275–295.

[Hig74] Graham Higman, *Finitely presented infinite simple groups*, Department of Pure Mathematics, Department of Mathematics, I.A.S. Australian National University, Canberra, 1974, Notes on Pure Mathematics, No. 8 (1974).

[IS10] Yutaka Ishii and John Smillie, *Homotopy shadowing*, Amer. J. Math. **132** (2010), no. 4, 987–1029.

[Law14] Mark V. Lawson, *On a class of countable Boolean inverse monoids and Matui's spatial realization theorem*, (preprint arxiv:1407.1473), 2014.

[Law15] ——, *Subgroups of the group of homeomorphisms of the Cantor space and a duality between a class of inverse monoids and a class of Hausdorff étale groupoids*, (preprint arxiv:1501.06824), 2015.

[Lea56] William G. Leavitt, *Modules over rings of words*, Proc. Amer. Math. Soc. **7** (1956), 188–193.

[Lea65] W. G. Leavitt, *The module type of homomorphic images*, Duke Math. J. **32** (1965), 305–311. MR 0178018 (31 #2276)

[Mat06] Hiroki Matui, *Some remarks on topological full groups of Cantor minimal systems*, Internat. J. Math. **17** (2006), no. 2, 231–251.

[Mat15] ——, *Topological full groups of one-sided shifts of finite type*, J. Reine Angew. Math. **705** (2015), 35–84.

[Med11] K. Medynets, *Reconstruction of orbits of Cantor systems from full groups*, Bull. Lond. Math. Soc. **43** (2011), no. 6, 1104–1110.

[Mil06] John Milnor, *Dynamics in one complex variable*, third ed., Annals of Mathematics Studies, vol. 160, Princeton University Press, Princeton, NJ, 2006.

[Nek03] Volodymyr Nekrashevych, *Hyperbolic spaces from self-similar group actions*, Algebra and Discrete Mathematics **1** (2003), no. 2, 68–77.

[Nek04] ——, *Cuntz-Pimsner algebras of group actions*, Journal of Operator Theory **52** (2004), no. 2, 223–249.

[Nek05] ——, *Self-similar groups*, Mathematical Surveys and Monographs, vol. 117, Amer. Math. Soc., Providence, RI, 2005.

[Nek08] ——, *Symbolic dynamics and self-similar groups*, Holomorphic dynamics and renormalization. A volume in honour of John Milnor's 75th birthday (Mikhail Lyubich and Michael Yampolsky, eds.), Fields Institute Communications, vol. 53, A.M.S., 2008, pp. 25–73.

[Nek09] ——, *C^*-algebras and self-similar groups*, Journal für die reine und angewandte Mathematik **630** (2009), 59–123.

[Nek11] ______, *Iterated monodromy groups*, Groups St. Andrews in Bath. Vol. 1, London Math. Soc. Lect. Note Ser., vol. 387, Cambridge Univ. Press, Cambridge, 2011, pp. 41–93.

[Nek14] ______, *Combinatorial models of expanding dynamical systems*, Ergodic Theory and Dynamical Systems **34** (2014), 938–985.

[Nek15] ______, *Hyperbolic groupoids and duality*, Memoirs of the A.M.S. **237** (2015), no. 1122, 108.

[Pat99] Alan L. T. Paterson, *Groupoids, inverse semigroups, and their operator algebras*, Birkhäuser Boston Inc., Boston, MA, 1999.

[Ren80] Jean Renault, *A groupoid approach to C^*-algebras*, Lecture Notes in Mathematics, vol. 793, Springer-Verlag, Berlin, Heidelberg, New York, 1980.

[Röv99] Claas E. Röver, *Constructing finitely presented simple groups that contain Grigorchuk groups*, J. Algebra **220** (1999), 284–313.

[Röv02] ______, *Commensurators of groups acting on rooted trees*, Geom. Dedicata **94** (2002), 45–61.

[Rub89] Matatyahu Rubin, *On the reconstruction of topological spaces from their groups of homeomorphisms*, Trans. Amer. Math. Soc. **312** (1989), no. 2, 487–538.

On characteristic modules of groups

Olympia Talelli

Abstract

We define a characteristic module for a group G to be a $\mathbb{Z}G$-module which is $\mathbb{Z}$-free, contains elements invariant under the action of G and has finite projective dimension over $\mathbb{Z}G$. We relate the existence of such a module to the Gorenstein dimension of G, the generalized cohomological dimension of G and proper actions of G. Moreover, we give a criterion, via complete cohomology, for the existence of characteristic modules.

1 Introduction

Let G be a group with periodic cohomology of period q after k steps, i.e there exist non negative integers q and k with $q \neq 0$ such that the functors $H^i(G, -)$ and $H^{i+q}(G, -)$ are naturally equivalent for all $i > k$, and let us assume that the periodicity isomorphisms are given by cup product with an element $g \in H^q(G, \mathbb{Z})$.

Examples of such groups are the countable groups which act freely and properly on $\mathbb{R}^n \times S^m$, or more generally the groups G which admit a finite dimensional G-CW-complex X, such that the action is free and X is homotopy equivalent to a sphere [23].

The q-fold extension representation of the element g, cup product with which induces the periodicity isomorphisms, is of the following form

$$0 \longrightarrow \mathbb{Z} \longrightarrow T \longrightarrow P_{q-2} \longrightarrow \cdots \longrightarrow P_0 \longrightarrow \mathbb{Z} \longrightarrow 0,$$

where $\mathbb{Z}$ is a $\mathbb{Z}G$-module with the trivial action, P_i are projective $\mathbb{Z}G$-modules and proj. $\dim_{\mathbb{Z}G} T \leq k$.

So, in particular the group G admits a $\mathbb{Z}G$-module T with the properties:

1. The module T is $\mathbb{Z}$-free
2. There is a $\mathbb{Z}$-split $\mathbb{Z}G$-monomorphism $\mathbb{Z} \longrightarrow T$
3. proj. $\dim_{\mathbb{Z}G} T \leq k$.

The existence of such a module leads to restrictions on the structure of G, for example if proj. $\dim_{\mathbb{Z}G} T \leq 1$ then G acts on a tree with finite vertex stabilizers [24] (The proof of this result uses the almost stability theorem of Dicks and Dunwoody [7]).

We call a $\mathbb{Z}G$-module with properties 1–3 a *characteristic module* for the group G.

Note that if T is a characteristic module for G and K a finite subgroup of G then by Rim's Theorem it follows that $\mathrm{Res}_K T$ is a projective $\mathbb{Z}K$-module.

For every group G there is a module with properties 1 and 2 such that when restricted to every finite subgroup of G becomes projective, e.g the module $B(G, \mathbb{Z})$ of bounded functions from G to $\mathbb{Z}$ [18]. However, not every group admits a characteristic module. We shall see, for example, that every group which contains a free abelian subgroup of infinite rank does not admit a characteristic module.

This survey consists of three sections. In the first section we show that if a group G admits a characteristic module, then any two such have the same projective dimension which coincides with the Gorenstein dimension of the group G which is the same as the generalized cohomological dimension defined by Ikenaga [14]. For the proof of this we will use the algebraic invariant $\mathrm{spli}\mathbb{Z}G$, the supremum of the projective lengths of the injective $\mathbb{Z}G$-modules. In the second section we give examples of groups which admit characteristic modules and discuss the conjecture relating the existence of a characteristic module for a group G to the existence of a finite dimensional contractible G-CW-complex with finite stabilizers [3]. If that conjecture were true then it would imply, in particular, that if a torsion free group G has periodic cohomology after some steps and the periodicity isomorphisms are induced by cup product with an element in $H^*(G, \mathbb{Z})$, then the group G has finite cohomological dimension. In the last section we give a criterion,via complete cohomology, for the existence of a characteristic module.

2 Gorenstein dimension, generalized cohomological dimension and characteristic modules

The Gorenstein projective dimension of a $\mathbb{Z}G$-module M, $\mathrm{Gpd}_{\mathbb{Z}G} M$, is a refinement of the projective dimension of M over $\mathbb{Z}G$, and was defined by Enochs and Jenda in [10]. This concept goes back to the concept of relative homological dimension, called the G-dimension, which was defined by Auslander in [2] for finitely generated modules over commutative Noetherian rings and provided a characterization of the Gorenstein rings.

The definition of the Gorenstein projective dimension is given as follows.

A $\mathbb{Z}G$-module M is said to admit a complete resolution $(\mathcal{F}, \mathcal{P}, n)$ if there is an acyclic complex $\mathcal{F} = \{(F_i, \vartheta_i)| \ i \in \mathbb{Z}\}$ of projective modules, and a projective resolution $\mathcal{P} = \{(P_i, d_i) | \ i \in \mathbb{Z}, i \geq 0\}$ of M such that $\mathcal{F}$ and $\mathcal{P}$ coincide in dimensions greater than n

$$\begin{array}{llllllll}
\mathcal{F}: & \cdots \to F_{n+1} \to & F_n \overset{\vartheta_n}{\to} F_{n-1} \to & \cdots & \to F_0 \to F_{-1} \to \cdots \\
& \quad\quad \| & \| & & \\
\mathcal{P}: & \cdots \to P_{n+1} \to & P_n \overset{d_n}{\to} P_{n-1} \to & \cdots & \to P_0 \to M \to 0
\end{array}$$

The number n is called the coincidence index of the complete resolution.

A $\mathbb{Z}G$-module M is said to admit a complete resolution in the strong sense if there is a complete resolution $(\mathcal{F}, \mathcal{P}, n)$ with $\mathrm{Hom}_{\mathbb{Z}G}(\mathcal{F}, Q)$ acyclic for every $\mathbb{Z}G$-projective module Q. It is not known whether a $\mathbb{Z}G$-module M that admits a complete resolution, also admits a complete resolution in the strong sense. However, if G is an $\mathbf{LH}\mathfrak{F}$ group, then every complete resolution of a $\mathbb{Z}G$-module M is a complete resolution in the strong sense [6]. The class $\mathbf{LH}\mathfrak{F}$ was defined by Kropholler in [15] as the class consisting of those groups G with the property that every finitely generated subgroup of G is in $\mathbf{H}\mathfrak{F}$. This in turn is defined as the smallest class of groups which contains the class of finite groups and whenever a group G admits a finite dimensional contractible G-CW-complex with stabilizers in $\mathbf{H}\mathfrak{F}$, then G is in $\mathbf{H}\mathfrak{F}$. In particular $H_1\mathfrak{F}$ is the class of groups which admit a finite dimensional contractible G-CW-complex with finite stabilizers.

Definition 2.1 A $\mathbb{Z}G$-module M is called Gorenstein projective if it admits a complete resolution in the strong sense of coincidence index 0; i.e., M is a syzygy of a doubly infinite acyclic complex of projective $\mathbb{Z}G$-modules $(\mathcal{P}_*)_{*\in\mathbb{Z}}$, which remains acyclic when applying the functor $\mathrm{Hom}_{\mathbb{Z}G}(-, P)$ for any projective $\mathbb{Z}G$-module P.

Examples of Gorenstein projective modules are [3]:

1. Every projective module is Gorenstein projective.

2. If F is a finite group then every $\mathbb{Z}F$-module which is $\mathbb{Z}$-free is Gorenstein projective.

3. If K is a subgroup of a group G and M a Gorenstein projective $\mathbb{Z}K$-module, then the induced module $\mathbb{Z}G \otimes_{\mathbb{Z}K} M$ is a Gorenstein projective $\mathbb{Z}G$-module.

4. If G has periodic cohomology after some steps and G is in $\mathbf{LH}\mathfrak{F}$, then G admits a complete resolution in the strong sense. Clearly, every kernel of this resolution is Gorenstein projective.

5. If $0 \to M' \to M \to M'' \to 0$ is a short exact sequence of $\mathbb{Z}G$-modules and M'' is Gorenstein projective, then M' is Gorenstein projective if and only

if M is Gorenstein projective. Moreover, direct sums and direct summands of Gorenstein projective modules are Gorenstein projective [13, theorem 2.5].

Definition 2.2 If A is an $\mathbb{Z}G$-module and n a non-negative integer, then the Gorenstein projective dimension of A is less than or equal to n, $\mathrm{Gpd}_{\mathbb{Z}G}A \leq n$, if there is a resolution of A by Gorenstein projective $\mathbb{Z}G$-modules of length n.

The Gorenstein dimension of a group G over $\mathbb{Z}$, $\mathrm{Gcd}_{\mathbb{Z}}G$, is the Gorenstein projective dimension of $\mathbb{Z}$, as a trivial $\mathbb{Z}G$-module.

It turns out that $\mathrm{Gcd}_{\mathbb{Z}}G$ coincides with the cohomological dimension of G over $\mathbb{Z}$, $\mathrm{cd}_{\mathbb{Z}}G$, when this is finite, and the vanishing of $\mathrm{Gcd}_{\mathbb{Z}}G$ characterizes the finite groups. I.e., $\mathrm{Gcd}_{\mathbb{Z}}G = 0$ if and only if the group G is finite. Moreover, $\mathrm{Gcd}_{\mathbb{Z}}G$ is related [3] to the algebraic invariants $\mathrm{spli}\mathbb{Z}G$, $\mathrm{silp}\mathbb{Z}G$ and the generalized cohomological dimension of G, $\underline{\mathrm{cd}}G$, where $\mathrm{spli}\mathbb{Z}G$ is the supremum of the projective dimensions of the injective $\mathbb{Z}G$-modules, $\mathrm{silp}\mathbb{Z}G$ is the supremum of the injective dimensions of the projective $\mathbb{Z}G$-modules, and

$$\underline{\mathrm{cd}}G = \sup\{n \in \mathbb{N} \mid \exists\, M, P : \mathrm{Ext}^n_{\mathbb{Z}G}(M,P) \neq 0,\ M\ \mathbb{Z}\text{-free},\ P\ \mathbb{Z}G\text{-projective}\}.$$

The invariants $\mathrm{spli}\mathbb{Z}G$ and $\mathrm{silp}\mathbb{Z}G$ were considered by Gedrich-Gruenberg [11] and the generalized cohomological dimension of G, $\underline{\mathrm{cd}}G$ by Ikenaga [14] in their study of extending the Farrell-Tate cohomology, which is defined for the class of groups of finite virtual cohomological dimension, to a larger class of groups.

The following theorem relates the finiteness of the Gorenstein dimension of a group G to the existence of a characteristic module for the group G.

Theorem 2.3 *The following statements are equivalent for a group G.*

(*i*) $\mathrm{Gcd}_{\mathbb{Z}}G < \infty$

(*ii*) $\mathrm{spli}\mathbb{Z}G < \infty$

(*iii*) *The group G admits a characteristic module.*

Moreover, if T is a characteristic module for G then $\text{proj. dim}_{\mathbb{Z}G}T = \mathrm{Gcd}_{\mathbb{Z}}G$.

The proof of (*i*) iff (*ii*) is in Remark 2.10 in [1] and the proof of (*ii*) iff (*iii*) is in [25, thm. 2.2]. Both proofs are based on the fact that $\mathrm{spli}\mathbb{Z}G < \infty$ iff G admits a complete resolution in the strong sense.

The proof that $\text{proj. dim}_{\mathbb{Z}G}T = \mathrm{Gcd}_{\mathbb{Z}}G$ is in [3, thm 2.7] and it uses the characterization of finite Gorenstein projective dimension given in [13].

It is clear that from Theorem 2.3 we obtain the following.

Corollary 2.4 *If a group G admits a characteristic module, then any two such have the same projective dimension which equals the Gorenstein dimension of G.*

Using the characterization of the Gorenstein dimension of groups provided by Theorem 2.3 we can prove that the Gorenstein dimension of groups enjoys the following properties:

Theorem 2.5 *([13, thm. 2.8])*

(1) If $H \leq G$ then $\mathrm{Gcd}_{\mathbb{Z}}H \leq \mathrm{Gcd}_{\mathbb{Z}}G$.

(2) If $1 \to N \to G \to K \to 1$ is an extension of groups, then $\mathrm{Gcd}_{\mathbb{Z}}G \leq \mathrm{Gcd}_{\mathbb{Z}}N + \mathrm{Gcd}_{\mathbb{Z}}K$.

(3) If F is a finite subgroup of G, and $N_G(F)$ its normalizer in G, then $\mathrm{Gcd}_{\mathbb{Z}}(N_G(F)/F) \leq \mathrm{Gcd}_{\mathbb{Z}}G$.

(4) If $1 \to N \to G \to K \to 1$ is an extension of groups with $|N| < \infty$ then $\mathrm{Gcd}_{\mathbb{Z}}G = \mathrm{Gcd}_{\mathbb{Z}}K$.

(5) If $1 \to N \to G \to K \to 1$ is an extension of groups with $|K| < \infty$ then $\mathrm{Gcd}_{\mathbb{Z}}G = \mathrm{Gcd}_{\mathbb{Z}}N$.

The following theorem relates the Gorenstein dimension to the generalized cohomological dimension

Theorem 2.6 *([3, thm. 2.5]) For any group G, we have that $\mathrm{Gcd}_{\mathbb{Z}}G = \underline{\mathrm{cd}}G$. Consequently, if $\underline{\mathrm{cd}}G < \infty$, then*

$$\underline{\mathrm{cd}}G = \sup\{n \in \mathbb{N} \mid H^n(G, P) \neq 0, \;\; P \textit{ is } \mathbb{Z}G\textit{-projective}\}.$$

The proof of this Theorem uses the fact that $\mathrm{silp}\mathbb{Z}G = \mathrm{spli}\mathbb{Z}G$. The fact that $\mathrm{silp}\mathbb{Z}G \leq \mathrm{spli}\mathbb{Z}G$ was proved by Gedrich and Gruenberg in [11] and the fact that $\mathrm{silp}\mathbb{Z}G$ finite implies $\mathrm{spli}\mathbb{Z}G$ finite by Emmanouil in [8]. It is easy to see that if both $\mathrm{silp}\mathbb{Z}G$ and $\mathrm{spli}\mathbb{Z}G$ are finite then they are equal.

Combining Theorem 2.3 with Theorem 2.6 we obtain:

Theorem 2.7. *The following are equivalent for a group G*
(i) G admits a characteristic module T
(ii) $\mathrm{Gcd}_{\mathbb{Z}}G < \infty$
(iii) The supremum of the following set is finite:

$$\{n \in \mathbb{N} \mid \exists M, P : \mathrm{Ext}^n_{\mathbb{Z}G}(M, P) \neq 0, M \; \mathbb{Z}\textit{-free and } P \; \mathbb{Z}G\textit{-projective}\}.$$

Moreover, if T is a characteristic module for G, then

$$\begin{aligned}\mathrm{Gcd}_{\mathbb{Z}}G &= \sup\{n \in \mathbb{N} \mid H^n(G, P) \neq 0 \textit{ and } P \; \mathbb{Z}G\textit{-projective}\} \\ &= \mathrm{proj.\ dim}_{\mathbb{Z}G} T.\end{aligned}$$

Remark 2.8 If fin.dim $\mathbb{Z}G$ denotes the finitistic dimension of $\mathbb{Z}G$, i.e the supremum of the projective dimensions of the modules which have finite projective dimension,then it is easy to see that fin.dim $\mathbb{Z}G \leq \text{silp}\mathbb{Z}G$. From this follows easily that if a group G contains a free abelian subgroup of infinite rank, then fin.dim $\mathbb{Z}G$ is not finite hence G does not admit a characteristic module.

3 Characteristic modules and proper actions

Proposition 3.1 *If a group G is in $\mathbf{H}_1\mathfrak{F}$, i.e., G admits a contractible finite dimensional G-CW-complex X with finite stabilizers, then G admits a characteristic module.*

Proof. If we tensor the cellular chain complex of X with $B(G, \mathbb{Z})$, the module of bounded functions from G to $\mathbb{Z}$, then as X is contractible, finite dimensional and the cell stabilizers are finite we obtain a resolution of $B(G, \mathbb{Z})$ of finite length by modules of the form $\oplus\mathbb{Z}(G/G_i) \otimes B(G, \mathbb{Z})$ (diagonal action), with G_i finite subgroups of G. Since restriction of $B(G, \mathbb{Z})$ to finite subgroups is projective we obtain a $\mathbb{Z}G$-projective resolution of $B(G, \mathbb{Z})$ of finite length. □

The converse was proved by Kropholler and Mislin in the case where G is in $\mathbf{H}\mathfrak{F}$ and there is a bound on the orders of the finite subgroups:

Theorem 3.2 *([17, thm. B]) Let G be an $\mathbf{H}\mathfrak{F}$-group and assume that there is a bound on the orders of the finite subgroups. If* proj. $\dim_{\mathbb{Z}G} B(G, \mathbb{Z})$ *is finite then G is in $\mathbf{H}_1\mathfrak{F}$.*

(Actually Theorem 3.2 holds for G in $\mathbf{LH}\mathfrak{F}$ [27].)

The proof of [17, theorem B] uses also essentially the following:

Theorem 3.3 *([24, 3]) For any group G, $\text{Gcd}_{\mathbb{Z}}G \leq 1$ if and only if G acts on a tree with finite vertex stabilizers.*

Theorems 3.2 and 3.3 are special cases of the following conjecture:

Conjecture 3.4 *The following statements are equivalent for a group G*

(i) *G is in $\mathbf{H}_1\mathfrak{F}$.*

(ii) *G admits a characteristic module.*

(iii) $\text{Gcd}_{\mathbb{Z}}G < \infty$.

Remarks 3.5

(i) It is worth mentioning the following paragraph from Kropholler's 1993 article in [16, p. 211]

...there is a perfectly sensible algebraic dimension for groups in the class $\mathbf{H}_1\mathfrak{F}$, defined to be the greatest integer n for which there is a projective $\mathbb{Z}G$ module P such that $H^n(G, P)$ is non zero...

This is indeed the Gorenstein dimension of G. Note that if G is in $\mathbf{H}_1\mathfrak{F}$ then

$\mathrm{Gcd}_{\mathbb{Z}}G$ is finite since the cellular chain complex of the corresponding G-CW-complex is a resolution of G by Gorenstein projective $\mathbb{Z}G$-modules.

(ii) If the group G is in $\mathbf{LH}\mathfrak{F}$and G admits a characteristic module then $B(G, \mathbb{Z})$ is also a characteristic module for G [27].

(iii) The following results of Lück [20] and Martinez-Perez [21] relate the minimal dimension of the classifying space for proper actions of a group G to the Gorenstein dimension of G.

Theorem 3.6 *([20, thm. 1.10]) If G is a group such that the dimension of the G-simplicial complex determined by the poset of the non-trivial finite subgroups of G is finite and G satisfies $B(n)$, for some non negative integer n, then G admits a model for $\underline{E}G$ of dimension at most* $\max 3,\ n + \lambda(n+1)$.

A group G is said to satisfy $B(n)$ if for every finite subgroup F of G the group $W(F) = N_G(F)/F$ has the property that a $\mathbb{Z}W(F)$-module has finite projective dimension over $\mathbb{Z}W(F)$ if and only if it has finite projective dimension over every finite subgroup of $W(F)$.

Theorem 3.7 *([21, thm. 3.10]) If a group G admits a finite dimensional model for $\underline{E}G$ and*

$$\lambda := \sup\{m|\ H_0 < H_1 < \cdots < H_m,\ H_i \textit{ non-trivial finite subgroups of } G\}$$

is finite, then the minimal dimension for a model of $\underline{E}G$ is bounded above by $\lambda + \text{proj. dim}_{\mathbb{Z}G} B(G, \mathbb{Z})$.

The integer λ is essentially the same in both Theorems. It turns out ([26]) that if G satisfies $B(n)$ then G satisfies $B(n_0)$, where $n_0 = \text{proj. dim}_{\mathbb{Z}G} B(G, \mathbb{Z}) = \mathrm{Gcd}_{\mathbb{Z}}G$ and $n_0 \le n$.

One does not expect the Gorenstein dimension of G over $\mathbb{Z}$ to be the minimal dimension of a model for the classifying space for proper actions of G, since there is a group G, with $\mathrm{vcd}\ G = 3 = \mathrm{Gcd}_{\mathbb{Z}}G$, and with any model for its classifying space for proper actions having dimension at least four [19].

(iv) It may well be that if $\mathrm{Gcd}_{\mathbb{Z}}G$ is finite then

$$\mathrm{Gcd}_{\mathbb{Z}}G = \sup\{n : H^n(G, \mathbb{Z}G) \neq 0\}.$$

4 Characteristic modules and complete cohomology

The existence of terminal completions of the ordinary Ext functors has been studied by Gedrich and Gruenberg in [11]. Using an approach that involves satellites, Mislin defined in [22] for any group G, complete cohomology functors $\widehat{H}^*(G, -)$, $* \in \mathbb{Z}$ and a natural transformation $\tau : H^*(G, -) \longrightarrow \widehat{H}^*(G, -)$ as the projective completion of $H^*(G, -)$. An immediate consequence of the definition is that the complete cohomology groups $\widehat{H}^*(G, T)$

vanish if proj. $\dim_{\mathbb{Z}G} T < \infty$. Equivalent definitions of the complete cohomology functors have been independently given by Vogel in [12] (using a hypercohomology approach) and by Benson and Carlson in [4] (using projective resolutions).

Here we present a criterion, via complete cohomology, for a group G to admit a characteristic module.

Theorem 4.1 *[9, thm. 6.4] The following statements are equivalent for a group G*

1. *There is a $\mathbb{Z}$-free $\mathbb{Z}G$-module A that admits a $\mathbb{Z}$-split $\mathbb{Z}G$-monomorphism $\iota : Z \longrightarrow A$ such that the image of $\iota \in \mathrm{Hom}_{\mathbb{Z}G}(\mathbb{Z}, A) = H^0(G, A)$ in $\widehat{H}^0(G, A)$, under the natural transformation τ, vanishes.*

2. *The group G admits a characteristic module.*

Proof. Clearly it is enough to show that (1)⇒(2).

(1)⇒(2): Let $P_* \longrightarrow \mathbb{Z} \longrightarrow 0$ be a projective resolution of the trivial $\mathbb{Z}G$-module $\mathbb{Z}$ and denote by K_i, $i \geq 0$, the corresponding syzygy modules. Then, $P_* \otimes A \longrightarrow \mathbb{Z} \otimes A \longrightarrow 0$ is a projective resolution of $\mathbb{Z} \otimes A = A$ and $K_i \otimes A$, $i \geq 0$, are the corresponding syzygy modules. A lifting of the $\mathbb{Z}G$-linear map ι is then provided by the chain map $1 \otimes \iota : P_* \longrightarrow P_* \otimes A$. Using the Benson-Carlson approach to complete cohomology, we conclude that the vanishing of the image of ι under the canonical map $H^0(G, A) \longrightarrow \widehat{H}^0(G, A)$ implies that there exists a non-negative integer s, such that the map $1 \otimes \iota : K_s \longrightarrow K_s \otimes A$ factors through a projective $\mathbb{Z}G$-module P. We note that for any $\mathbb{Z}G$-module B the map $1 \otimes \iota \otimes 1 : K_s \otimes B \longrightarrow K_s \otimes A \otimes B$ factors through the $\mathbb{Z}G$-module $P \otimes B$.

We now consider the direct system of $\mathbb{Z}G$-modules $(A^{\otimes n})_n$, with structural maps $A^{\otimes n} \longrightarrow A^{\otimes n+1}$ given by $\iota \otimes 1_{A^{\otimes n}}$ for all $n \geq 0$, and let T be its direct limit. Then it is easy to see that the $\mathbb{Z}G$-module T is $\mathbb{Z}$-projective and the natural $\mathbb{Z}G$-linear map $\iota' : \mathbb{Z} \longrightarrow T$ is a $\mathbb{Z}$-split monomorphism. In view of the very definition of T, it follows that there exists an isomorphism of $\mathbb{Z}G$-modules $\sigma : A \otimes T \longrightarrow T$, which is such that the composition $T \xrightarrow{\iota \otimes 1} A \otimes T \xrightarrow{\sigma} T$ is the identity map of T. Since the map $1 \otimes \iota \otimes 1 : K_s \otimes T \longrightarrow K_s \otimes A \otimes T$ factors through the $\mathbb{Z}G$-module $P \otimes T$, we conclude that the identity map of $K_s \otimes T$ factors through $P \otimes T$ as well. Since T is $\mathbb{Z}$-projective, the $\mathbb{Z}G$-module $P \otimes T$ is projective; hence, being a direct summand of it, the $\mathbb{Z}G$-module $K_s \otimes T$ is projective as well. We also note that the $\mathbb{Z}G$-modules $P_j \otimes T$ are projective for all $j \geq 0$. Then, the exact sequence

$$0 \longrightarrow K_s \otimes T \longrightarrow P_{s-1} \otimes T \longrightarrow \cdots \longrightarrow P_0 \otimes T \longrightarrow \mathbb{Z} \otimes T \longrightarrow 0$$

provides us with a projective resolution of the $\mathbb{Z}G$-module $\mathbb{Z} \otimes T = T$ of length at most s and hence proj. $\dim_{\mathbb{Z}G} T \leq s < \infty$. □

Remark 4.2 This result is closely related to the finiteness criterion proved by Cornick and Kropholler in [5, lemma 2.2 and remark 2.4]; we avoid any assumption about the existence of a multiplicative structure on A, at the expense of replacing A with T (which is still a $\mathbb{Z}$-projective $\mathbb{Z}G$-module that contains $\mathbb{Z}$ as a $\mathbb{Z}$-split $\mathbb{Z}G$-submodule).

References

[1] J. Asadollahi, A. Bahlekeh and Sh. Salarian, 'On the hierarchy of cohomological dimensions of groups', preprint 2007.

[2] M. Auslander, 'Anneaux de Gorenstein, et torsion en algèbre commutative', Secrétariat mathématique, Paris, 1967, Séminaire d'Algèbre Commutative dirigé par Pierre Samuel, 1966/67. Texte rédigé, d'après des exposés de Maurice Auslander, par Marquerite Mangeney, Christian Peskine et Lucien Szpiro. École Normale Supérieure de Jeunes Filles.

[3] A. Bahlekeh, F. Dembegioti and O. Talelli, 'Gorenstein dimension and proper actions', Bull.London Math.Soc.41 (2009), 859-871.

[4] D. J. Benson, J. F. Carlson, 'Products in negative cohomology', J. Pure Appl. Algebra 82 (1992), 107-129.

[5] J. Cornick and P. H. Kropholler, 'Homological finiteness conditions for modules over group algebras', J. London Math. Soc. 58 (1998), 19-62.

[6] F. Dembegioti and O. Talelli, 'A note on complete resolutions', to appear in the Proc. AMS.

[7] W. Dicks and M. J. Dunwoody, 'Groups acting on graphs', Cambridge University Press (1989).

[8] I. Emmanouil, 'On certain cohomological invariants of groups', Adv. Math. 225 (2010), 3446-3462.

[9] I. Emmanouil and O. Talelli, 'Finiteness Criteria in Gorenstein Homological Algebra', to appear in the Tran AMS.

[10] E. E. Enochs and O. M. G. Jenda, 'Gorenstein injective and projective modules', Math. Z 220 (4) (1995), 611-633.

[11] T.V. Gedrich and K.W. Gruenberg, 'Complete cohomological functors on groups', Topology and its Applications 25 (1987), 203-223.

[12] F. Goichot, 'Homologie de Tate-Vogel equivariant', J. Pure Appl. Algebra 82 (1992), 39-64.

[13] H. Holm, 'Gorenstein homological dimensions', J. Pure Appl. Algebra 189 (2004), 167-193.

[14] B. M. Ikenaga, 'Homological dimension and Farrell cohomology', J. Algebra 87 (1984), 422-457.

[15] P. H. Kropholler, 'On groups of type FP_∞', J. Pure Appl. Algebra 90 (1993), 55-67.

[16] P. H. Kropholler, 'Hierarchical decompositions, generalized Tate cohomology, and groups of type $(FP)_\infty$', Combinatorial and Geometric Group Theory, LMS, Lecture Note Series 204.

[17] P. H. Kropholler and G. Mislin, 'Groups acting on finite dimensional spaces with finite stabilizers', Comment. Math. Helv. 73 (1998), 122-136.

[18] P. H. Kropholler and O. Talelli, 'On a property of fundamental groups of graphs of finite groups', J. Pure Appl. Algebra 74 (191), 57-59.

[19] I. J. Leary and B. E. A. Nucinkis, 'Some groups of type VF', Inventiones Math. 151 (1) (2003), 135-165.

[20] W. Lück, 'The type of classifying space for a family of subgroups', J. Pure Appl. Algebra 149, (2000),177-203.

[21] C. Martinez-Perez, 'A bound for the Bredon cohomological dimension for groups', J.Group Theory 10 (2007), 731-747.

[22] G. Mislin, 'Tate cohomology for arbitrary groups via satellites', Topology and its Applications 56 (1994), 293-300.

[23] G. Mislin and O. Talelli, 'On groups which act freely and properly on finite dimensional homotopy spheres', in Computational and Geometric Aspects of Modern Algebra, M. Atkinson et al. (Eds.), London Math. Soc. Lecture Note Ser. 275, Cambridge Univ. Press (2000), 208-228.

[24] O. Talelli, 'On groups with $\mathrm{cd}_{\mathbb{Q}}G \leq 1$', J. Pure Appl. Algebra 88 (1993) 245-247.

[25] O. Talelli, 'Periodicity in group cohomology and complete resolutions', Bull. London Math. Soc. 37 (2005), 547-554.

[26] O. Talelli, 'On groups of type Φ', Archiv der Mathematik 89 (1) (2007) 24-32.

[27] O. Talelli, 'A characterization for cohomological dimension for a big class of groups', J.Algebra 326(2011), 238-244.

Controlled Algebra for Simplicial Rings and Algebraic K-theory

Mark Ullmann

Abstract

We develop a version of controlled algebra for simplicial rings. This generalizes the methods which lead to successful proofs of the algebraic K- theory isomorphism conjecture (Farrell-Jones Conjecture) for a large class of groups. This is the first step to prove the algebraic K-theory isomorphism conjecture for simplicial rings. We show that the category in question has the structure of a Waldhausen category and discuss its algebraic K-theory.

We lay emphasis on detailed proofs. Highlights include the discussion of a simplicial cylinder functor, the glueing lemma, a simplicial mapping telescope to split coherent homotopy idempotents, and a direct proof that a weak equivalence of simplicial rings induces an equivalence on their algebraic K-theory. Because we need a certain cofinality theorem for algebraic K-theory, we provide a proof and show that a certain assumption, sometimes omitted in the literature, is necessary. Last, we remark how our setup relates to ring spectra.

Contents

1 Introduction

Controlled algebra is a powerful tool to prove statements about the algebraic K-theory of a ring R. While early on it was used in [PW85] to construct a non-connective delooping of $K(R)$—a space such that $\pi_i(K(R)) = K_{i-1}(R)$—it is a crucial ingredient in recent progress of the so-called Farrell-Jones Conjecture. Our aim here is to construct for a simplicial ring R, and a so-called "control space" X, a category of "controlled simplicial R-modules over a X". It should be regarded as a generalization of controlled algebra from rings to simplicial rings.

The category of "controlled simplicial modules" supports a homotopy theory which is formally very similar to the homotopy theory of CW-complexes. In particular we have a "cylinder object" which yields a notion of homotopy and therefore the category has homotopy equivalences. Waldhausen nicely summarized a minimal set of axioms to do homotopy theory in his notion of a Waldhausen category, called "category with cofibrations and weak equivalences in [Wal85]. He did this to define algebraic K-theory of such a category. Our category satisfies Waldhausen's axioms, which is our main result:

Theorem. *Let X be a control space and R a simplicial ring. The category of controlled simplicial modules over X, $\mathcal{C}(X;R)$, together with the homotopy equivalences and a suitable class of cofibrations is a "category with cofibrations and weak equivalences" in the sense of Waldhausen ([Wal85]). Therefore Waldhausen's algebraic K-theory of $\mathcal{C}(X;R)$ is defined.*

The category has a cylinder functor and it satisfies Waldhausen's cylinder axiom, his saturation axiom and his extension axiom.

In fact, for G a (discrete) group, there is a G-equivariant version, $\mathcal{C}^G(X;R)$, of this theorem, which is crucial for applications to the Farrell-Jones Conjecture. Maybe surprisingly the G-equivariant version of the theorem is not more difficult to prove than its non-equivariant counterpart. It is stated in Section 3.1 as Theorem 3.1.4 and Section 6 is devoted completely to its proof.

It is well-known that if a category has infinite coproducts of one objects, its algebraic K-theory vanishes. As $\mathcal{C}^G(X;R)$ suffers from this, we define a full subcategory of *finite objects* $\mathcal{C}^G_f(X;R)$. It behaves from the homotopy theoretic point of view like finite CW-complexes. From the algebraic point of view it corresponds to finitely generated free modules. Corresponding to projective modules we define the full subcategory of *homotopy finitely dominated objects* $\mathcal{C}^G_{hfd}(X;R)$ of $\mathcal{C}^G(X;R)$. (An object X is homotopy finitely dominated if there is a finite object A and maps $r\colon A \to X$, $i\colon X \to A$ such that $r \circ i \simeq \mathrm{id}_X$.)

Theorem. *Both $\mathcal{C}^G_f(X;R)$ and $\mathcal{C}^G_{hfd}(X;R)$ are Waldhausen categories, with the inherited structure from $\mathcal{C}^G(X;R)$. They still have a cylinder functor and satisfy the saturation, extension and cylinder axiom.*

Further the inclusion $\mathcal{C}^G_f(X;R) \to \mathcal{C}^G_{hfd}(X;R)$ induces an isomorphism

$$K_i(\mathcal{C}^G_f(X;R)) \to K_i(\mathcal{C}^G_{hfd}(X;R))$$

for $i \geq 1$ and an injection for $i = 0$.

The category $\mathcal{C}^G_{hfd}(X;R)$ the closest analogy to the idempotent completion of an additive category. We show in the appendix that indeed idempotents and "coherent" homotopy idempotents split up to homotopy in $\mathcal{C}^G_{hfd}(X;R)$.

We think that both $\mathcal{C}^G_f(X;R)$ and $\mathcal{C}^G_{hfd}(X;R)$ are basic ingredients to attack the Farrell-Jones Conjecture for simplicial rings.

1.1 Results of independent interest

In this article we need to discuss several topics which might be of interest for readers who are not interested in our main theorems. Here is a guide for these topics.

1.1.1 Controlled algebra for discrete rings We explain in Subsection 9.1 that the constructions here specialize to a construction of a category of controlled modules over a (discrete) ring. Readers who are interested in controlled algebra for rings can read Sections 2.2, 2.3 and the relevant part of 2.5, as well as 9.1. This gives in very few pages a construction of a category of controlled modules. We think our category is technically nicer than the model described in [BFJR04], because it is e.g. functorial in the control space and has an obvious forgetful functor to free modules. Otherwise the categories are interchangeable.

1.1.2 Establishing a Waldhausen structure and the glueing lemma A basic result in the homotopy theory of topological spaces is the Glueing Lemma: Assume that D_i is pushout of $C_i \leftarrow A_i \rightarrowtail B_i$ for $i = 0, 1$, where $\rightarrowtail$ denotes a cofibration. Assume we have maps $\varphi_A\colon A_0 \to A_1$ etc., which form a map of pushout diagrams. If φ_A, φ_B, φ_C are homotopy equivalences, then φ_D is one. This is not obvious, as the homotopy inverse of φ_D is not induced by the homotopy inverses of the other maps.

Waldhausen made the Glueing Lemma into one of the axioms of a Waldhausen category (called a "category with cofibrations and weak equivalences" in [Wal85]). Proofs that a given category satisfies Waldhausen's axioms are usually omitted in the literature. Section 6 contains a detailed proof that our category $\mathcal{C}^G(X;R)$ satisfies Waldhausen's axioms. Because in $\mathcal{C}^G(X;R)$ the weak equivalences are homotopy equivalences, which one can define once one has a cylinder functor, the proofs might be helpful for readers who seek for proofs in related situation.

1.1.3 A Cofinality Theorem for algebraic K-theory Let $\mathcal{B}$ the category of finitely generated projective modules over a (discrete) ring R and $\mathcal{A}$ the subcategory of free modules. It is well-known that the algebraic K-theory of $\mathcal{B}$ differs from the one of $\mathcal{A}$ only in degree 0. A way to describe this is to say that

$$K(\mathcal{A}) \to K(\mathcal{B}) \to \text{“}K_0(\mathcal{B})/K_0(\mathcal{A})\text{”}$$

is a homotopy fiber sequence of connective spectra, where the last term is the Eilenberg-MacLane spectrum of the group $K_0(\mathcal{B})/K_0(\mathcal{A})$ in degree 0. There are statements in the literature providing such a homotopy fiber sequence when $\mathcal{A}$ and $\mathcal{B}$ satisfy a list of conditions, e.g. in [Wei13, TT90]. We show these miss an essential assumption and provide counterexamples, as well as a proof of such a cofinality theorem, in Subsection 8.2. (Note that the above example of free and projective modules is just an illustration. To apply the theorem to finite and projective modules we would need to replace them by suitable categories of chain complexes first, as they do not have mapping cylinders.)

1.1.4 A simplicial mapping telescope In topological spaces one can form a mapping telescope of a sequence $A_0 \to A_1 \to A_2 \ldots$ by glueing together the mapping cylinder of the individual maps. It can be used to show that a space which is dominated by a CW-complex is homotopy equivalent to a CW-complex, see e.g. [Hat02, Proposition A.11]. We need an analogue in our category $\mathcal{C}^G(X;R)$. Because it is a simplicial category, and the homotopies are simplicial, a lot more care is required. We construct a simplicial mapping telescope in Appendix A. For this we define an analogue of Moore homotopies and provide the necessary tools to deal with them. Our results are summarized as Theorem A.2.1. We also define what we call a coherent homotopy idempotent and use the mapping telescope to show these split up to homotopy in $\mathcal{C}^G(X;R)$. We only use a few formal properties of $\mathcal{C}^G(X;R)$ to derive that result. We expect this construction to work in other settings to split idempotents there. But because we have no further examples of such categories we refrained from providing an axiomatic framework in which the Theorem would hold.

1.1.5 Weak equivalences of simplicial rings and algebraic K-theory A map $f\colon R \to S$ of simplicial rings is a weak equivalence if it one on the geometric realization of the underlying simplicial sets. Such a map induces an equivalence $K(R) \to K(S)$ on algebraic K-theory. Usually this is proved by using a plus-construction description of $K(R)$ (e.g. in [Wal78, Proposition 1.1]). Here we provide a proof which only uses Waldhausen's Approximation Theorem. The proof shows that f induces a weak equivalence on the algebraic K-theory of the categories of controlled modules, for which we do not have a plus-construction description. Note however, that [Wal78, Proposition 1.1] provides the stronger statement that an n-connected map induces an $n+1$-connected map on K-theory. We currently have no analogue of this for controlled modules over simplicial rings.

1.2 The idea of control

Let us now sketch the construction of $\mathcal{C}^G(X;R)$. For simplification we assume that G is the trivial group and X arises from a metric space (X,d), for example from $\mathbb{R}^n$ with the euclidean metric. The complete and precise definitions can be found in Section 2.

As simplicial R-module M is generated by a set $\diamond_R M = \{e_i\}_{i\in I} \subseteq \coprod_n M_n$ if every R-submodule $M' \subseteq M$ which contains $\{e_i\}_{i\in I}$ is equal to M. The idea is now to label each of the chosen generators e_i of M by an element $\kappa(e_i)$ of X and require that maps respect the labeling "up to an $\alpha > 0$". More precisely, a *controlled simplicial R-module over X* is a simplicial R-module M, a set of generators $\diamond_R M$ of M and a map $\kappa^M \diamond_R M \to X$. A morphism $f\colon (M, \diamond_R M, \kappa^M) \to (N, \diamond_R N, \kappa^N)$ of controlled simplicial R-modules is a map $f\colon M \to N$ of simplicial R-modules such that there is an $\alpha \in \mathbb{R}_{>0}$ such that for each $e \in \diamond_R M$ we have that $f(e) \subseteq N$ is contained in an R-submodule generated by elements $e' \in \diamond_R N$ with $d(\kappa^N(e'), \kappa^M(e)) \leq \alpha$.

There are two problems with the objects here: First, we want to have the generators as few relations as possible. This is the case for cellular R-modules, when $\diamond_R M$ is a set cells of M. We define this notion in Section 2.1. Second, the boundary maps in M should behave well with respect to the labels in the control space. A quick way of requiring that is that $\mathrm{id}_M\colon (M, \diamond_R M, \kappa^M) \to (M, \diamond_R M, \kappa^M)$ should be controlled. This is a condition on $(M, \diamond_R M, \kappa^M)$. We restrict to such modules which are controlled. This defines $\mathcal{C}(X;R)$ for X a metric space. The general notion is carefully introduced in Section 2.

The category $\mathcal{C}(X;R)$ relates to the "ordinary" controlled modules of for example [PW85] or [BFJR04] like chain complexes of free modules relate to projective modules, or like CW-complexes relate to projective $\mathbb{Z}$-modules. For M a simplicial R-module and $\mathbb{Z}[\Delta^1]$ the free simplicial abelian group on the 1-simplex define $M[\Delta^1] = M \otimes_{\mathbb{Z}} \mathbb{Z}[\Delta^1]$. If $(M, \diamond_R M, \kappa^M)$ is a controlled simplicial R-modules, $M[\Delta^1]$ is also one, canonically. This is the cylinder which yields the homotopy theory in $\mathcal{C}(X;R)$.

1.3 Structure of this article

The proof of our main theorem is quite involved as we need to develop the homotopy theory in $\mathcal{C}^G(X;R)$ from scratch. We therefore split this article into two main parts. The first, Sections 2 to 4 provides only definitions without any proofs, such that we can state our main theorems as soon as possible. We hope that this makes it easier for the reader to grasp the main definitions of this article, compared to when the definitions would be scattered over the rather long proofs.

The second part, Sections 5 to 8 and the appendix provide the proofs and all intermediate definitions and theorems we need. There does not seem to exist an established way to verify the axioms of a Waldhausen category in the literature, apart from trivial cases, although they are surely well-known. Therefore we provide a reasonable level of detail. Most of the proofs are rather formal once we established the Relative Horn-Filling Lemma 6.2.1. We hope the level of detail is helpful in case one wants to transfer the proofs here to other settings.

Section 9 gives some applications. We briefly elaborate on the relation of this work to the Farrell-Jones Conjecture and to controlled algebra for (discrete) rings. We give a construction of non-connective delooping of the algebraic K-theory of a simplicial ring without any proofs. We add a remark on ring spectra. Appendix A constructs a simplicial mapping telescope and proves the main Theorem A.2.1 about them, which is used to analyse idempotents and coherent homotopy idempotents in $\mathcal{C}^G(X;R)$.

1.4 Contents of the Sections 2 to 8

In Section 2 we concisely review simplicial rings and simplicial modules, as well as the idea of control. We define the category $\mathcal{C}^G(X;R)$. Section 3 defines the Waldhausen structure on $\mathcal{C}^G(X;R)$. Then we introduce the finiteness conditions of finite, homotopy finite, and homotopy finitely dominated modules. Each of these gives us a full subcategory of $\mathcal{C}^G(X;R)$. We show that the full subcategories of these are naturally Waldhausen categories. In Section 4 we state that the category of finite and homotopy finite modules have the same algebraic K-theory, and the algebraic K-theory of the homotopy finitely dominated ones differ one at K_0. We also compare the K-theory of our category for weakly equivalent rings.

The second part, Sections 5 to 8, contains the proofs for the previous sections and some elaborations. First we state some initial results on simplicial modules and controlled maps between them in Section 5. We are rather brief there. The result provided should be enough to make it possible for the experienced reader to verify all statements we made in Section 3. (Most of the statements in Section 3 are definitions anyway.)

Section 6 verifies the axioms of a Waldhausen category for $\mathcal{C}^G(X;R)$ for the cofibrations and weak equivalences we defined in Section 3.1. We give careful and complete proofs. The key ingredient is the Relative Horn-Filling Lemma 6.2.1. Further important results are the establishing of a Cylinder Functor 6.1.3, the Glueing Lemma 6.4 and the Extension Axiom for the homotopy equivalences 6.5.

Section 7 discusses the different finiteness conditions. This proves the results of Section 3.2, i.e., it establishes that each of the full subcategories of finite, homotopy finite and homotopy finitely dominated modules are again Waldhausen categories which satisfy all the extra axioms we listed.

In Section 8 we switch to algebraic K-theory and prove comparison theorems of the algebraic K-theory of the aforementioned categories. As an important part we prove a cofinality theorem 8.2.1 for algebraic K-theory. It is stated as an exercise in [TT90] and as Corollary V.2.3.1 in [Wei13], but we show that a crucial assumption is missing there and prove the correct statement. Last, we give a direct proof that a weak equivalence of simplicial rings gives an equivalence on algebraic K-theory of controlled modules.

1.5 Previous results

In [PW85] Pedersen and Weibel first used controlled modules to construct a non-connective delooping of the algebraic K-theory space of a (discrete) ring. In [Vog90], Vogell used the idea of control to construct a category related to Waldhausen's algebraic K-theory of spaces $A(X)$, which is homotopically flavored. Unfortunately Vogell does not provide any details on why his category is a Waldhausen category. Later Weiss [Wei02] gave a quick construction of a category similar to Vogell's one, but he also does not give a proof of the Waldhausen structure. Weiss' definitions inspired the definitions we use here.

With regard to discrete rings Controlled algebra was developed with the applications to the Farrell-Jones Conjecture in mind. A fundamental result is in [CP97], which constructs a highly useful fiber sequence on algebraic K-theory spaces, arising solely from control spaces. The most recent incarnation of controlled algebra is described in [BFJR04] which describes the category which is used in the most recent approaches to the Farrell-Jones Conjecture. [Ped00] contains a nice survey of the area at the time of its writing.

1.6 Acknowledgements

This work grew from the PhD-thesis [Ull11] the author wrote at the Heinrich-Heine-Universität Düsseldorf under the advise of Holger Reich. Compared to the thesis the definition of "controlled module" has been generalized, so most subsequent results needed to be adjusted. This work was partially supported by the PhD-grant of the Graduiertenkolleg 1150 "Homotopy and Cohomology" and the SFB 647: "Space - Time - Matter. Analytic and Geometric Structures". The author would like to thank everybody who has supported him during the long time this work needed to get in its final shape. Finally, the author would like to thank Charles Weibel for answering a question which lead to the counterexample 8.2.5.

1.7 Conventions

We sometimes use the property that for a diagram in a category

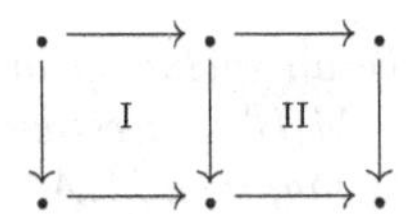

the whole diagram I + II is a pushout if I and II are pushouts and II is a pushout if I and I + II are pushouts. The dual version is proved in [Bor94, I.2.5.9], the third possible implication does not hold in general.

The set of natural numbers $\mathbb{N}$ contains zero. All rings have a unit.

2 Simplicial modules and control

2.1 Basic definitions

We assume familiarity with the theory of simplicial sets. A good reference is [GJ99].

2.1.1 Simplicial modules We recall the definition of simplicial modules. Δ is always the simplicial category $\{[n] \mid n \in \mathbb{N}\}$. A simplicial abelian group is a functor $\Delta^{\mathrm{op}} \to Ab$, similar a simplicial ring is a functor $\Delta^{\mathrm{op}} \to \mathcal{R}ings$. There are obvious generalizations of the notions of left and right modules and tensor products.

We introduce a notation. For a simplicial set A let $\mathbb{Z}[A]$ be the free simplicial abelian group on A. For M a simplicial left R-module define $M[A]$ as the simplicial left R-module $M \otimes_{\mathbb{Z}} \mathbb{Z}[A]$. For M, N simplicial left R-modules define $\mathrm{HOM}_R(M, N)$ as the simplicial abelian group $[n] \mapsto \mathrm{Hom}_R(M[\Delta^n], N)$.

2.1.2 Cellular modules We call the simplicial left R-module $R[\Delta^n]$ an n-cell and $R[\partial\Delta^n]$ the boundary of an n-cell. We say M arises from M' by attaching an n-cell if M is the isomorphic to the pushout $M' \cup_{R[\partial\Delta^n]} R[\Delta^n]$. Like a CW-complex in topological spaces, a cellular R-module relative to a submodule A is a module M together with a filtration of R-submodules $M^i, i \geq -1$ with $M^{-1} = A$ and $\bigcup M^i = M$ such that M^i arises from M^{i-1} by attaching i-cells. We call the map $A \to M$ a cellular inclusion. The composition of two cellular inclusions is again a cellular inclusion. A simplicial left R-module is called cellular if $* \to M$ is a cellular inclusion. $\rightarrowtail$ denotes cellular inclusions.

For our setting we will always remember the attaching maps of the cells to M and call this a *cellular structure* on M. This gives and can be reconstructed from an element $e_n \in M_n$ for each n-cell of M, where M_n denotes the set of n-simplices of M. This gives a set $\diamond_R M \subseteq \bigcup_n M_n$ to which we refer as *the cells of* M. As R-module, M can have many different cellular structures and we do not require maps to respect them.

Lemma 2.1.3. *Let $A \rightarrowtail B$ be an inclusion of simplicial sets. Let M be a cellular module. Then $M[A] \to M[B]$ is a cellular inclusion.*

If $M \rightarrowtail N$ is a cellular inclusion, then $M[A] \to N[A]$ is a cellular inclusion.

2.1.4 Finiteness conditions Similar to the case of CW-complexes or simplicial sets we call a cellular module *finite* if it has only finitely many cells and *finite-dimensional* if it has only cells of finitely many dimensions.

2.1.5 Dictionary We compare the notions introduced in this section to the corresponding notions of "ordinary", or "discrete" rings.

simplicial R-modules, R a simplicial ring	discrete R-modules, R a discrete ring
cellular module	free module
cellular structure	choice of a basis
cellular inclusion	direct summand with a free complement
$M[A]$ (A a simplicial set)	$\bigoplus_{a\in A} M$ (A a set)
$\coprod_I R[\Delta^n]$	$\bigoplus_I R$
of finite dimension	—
finite	finite dimensional

Table 1: Dictionary simplicial rings and modules.

2.2 Control spaces

Definition 2.2.1. Let X be a topological Hausdorff space. A *morphism control structure on* X consists of a set $\mathcal{E}$ of subsets E of $X \times X$ (i.e., relations on X), called the *morphism control conditions.* We require:

1. For $E, E' \in \mathcal{E}$ there is an $\overline{E} \in \mathcal{E}$ such that $E \circ E' \subseteq \overline{E}$ where "$\circ$" is the composition of relations.
2. For $E, E' \in \mathcal{E}$ there is an $E'' \in \mathcal{E}$ such that $E \cup E' \subseteq E''$.

3. Each $E \in \mathcal{E}$ is symmetric, i.e., $(x, y) \in E \Leftrightarrow (y, x) \in E$.
4. The diagonal $\Delta \subseteq X \times X$ is a subset of each $E \in \mathcal{E}$.

The topology is only relevant for finiteness conditions later.

2.2.2 Thickenings For $U \subseteq X$ and $E \in \mathcal{E}$ we call $U^E = \{x \in X \mid \exists y \in U \colon (x, y) \in E\}$ the *E-thickening* of X.

Definition 2.2.3. Given X and a morphism control structure $\mathcal{E}$ on X. An *object support structure on* $(X, \mathcal{E})$ is a set $\mathcal{F}$ of subsets F of X, called the *object support conditions.* We require:

1. For $F, F' \in \mathcal{F}$ there is an $F'' \in \mathcal{F}$ such that $F \cup F' \subseteq F''$.
2. For $F \in \mathcal{F}$ and $E \in \mathcal{E}$ there is an $F''' \in \mathcal{F}$ such that $F^E \subseteq F'''$.

2.2.4 In all applications we can close up both conditions under taking subsets, i.e., require if $E \in \mathcal{E}$ and $\Delta \subseteq E' \subseteq E$ then $E' \in \mathcal{E}$, and if $F' \subseteq F$ and $F \in \mathcal{F}$ then $F' \in \mathcal{F}$. We call the triple $(X, \mathcal{E}, \mathcal{F})$ a *control space.* If $\mathcal{F} = \{X\}$ we often leave it out of the notation.

2.2.5 Maps A map of control space $(X_1, \mathcal{E}_1, \mathcal{F}_1) \to (X_2, \mathcal{E}_2, \mathcal{F}_2)$ is a (not necessarily continuous) map $f \colon X_1 \to X_2$ such that for each $E_1 \in \mathcal{E}_1$ and $F_1 \in \mathcal{F}_1$ there are $E_2 \in \mathcal{E}_2$ and $F_2 \in \mathcal{F}_2$ with $(f \times f)(E_1) \subseteq E_2$ and $f(F_1) \subseteq F_2$.

We give the most important examples, see [BFJR04, Section 2.3] for more.

Example 2.2.6 (metric control). Let X have a metric d. Then

$$\mathcal{E}_d = \{E \mid \text{there is an } \alpha \text{ such that } E = \{(x, y) \mid d(x, y) \leq \alpha\}\}$$

is a morphism control structure on X.

Example 2.2.7 (continuous control). Let Z be a topological space and $[1, \infty)$ the half-open interval with closure $[1, \infty]$. Define a morphism control structure $\mathcal{E}_{cc}$ on $X := Z \times [1, \infty)$ as follows. E is in $\mathcal{E}_{cc}$ if it is symmetric and

1. For every $x \in Z$ and each neighborhood U of $x \times \infty$ in $Z \times [1, \infty]$ there is a neighborhood $V \subseteq U$ of $x \times \infty$ in $Z \times [1, \infty]$ such that $E \cap ((X \smallsetminus U) \times V) = \varnothing$.
2. $p_{[1,\infty)} \times p_{[1,\infty)}(E) \in \mathcal{E}_d([1, \infty))$, where d is the standard euclidean metric on $[1, \infty)$ and $p_{[1,\infty)}$ is the projection to $[1, \infty)$.

Example 2.2.8 (compact support). Let set $F \subseteq X$ be in $\mathcal{F}_c$ if it is compact. These are the compact object support conditions. They are object support conditions for $(X, \mathcal{E}_d)$ where X is a proper metric space (closed balls are compact) or for the continuous control conditions $\mathcal{E}_{cc}$ on $Z \times [1, \infty)$.

2.3 Controlled simplicial modules

2.3.1 Cellular submodules We required cellular R-modules to come with a chosen cellular structure $\diamond_R M$. A *cellular submodule* is an R-submodule M' of M which is generated by a subset of $\diamond_R M$. In particular we have an inclusion $\diamond_R M' \subseteq \diamond_R M$ induced by $M' \hookrightarrow M$.

Definition 2.3.2. For a set of simplices $Q \subseteq \bigcup M_n$ define $\langle C \rangle_M$ as the smallest cellular submodule of M containing Q.

We abbreviate $\langle \{e\} \rangle_M$ by $\langle e \rangle_M$ or $\langle e \rangle$.

2.3.3 Modules over a space For a control space $(X, \mathcal{E}, \mathcal{F})$ define a *general module over* X to be a cellular module $(M, \diamond_R M)$ together with a map $\kappa_R \colon \diamond_R M \to X$. (We followed [Wei02] in the notation.)

Definition 2.3.4 (Controlled module)**.** A controlled R-module over X is a general R-module $(M, \diamond_R M, \kappa_R)$ over X such that there are $E \in \mathcal{E}$, $F \in \mathcal{F}$ with:

1. For all $e \in \diamond_R M$ and $e' \in \langle e \rangle_M$ we have $(\kappa_R(e), \kappa_R(e')) \in E$.

2. $\kappa_R(\diamond_R M) \subseteq F$.

We say (M, κ_R) is E-controlled and has support in F, and often leave κ_R understood.

Definition 2.3.5 (Controlled maps)**.** A map $(M, \kappa_R^M) \to (N, \kappa_R^N)$ of controlled modules is a map $f \colon M \to N$ of simplicial R-modules such that there is an $E \in \mathcal{E}$ and for all $e \in \diamond_R M$, $e' \in \diamond_R \langle f(e) \rangle_N$ we have $(\kappa_R^M(e), \kappa_R^N(e')) \in E$.

We say f is *E-controlled.* We just say f is *controlled* if we do not want to specify the E.

2.3.6 Composition If $f, f_1, f_2 \colon M \to M'$ and $g \colon M' \to M''$ are controlled maps (of controlled modules), then $g \circ f$ and $f_1 + f_2$ are controlled.

2.3.7 The category of controlled modules For $(X, \mathcal{E}, \mathcal{F})$ a control space the controlled R-modules over X together with the controlled maps between them form a category which we denote by $\mathcal{C}(X, \mathcal{E}, \mathcal{F}; R)$. We will usually abbreviate it by $\mathcal{C}(X; R)$, $\mathcal{C}(X)$, $\mathcal{C}(X, \mathcal{E}, \mathcal{F})$ or $\mathcal{C}$.

If M is a controlled module over X and A a simplicial set then $M[A]$ is canonically a controlled module over X. This is functorial in A and M.

2.3.8 Cellular inclusion of controlled modules Define a cellular inclusion of controlled modules to be a map $(M, \kappa_R^M) \to (N, \kappa_R^N)$ such that $(M, \diamond_R M) \to (N, \diamond_R N)$ is a cellular inclusion of simplicial R-modules and the inclusion $i\colon \diamond_R M \hookrightarrow \diamond_R N$ satisfies $\kappa_R^M = \kappa_R^N \circ i$. This is the right notion of a subobject in $\mathcal{C}$.

If $A \rightarrowtail B$ is an inclusion of simplicial sets, then $M[A] \to M[B]$ is a cellular inclusion of controlled modules. If $M \to N$ is a cellular inclusion of controlled modules, then $M[A] \to N[A]$ is one.

2.4 A kind of an adjunction

2.4.1 Controlled filtration on the HOM-space Let $M, N \in \mathcal{C}(X, \mathcal{E}, \mathcal{F})$, $E \in \mathcal{E}$. Define $\mathrm{Hom}_R^E(M, N)$ as the subset of maps $f\colon M \to N$ in $\mathcal{C}(X)$ which are E-controlled. Similar define $\mathrm{HOM}_R^E(M, N)$ as the sub-simplicial set of E-controlled maps $M[\Delta^n] \to N$. Boundaries and degeneracies respect E, so this is a well-defined simplicial subset of $\mathrm{HOM}_R(M, N)$. Define $\mathrm{HOM}_R^{\mathcal{E}}(M, N)$ as $\bigcup_{E \in \mathcal{E}} \mathrm{HOM}_R^E(M, N)$.

2.4.2 Uncontrolled adjunction and its controlled counterparts If A is a simplicial set, we have an adjunction

$$\mathrm{Hom}_R(M[A], N) \cong \mathrm{Hom}_{sSet}(A, \mathrm{HOM}_R(M, N))$$

in simplicial R-modules. This restricts to a bijection

$$\mathrm{Hom}_R^E(M[A], N) \cong \mathrm{Hom}_{sSet}(A, \mathrm{HOM}_R^E(M, N)).$$

If A is a finite simplicial set, we have a bijection $\mathrm{Hom}_R^{\mathcal{E}}(M[A], N) \cong \mathrm{Hom}_{sSet}(A, \mathrm{HOM}_R^{\mathcal{E}}(M, N))$ which is natural in A, M and N.

2.5 G-equivariance

Let G be a (discrete) group. All notions above generalize in a straightforward way to G-equivariant versions:

2.5.1 G-equivariant cellular modules An action of G on a simplicial R-module M is a group homomorphism $\rho\colon G \to \mathrm{Aut}_R(M)$. The action is a called *cell-permuting* if it induces an action on $\diamond_R M$. An action is free if it is cell-permuting and the action on $\diamond_R M$ is free. If M, N are simplicial R-modules with G-actions ρ_M, ρ_N a map $f\colon M \to N$ of simplicial R-modules is G-equivariant if for each $g \in G$ the diagram

$$\begin{array}{ccc} M & \xrightarrow{f} & N \\ \downarrow{\scriptstyle \rho_M(g)} & & \downarrow{\scriptstyle \rho_N(g)} \\ M & \xrightarrow{f} & N \end{array}$$

commutes. A G-equivariant map $L \to M$ is a cellular inclusion, if it is one after forgetting the G-action.

If M is a simplicial R-module with G-action and A a simplicial set then $M[A]$ has a G-action by the functoriality in M.

2.5.2 ***G*-equivariant control spaces** A control space $(X, \mathcal{E}, \mathcal{F})$ is G-equivariant if X has a continuous G-action such that $gE = E$ (diagonal action) and $gF = F$ for all $g \in G$, $E \in \mathcal{E}$, $F \in \mathcal{F}$. A free control space is one where the action of G on X is free. The examples of control spaces in Section 2.2 have G-equivariant analogues, see [BFJR04, 2.7, 2.9, 3.1, 3.2].

2.5.3 The category of *G*-equivariant controlled modules We now let $(X, \mathcal{E}, \mathcal{F})$ be a free G-equivariant control space. A controlled simplicial R-module with G-action over X is a controlled module $(M, \diamond_R M, \kappa_R)$ with cell-permuting G-action such that κ_R is G-equivariant. A morphism $(M, \kappa_R) \to (N, \kappa_R)$ of such modules is a G-equivariant morphism $M \to N$ which is controlled over X. Denote the category of these as $\mathcal{C}^G(X, \mathcal{E}, \mathcal{F}; R)$. We use abbreviations like $\mathcal{C}^G$, etc. All further definitions of Section 2.3 transfer to $\mathcal{C}^G$.

2.5.4 The *G*-equivariant kind of adjunction The adjunction between $M[-]$ and $\mathrm{HOM}_R(M, -)$ and its controlled counterparts generalizes to the G-equivariant setting. Denote by $\mathrm{Hom}_R(M, N)^G$ and $\mathrm{HOM}_R(M, N)^G$ the subset of $\mathrm{Hom}_R(M, N)$, resp. subspace of $\mathrm{HOM}_R(M, N)$, of G-equivariant maps. The adjunctions of 2.4.2 restrict to adjunctions

$$\mathrm{Hom}_R(M[A], N)^G \cong \mathrm{Hom}_{sSet}(A, \mathrm{HOM}_R(M, N)^G)$$

and

$$\mathrm{Hom}_R^E(M[A], N)^G \cong \mathrm{Hom}_{sSet}(A, \mathrm{HOM}_R^E(M, N)^G).$$

Similarly, if A is a finite simplicial set there is a bijection $\mathrm{Hom}_R^{\mathcal{E}}(M[A], N)^G \cong \mathrm{Hom}_{sSet}(A, \mathrm{HOM}_R^{\mathcal{E}}(M, N)^G)$.

3 Waldhausen categories of controlled modules

In the following $(X, \mathcal{E}, \mathcal{F})$ is always a free G-equivariant control space which we abbreviate as X. We choose a simplicial ring R for this section. We put some additional structure on $\mathcal{C}^G(X; R)$. We will always work in this category in this section.

3.1 $\mathcal{C}^G(X, \mathcal{E}, \mathcal{F}; R)$ as a Waldhausen category

First we make $\mathcal{C}^G(X; R)$ into a Waldhausen category, called "category with cofibrations and weak equivalences" in [Wal85]. We will use the definitions of category with cofibrations, category with weak equivalences, cylinder functor and the saturation, cylinder and extension axiom from there. We will give detailed proofs of the statements below in Section 6.

3.1.1 Cofibrations Define a map $f\colon M \to N$ in $\mathcal{C}^G(X)$ to be a *cofibration* if there are isomorphisms $\alpha\colon M' \to M$ and $\beta\colon N \to N'$ in $\mathcal{C}^G(X)$ such that $\beta \circ f \circ \alpha$ is a cellular inclusion. Note that α, β do not need to preserve the cellular structures, so the notion is independent of chosen cellular structures. The compositions of cofibrations is a cofibration. We also denote cofibrations by $\rightarrowtail$. If $A \rightarrowtail B$ is a cofibration and $A \to C$ a map then the pushout $B \cup_A C$ exists and $C \to B \cup_A C$ is a cofibration. If $A \to C$ has been a cellular inclusion, then this pushout can be chosen canonically, in particular functorially. This makes $\mathcal{C}^G(X;R)$ into a category with cofibrations.

3.1.2 Cylinders Consider the simplicial set Δ^1, call it the *interval.* It comes with inclusions $i_0, i_1\colon \mathrm{pt} \to \Delta^1$ and a projection $p\colon \Delta^1 \to \mathrm{pt}$. For M in $\mathcal{C}^G(X;R)$ this induces the corresponding cellular inclusions $M \to M[\Delta^1]$ and a projection $M[\Delta^1] \to M$ which makes $M[\Delta^1]$ into a *cylinder* object. For a map $f\colon A \to B$ define $T(f)$ as $A[\Delta^1] \cup_{i_{1*}} B$. This is functorial in the arrow category and therefore gives cylinder functor on $\mathcal{C}^G(X;R)$ in the sense of Waldhausen [Wal85, 1.6].

3.1.3 Weak equivalences Two maps $f, g\colon A \to B$ are homotopic if there is a *homotopy* $H\colon A[I] \to B$ such that $H \circ i_{0*} = f$ and $H \circ i_{1*} = g$. This gives rise to the obvious notion of homotopy equivalence.

Theorem 3.1.4. *The subcategory of homotopy equivalences in $\mathcal{C}^G(X;R)$ forms a category of weak equivalences, in particular it satisfies the glueing lemma. It also satisfies the saturation axiom and the extension axiom. The cylinder functor satisfies the cylinder axiom with respect to these weak equivalences.*

Note that this category is too big, it has an Eilenberg-Swindle. But it contains interesting full subcategories, which we discuss next.

3.1.5 A remark on the proofs Let us interrupt for a remark on the proofs. The main tool for the proofs are the adjunctions of 2.5.4 and a careful analysis of the control conditions in the settings, often accompanied by an induction over the cells. Here is a prototype of such a proof. We need to show that we have horn-filling in our category. In particular the following (simplified) lemma should hold.

Lemma. *Given a map $M[\Lambda^n_i] \to P$. Then there is an extension to a map $M[\Delta^n] \to P$.*

Proof. By the adjunction the situation is equivalent to finding a (dotted) lift in the diagram

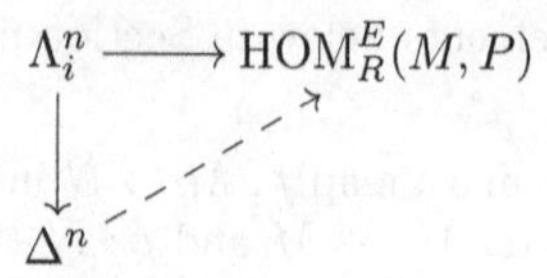

But the simplicial set $\mathrm{HOM}^E_R(M,P)$ is in fact a simplicial abelian group, hence the lift exists by the Kan-property, i.e., it is fibrant. □

The general proofs are considerably more complicated and quite long if carried out in detail. We will devote Section 6 to them.

3.2 Finiteness conditions

We define full subcategories of $\mathcal{C}^G(X;R)$ by specifying conditions on the objects.

3.2.1 (Locally) finite controlled modules Here we use the topology on X. Let $(M, \diamond_R M, \kappa_R)$ be a controlled module over X. We say that M is *locally finite* if for each $x \in X$ there is a neighborhood U of x in X such that $\kappa_R^{-1}(U)$ is a finite subset of $\diamond_R M$. Then M is called *finite* if it is locally finite and finite-dimensional. Denote the full subcategory of finite module by $\mathcal{C}^G_f(X;R)$. It inherits the structure of a Waldhausen category from $\mathcal{C}^G(X;R)$. It satisfies the saturation and extension axiom and has a cylinder functor satisfying the cylinder axiom. Also, it turns out that cofibrations are isomorphic in $\mathcal{C}^G_f(X;R)$ to cellular inclusions. The proof needs the Hausdorff-property of X.

3.2.2 Homotopy finite controlled modules An object $M \in \mathcal{C}^G(X;R)$ is *homotopy finite* if there is a homotopy equivalence $M \xrightarrow{\sim} M'$ such that M' is a finite module. We denote the full subcategory of homotopy finite modules by $\mathcal{C}^G_{hf}(X;R)$. Similar to $\mathcal{C}^G_f(X;R)$ it inherits the structure of a Waldhausen category. It has a cylinder functor satisfying the cylinder axiom. The saturation and extension axiom hold.

3.2.3 Homotopy finitely dominated controlled modules An object $M \in \mathcal{C}^G(X;R)$ is *homotopy finitely dominated* if it is a strict retract of a homotopy finite object. We denote the full subcategory of homotopy finitely dominated modules by $\mathcal{C}^G_{hfd}(X;R)$. Similar to $\mathcal{C}^G_f(X;R)$ it inherits the structure of a Waldhausen category. It has a cylinder functor satisfying the cylinder axiom. The saturation and extension axiom hold. Homotopy finitely dominated modules can equivalently be characterized by being a retract up to homotopy of a finite object.

4 Algebraic K-theory of controlled modules

4.1 Connective K-theory

Let us make explicit that we can get algebraic K-theory out of the defined Waldhausen categories.

Definition 4.1.1 (Algebraic K-theory of categories of controlled modules). Let G be a group, $(X, \mathcal{E}, \mathcal{F})$ be a free G-equivariant control space. Let R be a simplicial ring. Define the algebraic K-theory spectrum of the category with cofibrations and weak equivalences $\mathcal{C}_f^G(X, R, \mathcal{E}, \mathcal{F})$ as the connective spectrum

$$K(w\mathcal{C}_f^G(X, R, \mathcal{E}, \mathcal{F}))$$

where K is Waldhausen's algebraic K-theory of spaces [Wal85]. We define similar the algebraic K-theory of $\mathcal{C}_{hf}^G$ and $\mathcal{C}_{hfd}^G$.

Remark 4.1.2. In [Wal85] Waldhausen defines the K-theory as a space and then constructs a delooping, i.e., an Ω-spectrum. This is what we use here, because for the Cofinality Theorem it is more convenient to work with spectra. See also [TT90, 1.5.3].

Theorem 4.1.3 (Different finiteness conditions). *Let $(X, \mathcal{E}, \mathcal{F})$ be a control space and R a simplicial ring.*

1. *The inclusion $\mathcal{C}_f^G(X, R, \mathcal{E}, \mathcal{F}) \to \mathcal{C}_{hf}^G(X, R, \mathcal{E}, \mathcal{F})$ is exact and induces a homotopy equivalence on K-Theory.*

2. *The inclusion $\mathcal{C}_{hf}^G(X, R, \mathcal{E}, \mathcal{F}) \to \mathcal{C}_{hfd}^G(X, R, \mathcal{E}, \mathcal{F})$ is exact and induces an isomorphism on K_n for $n \geq 1$ and an injection $K_0(\mathcal{C}_{hf}^G) \to K_0(\mathcal{C}_{hfd}^G)$.*

4.1.4 Separations of variables The categories $\mathcal{C}^G(G/1, \{G \times G\}, \{G\}; R)$ and $\mathcal{C}(\text{pt}, \{\text{pt}\}, \{\text{pt}\}; R[G])$ are equivalent. Both are equivalent to the category of cellular $R[G]$-modules. The equivalences respect the finiteness conditions f, hf and hfd.

Corollary 4.1.5. *The algebraic K-theory of $\mathcal{C}_{hfd}^G(G/1, \{G \times G\}, \{G\}; R)$ is homotopy equivalent to the algebraic K-theory of the simplicial ring $R[G]$.*

4.1.6 Change of rings The constructions are functorial in change of ring maps $f \colon R \to S$. If f is a weak equivalence, then the induced map $\mathcal{C}_?^G(X; R) \to \mathcal{C}_?^G(X; S)$ is an equivalence on algebraic K-theory. Here ? can be f, hf, or hfd.

5 Proofs I: About control

Section 2 introduced our basic categories of controlled simplicial modules. Most results are straightforward to check. Therefore we discuss only a few important lemmas and leave the rest to the reader. For the structure of a Waldhausen category which we introduced in Section 3 we will provide much more detailed proofs in Section 6.

5.1 Cellular structure

5.1.1 Proof of Lemma 2.1.3

Lemma 5.1.2 (Lemma 2.1.3). *Let $A \rightarrowtail B$ be an inclusion of simplicial sets. Let M be a cellular module. Then $M[A] \to M[B]$ is a cellular inclusion.*

If $M \rightarrowtail N$ is a cellular inclusion, then $M[A] \to N[A]$ is a cellular inclusion.

Proof. We need to show that $M[A]$ respects cellular inclusion in both variables. First, the pushout of a cellular inclusion is a cellular inclusion. Also sequential colimits and coproducts of cellular inclusions are cellular inclusions. Further, $M[-]$ commutes with colimits as it is the composition of two left adjoint functors. Therefore it suffices to consider the case $M[\partial\Delta^n] \to M[\Delta^n]$. This map factors over $M[\partial\Delta^n] \cup_{L_i[\partial\Delta^n]} L_i[\Delta^n]$ for each i, where L_i is the i-skeleton of M. Using that $-[\Delta^n]$ commutes with pushouts one can show that the map $M[\partial\Delta^n] \cup_{L_i[\partial\Delta^n]} L_i[\Delta^n] \to M[\partial\Delta^n] \cup_{L_{i+1}[\partial\Delta^n]} L_{i+1}[\Delta^n]$ is a cellular inclusion. Then $M[\partial\Delta^n] \to M[\Delta^n]$ is the sequential colimit of these maps.

For the second part it suffices to consider the case of attaching one R-cell. Then it follows from $R[\partial\Delta^n][A] \to R[\Delta^n][A]$ being a cellular inclusion. Namely applying $-[A]$ to the pushout "attaching an n-cell" gives again a pushout with desired map being the pushout of the map above. □

5.2 Controlled maps

We denote the support of a controlled module by $\operatorname{supp}(M)$. We have the following more precise statement about the control of maps.

Lemma 5.2.1.

1. *If $f\colon M \to M'$ is E-controlled and $g\colon M' \to M''$ is E'-controlled, then $g \circ f$ is $E' \circ E$-controlled.*

2. *If $f_1, f_2\colon M \to M'$ are E_1-, resp. E_2-controlled and $E_1 \cup E_2 \subseteq E_3$, then $f_1 + f_2$ is E_3-controlled.*

3. *If M is an E-controlled module, then id_M is E-controlled.*

Proof. Let $e \in \diamond_R M$. Then $\operatorname{supp}(\langle f(e)\rangle) \subseteq \{\kappa^M(e)\}^E$. Further for $e' \in \diamond_R \langle f(e)\rangle$ we have $\operatorname{supp}(\langle g(e')\rangle) \subseteq \{\kappa^{M'}(e')\}^{E'}$. By minimality it follows that $\langle (g \circ f)(e))\rangle_{M''} \subseteq \langle g(\langle f(e)\rangle_{M'})\rangle_{M''}$, so its support is contained in $\{\kappa^M(e)\}^{E' \circ E}$.

Further $\langle (f_1 + f_2)(e)\rangle \subseteq \langle (f_1)(e)\rangle \cup \langle (f_2)(e)\rangle$. So its support is contained in $\{\kappa^M(e)\}^{E_1} \cup \{\kappa^M(e)\}^{E_2} \subseteq \{\kappa^M(e)\}^{E_3}$. The third part is clear. □

We say that a map $f\colon M \to N$ of cellular R-modules is *0-controlled* if it induces a map $\diamond_R M \to \diamond_R N$ and $\kappa_R^M = \kappa_R^N \circ f$. Cellular inclusions are 0-controlled. The name 0-controlled is a misuse of notation, such a map has the control of its image. However, $g \circ f$ has the control of g if f is 0-controlled.

Lemma 5.2.2. *Let (M, κ_R) be an E-controlled R-module. Let A be a simplicial set. Then $M[A]$ can be made canonically into an E-controlled R-module.*

Further, each map $A \to B$ of simplicial sets induces a 0-controlled map $M[A] \to M[B]$.

Proof. From Lemma 2.1.3 it follows that each cell e of $M[A]$ arises from exactly on cell $p(e)$ of M. Define $\kappa^{M[A]}(e) := \kappa^M(e)$. It makes $M[A]$ into an E-controlled module: For $e \in M[A]$ we have $p(e) \in \diamond_R M$. Then $e \in \langle p(e)\rangle [A] \subseteq M[A]$, and $\langle p(e)\rangle [A]$ is supported on $\{\kappa(e)\}^E$. This shows the first part.

Another way to describe the control map is to note that the map $M[A] \to M[\mathrm{pt}] \cong M$ is 0-controlled. A cell of M is given by a map $R[\Delta^n] \to M$. Hence for each cell of M we get a commutative diagram

$$\begin{array}{ccc} M[A] & \xrightarrow{f} & M[B] \\ \uparrow & & \uparrow \\ R[\Delta^n][A] & \longrightarrow & R[\Delta^n][B] \end{array}$$

which shows that f maps cells to cells. As f commutes with the map to $M[\mathrm{pt}]$ this shows that f is 0-controlled. □

6 Proofs II: Controlled simplicial modules as a Waldhausen category

In this section we establish that $\mathcal{C}^G$ is indeed a Waldhausen category, thus proving Subsection 3.1. We will use the definitions from 3.1 without further notice.

It would have been great if we could have followed an established pattern to show that $\mathcal{C}^G(X; R)$ is a Waldhausen category, but the author is not aware of worked-out proofs of the structure of a Waldhausen category in the literature in elementary terms. If the category in question is a subcategory of cofibrant objects of a Quillen model category, one basically gets the structure of a Waldhausen category for free, and lots of examples in the literature are of that form. But unfortunately there does not seem to be a suitable Quillen model category which contains our category of controlled modules. In fact, neither general pushouts, nor infinite unions exists in general in any of our categories of controlled modules.

Therefore we prove all results directly. The hardest part is to prove the glueing lemma. We follow a strategy the author learned from a proof of Waldhausen of the proof of the glueing lemma for topological spaces.

Sometimes we are brief or do not comment on statements which are easy to prove. We expect the experienced reader to be able to fill the gaps easily, but otherwise refer to the author's thesis [Ull11] which has even more details. Note that the thesis works with a slightly different definition of controlled module.

6.1 The Waldhausen structure

We assume familiarity with Section 1.1 to 1.6 of [Wal85] and use the language from there freely.

Lemma 6.1.1 (Pushouts along cellular inclusions). *Let $(A, \kappa_R^A) \to (B, \kappa_R^B)$ be a cellular inclusion in $\mathcal{C}^G$, let $f\colon (A, \kappa_R^A) \to (C, \kappa_R^C)$ be any controlled map in $\mathcal{C}^G$. The pushout $D := C \cup_A B$,*

$$\begin{array}{ccc} A & \rightarrowtail & B \\ \downarrow{\scriptstyle f} & & \downarrow \\ C & \rightarrowtail & D \end{array}$$

of simplicial R-modules can be chosen canonically and further it has a canonical structure of an object (D, κ_R^D) in $\mathcal{C}^G$. Further $(C, \kappa_R^C) \to (D, \kappa_R^D)$ is a cellular inclusion.

Hence $\mathcal{C}^G$ has canonical pushouts along cellular inclusions.

Proof. The category of simplicial R-modules can be equipped with canonical pushouts. For example, in the above situation one could take the coproduct of B and C and divide out the relations from A. We discuss the cellular structure.

Let e be a cell in B not in A with attaching map α. This gives a cell in D with attaching map $f \circ \alpha$. This way one shows that $C \to D$ is cellular, and therefore so is D. Hence there is a canonical isomorphism $\diamond_R D \cong \diamond_R C \cup (\diamond_R B \setminus \diamond_R A)$. Define $\kappa_R^D\colon \diamond_R D \to X$ via that isomorphism. We get the control conditions of Table 2. As φ is G-equivariant, D is a controlled G-equivariant cellular R-module. □

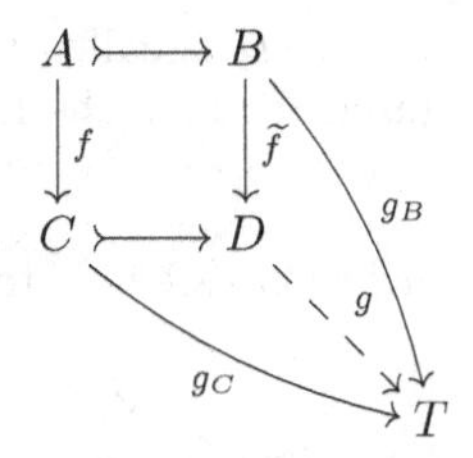

control conditions we have	
A, B	E_B-controlled
C	E_C-controlled
f	E_f-controlled
g_C, g_B	E-controlled

control conditions we get	
D	$E_C \cup E_f \circ E_B$-controlled
$\widetilde{f}$	$E_f \circ E_B$-controlled
g	E-controlled

Table 2: Control conditions on pushouts along cellular inclusions in $\mathcal{C}^G$.

Having a canonical pushout is import for the functoriality of the cylinder functor, which we will introduce later. There are no canonical pushouts along cofibrations in $\mathcal{C}^G$, but as cofibrations are isomorphic to cellular inclusions, pushouts along cofibrations also exits in $\mathcal{C}^G$.

6.1.2 The subcategory of cofibrations One can use the lemma above to show that the composition of cofibrations is again a cofibration. Isomorphisms are cofibrations, the map $* \to M$ from the trivial module to any controlled module M is a cofibration. Lemma 6.1.1 immediately implies that pushouts along cofibrations exist. These are the axioms of a category with cofibrations in the sense of [Wal85, 1.1], which are therefore satisfied by $\mathcal{C}^G$.

Note that in our setting a retract of a cofibration can not be a cofibration in general, as pushouts along such maps do not need to be cellular. This is already true for discrete rings and free modules.

6.1.3 The Cylinder Functor We defined for a map $f\colon M \to N$ in $\mathcal{C}^G$ the cylinder functor as $T(f) := A[\Delta^1] \cup_{i_{1*}} B$. We will outline how to verify that this indeed gives a cylinder functor in the sense of [Wal85, 1.6]. If we need to refer to the object $T(f)$, we call it the mapping cylinder of f.

T gives a functor from $\mathrm{Ar}\mathcal{C}^G$ into diagrams in $\mathcal{C}^G$, taking $f\colon A \to B$ to a commutative diagram

$$\begin{array}{ccccc} A & \xrightarrow{\iota_0} & T(f) & \xleftarrow{\iota_1} & B \\ & {\scriptstyle f}\searrow & \downarrow{\scriptstyle p} & \swarrow{\scriptstyle \mathrm{id}} & \\ & & B & & \end{array} \qquad (1)$$

Here ι_0 is called the *front inclusion*, ι_1 is called the *back inclusion* and p is called the *projection*. Waldhausen requires the following two axioms to be satisfied.

1. (Cyl 1) Front and back inclusion assemble to an exact functor

$$\mathrm{Ar}\mathcal{C} \longrightarrow \mathcal{F}_1\mathcal{C}$$
$$f \mapsto \big(\iota_0 \vee \iota_1 \colon A \vee B \rightarrowtail T(f)\big).$$

2. (Cyl 2) $T(* \to A) = A$ for every $A \in \mathcal{C}$ and the projection and the back inclusion are the identity map on A.

Here $\mathrm{Ar}\mathcal{C}$ is the arrow category of $\mathcal{C}$ and $\mathcal{F}_1\mathcal{C}$ is the full subcategory of $\mathrm{Ar}\mathcal{C}$ with objects the cofibrations. Both can be made into categories with cofibrations, with the cofibrations of $\mathcal{F}_1\mathcal{C}$ being slightly non-obvious. See [Wal85, 1.1] for the precise definitions and the notion of an exact functor. We use the notion "$A \vee B$" from [Wal85] for the coproduct of A and B. We will first exclude the weak equivalences from the discussion.

(Cyl 2) is directly verified using $*[\Delta^1] = *$ and choosing the right canonical pushouts along $* \to *$.

Lemma 6.1.4. *Front and back inclusion give a functor* $\mathrm{Ar}\mathcal{C}^G \to \mathcal{F}_1\mathcal{C}^G$,

$$f \mapsto (A \vee B \rightarrowtail T(f)).$$

Proof. The only thing to show is that $A \vee B \to T(f)$ is a cellular inclusion. Consider the diagram

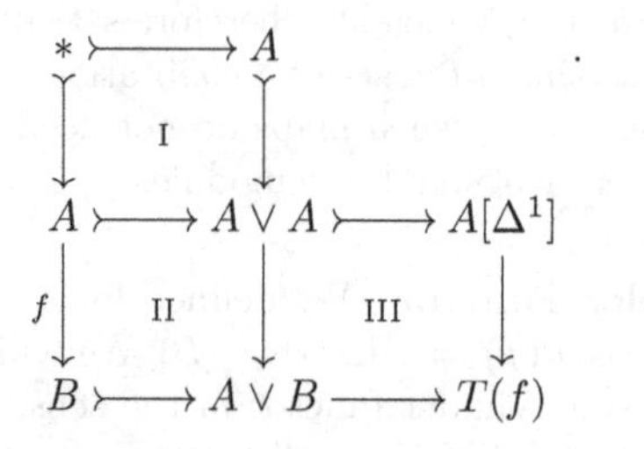

Here $A \vee A \rightarrowtail A[\Delta^1]$ is the cellular inclusion $\iota_0 \vee \iota_1 \colon A[0] \vee A[1] = A[0 \amalg 1] \rightarrowtail A[\Delta^1]$. We claim that every possible square is a pushout along a cellular inclusion. I is a pushout square by definition, as well as I+II. It follows that II is one. Further II + III is a pushout square by definition of $T(f)$, so III is one. Hence the lower map $A \vee B \to T(f)$ is a cellular inclusion by Lemma 6.1.1. □

Lemma 6.1.5. *The functor of Lemma 6.1.4 is exact.*

Proof. We need to show that the functor respects the structure of a category with cofibrations. Pushouts and the zero object are defined pointwise in $\mathrm{Ar}\mathcal{C}^G$ and $\mathcal{F}_1\mathcal{C}^G$. Therefore $A \vee B$ and $T(f)$ commute with pushouts. So we only have to show that the functor maps cofibrations to cofibrations.

Let us briefly recall the cofibrations in $\mathrm{Ar}\mathcal{C}^G$ and $\mathcal{F}_1\mathcal{C}^G$, cf. [Wal85, Lemma 1.1.1]. For notation let

$$\begin{array}{ccc} A & \longrightarrow & A' \\ \downarrow{\scriptstyle f} & & \downarrow{\scriptstyle f'} \\ B & \longrightarrow & B' \end{array} \tag{2}$$

be a map in $\mathrm{Ar}\mathcal{C}^G$ from $A \to B$ to $A' \to B'$. It is a cofibration in $\mathrm{Ar}\mathcal{C}^G$ if both horizontal maps are cofibrations. The category $\mathcal{F}_1\mathcal{C}^G$ is the full subcategory of $\mathrm{Ar}\mathcal{C}^G$ with objects being the cofibrations in $\mathcal{C}^G$. Hence if f and f' are cofibrations, the diagram also shows a map in $\mathcal{F}_1\mathcal{C}^G$. It is a cofibration in $\mathcal{F}_1\mathcal{C}^G$ if $A \to A'$ and $A' \cup_A B \to B'$ are cofibrations in $\mathcal{C}^G$. (It follows that $B \to B'$ is a cofibration.) See [Wal85, Lemma 1.1.1] for details and a proof that the composition of cofibrations in $\mathcal{F}_1\mathcal{C}^G$ is again a cofibration.

We have to show that for a map (2) which is a cofibration in $\mathrm{Ar}\mathcal{C}^G$ the maps $A \vee B \to A' \vee B'$ and $(A' \vee B') \cup_{A \vee B} T(f) \to T(f')$ are cofibrations in $\mathcal{C}^G$. As functors respect isomorphisms we can assume that all cofibrations are cellular inclusions.

So assume we have a diagram (2) where the vertical maps are cellular inclusions. We can factor (2) into

$$\begin{array}{ccccc} A & \overset{\mathrm{id}}{=\!=\!=} & A & \longrightarrow & A' \\ \downarrow{\scriptstyle f} & & \downarrow{\scriptstyle f^*} & & \downarrow{\scriptstyle f'} \\ B & \longrightarrow & B' & \overset{\mathrm{id}}{=\!=\!=} & B' \end{array} \ .$$

It suffices to check each map individually. The map $A \vee B \to A \vee B'$ is a pushout along the cofibration $B \to B'$, similar for $A \vee B' \to A' \vee B'$. Hence both are cofibrations.

Recalling that by Definition $T(f)$ is the pushout $B \cup_f A[\Delta^1]$

$$\begin{array}{ccc} A & \longrightarrow & A[\Delta^1] \\ \downarrow{\scriptstyle f} & & \downarrow \\ B & \longrightarrow & T(f) \end{array}$$

we see that $T(f^*)$ is the pushout

$$\begin{array}{ccc} B & \longrightarrow & T(f) \\ \downarrow & & \downarrow \\ B' & \longrightarrow & T(f^*) \end{array}$$

and hence (by "canceling A" by a similar pushout argument as in the proof of Lemma 6.1.4) it is the pushout

$$\begin{array}{ccc} A \vee B & \longrightarrow & T(f) \\ \downarrow & & \downarrow \\ A \vee B' & \longrightarrow & T(f^*) \end{array}$$

so the map $(A \vee B') \cup_{A \vee B} T(f) \to T(f^*)$ is an isomorphism and therefore a cellular inclusion. Using the canceling argument for B' we can write the other map

$$(A' \vee B') \cup_{A \vee B'} T(f^*) \to T(f')$$

as

$$A' \cup_{A[0]} A[\Delta^1] \cup_{f^*} B' \to A'[\Delta^1] \cup_{f'} B'. \tag{3}$$

Here the first object is a cylinder where we glued in spaces at both sides. But because $A \rightarrowtail A'$ is a cellular inclusion so is $A'[0] \cup_{A[0]} A[\Delta^1] \cup_{A[1]} A'[1] \rightarrowtail A'[\Delta^1]$. We have the commutative diagram

$$\begin{array}{ccccc} A[1] & \rightarrowtail & A'[0] \cup_{A[0]} A[\Delta^1] & & \\ \downarrow & & \downarrow & & \\ A'[1] & \rightarrowtail & A'[0] \cup_{A[0]} A[\Delta^1] \cup_{A[1]} A'[1] & \rightarrowtail & A'[\Delta^1] \\ \downarrow f' & & \downarrow & & \downarrow \\ B' & \rightarrowtail & A'[0] \cup_{A[0]} A[\Delta^1] \cup_{f^*} B' & \longrightarrow & A'[\Delta^1] \cup_{f'} B' \end{array}$$

(with the curved arrow $f^*: A[1] \to B'$)

where every square and in particular the lower right one is a pushout (by the same reasoning as in the proof of Lemma 6.1.4). The lower right horizontal map is the map (3). Hence using Lemma 6.1.1 one last time it follows that the map (3) is a cellular inclusion. □

Weak equivalences in $\mathrm{Ar}\mathcal{C}^G$ and $\mathcal{F}_1\mathcal{C}^G$ are defined pointwise. As $A \to A[\Delta^1]$ respects homotopy equivalences, the cylinder functor respects homotopy equivalences, too.

6.2 The homotopy extension property and the glueing lemma

We want to prove the glueing lemma for the homotopy equivalences in $\mathcal{C}^G(X;R)$. The main ingredient is the relative homotopy extension property, which we will prove first. The glueing lemma is then a relatively formal consequence, but the proof is a bit lengthy and in the end relies on the fact that the glueing lemma holds in a category of cofibrant objects. So we will show that $\mathcal{C}^G(X;R)$ is a category of cofibrant object in the sense of [GJ99, II.8].

Recall that the ith horn $\Lambda_i^n \subseteq \Delta^n$ is $\partial\Delta^n$ minus the ith face, see e.g. [GJ99, I.1, p. 6].

Lemma 6.2.1 (Relative Horn-Filling). *Let $M, P \in \mathcal{C}^G$. Let A be a cellular submodule of M, let $\Lambda_i^n \subseteq \Delta^n$ be a horn. Any controlled maps $A[\Delta^n] \to P$ and $M[\Lambda_i^n] \to P$ which agree on $A[\Lambda_i^n]$ can be extended to a controlled map $M[\Delta^n] \to P$.*

If M is E_M-controlled and both maps to P are E_f-controlled, then the extended map can be chosen to be $E_f \circ E_M$-controlled.

Proof. First we prove the claim of the lemma if $G = \{1\}$. It suffices to produce a controlled retraction $r\colon M[\Delta^n] \to M[\Lambda_i^n] \cup_{A[\Lambda_i^n]} A[\Delta^n]$ with section the inclusion.

Let $B_k := A \cup M_k$, where M_k is the submodule of M generated by all cells of dimension $\leq k$. We do induction over k. We assume the following induction hypothesis:

1. There is a retraction $g_k\colon M[\Lambda_i^n] \cup_{B_k[\Lambda_i^n]} B_k[\Delta^n] \to M[\Lambda_i^n] \cup_{A[\Lambda_i^n]} A[\Delta^n]$.

2. For each $e_0 \in \diamond_R M$ the map g_k restricts to

$$\langle e_0\rangle_M [\Delta^n] \cap \left(M[\Lambda_i^n] \cup_{B_k[\Lambda_i^n]} B_k[\Delta^n]\right) \xrightarrow{g_k} \langle e_0\rangle_M [\Delta^n] \cap \left(M[\Lambda_i^n] \cup_{A[\Lambda_i^n]} A[\Delta^n]\right) \quad (4)$$

The second condition is needed to gain enough control. It is important that the condition holds for all cells of M and not only the ones from B_k. We abbreviate $N_k := \left(M[\Lambda_i^n] \cup_{B_k[\Lambda_i^n]} B_k[\Delta^n]\right)$. Hence g_k is a map $N_k \to N_{-1}$.

For $k = -1$ the induction hypothesis is satisfied because $g_{-1} = \mathrm{id}$.

So assume the induction hypothesis holds for $k-1$. As B_k arises from B_{k-1} by attaching cells of dimension k it suffices to treat the case of attaching one cell e, as cells of the same dimension can be attached independently.

We get a diagram

$$\begin{array}{ccccc} \underline{R}[\Delta^k \times \Lambda^n_i \cup \partial\Delta^k \times \Delta^n] & \xrightarrow{\partial e_*} & B_k[\Lambda^n_i] \cup_{B_{k-1}[\Lambda^n_i]} B_{k-1}[\Delta^n] & \xrightarrow{g_{k-1}} & N_{-1} \\ \downarrow & & \downarrow & \nearrow & \\ \underline{R}[\Delta^k \times \Delta^n] & \xrightarrow{e_*} & B_k[\Delta^n] & & \end{array}, \tag{5}$$

where we want to find the dashed lift to get g_k. Here $\underline{R}$ is the module $R[\Delta^0]$ over $\kappa^M(e)$. But as the square is a pushout, as one checks, it suffices to find a lift $R[\Delta^k \times \Delta^n] \to N_{-1}$. The lower horizontal map e_* of course factors over $\langle e \rangle_M [\Delta^n]$, so does the upper horizontal one. By the induction hypothesis then $g_{k-1} \circ \partial e_* \subseteq \langle e \rangle_M [\Delta^n] \cap N_{-1}$. We want to find a lift $R[\Delta^k \times \Delta^n] \to \langle e \rangle_M [\Delta^n] \cap N_{-1}$.

By the adjunction from section 2.5.4 it suffices to find a lift in the diagram of simplicial sets

$$\begin{array}{ccc} \Delta^k \times \Lambda^n_i \cup \partial\Delta^k \times \Delta^n & \longrightarrow & \mathrm{HOM}^{\mathcal{E}}_R(\underline{R}, \langle e \rangle_M [\Delta^n] \cap N_{-1}) \, . \\ \downarrow & \nearrow & \\ \Delta^k \times \Delta^n & & \end{array}$$

But

$$\mathrm{HOM}^{\mathcal{E}}_R(\underline{R}, \langle e \rangle_M [\Delta^n] \cap N_{-1}) = \langle e \rangle_M [\Delta^n] \cap N_{-1}$$

as $\langle e \rangle_M$ has bounded support. Such a lift exists as the vertical inclusion arises by repeated horn-filling (cf. [GJ99, p. 18/19]) and $\langle e \rangle_M [\Delta^n] \cap N_{-1}$ is an abelian group and hence Kan. This gives a lift $R[\Delta^k \times \Delta^n] \to \langle e \rangle_M [\Delta^n] \cap N_{-1}$, which extends g_{k-1} to $g_k \colon N_k \to N_{-1}$.

Note that g_k restricts to a map

$$\langle e \rangle_M [\Delta^n] \to \langle e \rangle_M [\Delta^n] \cap N_{-1} \tag{6}$$

which is exactly the second condition of the induction hypothesis for k if $e_0 = e \in \diamond_R M$. For general $e_0 \in \diamond_R M$ consider first the case $e \notin \diamond_R \langle e_0 \rangle_M$. Then $\langle e_0 \rangle_M [\Delta^n] \cap N_k \subseteq \langle e_0 \rangle_M [\Delta^n] \cap N_{k-1}$ and the induction hypothesis follows from the induction hypothesis for $k-1$. Otherwise $\langle e \rangle_M \subseteq \langle e_0 \rangle_M$ and then g_k restricts to

$$\begin{aligned} \langle e_0 \rangle_M [\Delta^n] \cap N_k &= (\langle e_0 \rangle_M [\Delta^n] \cap N_{k-1}) \cup \langle e \rangle_M [\Delta^n] \\ &\longrightarrow (\langle e_0 \rangle_M [\Delta^n] \cap N_{-1}) \cup (\langle e \rangle_M [\Delta^n] \cap N_{-1}) \\ &\subseteq \langle e_0 \rangle_M [\Delta^n] \cap N_{-1} \end{aligned}$$

Setting $r := \mathrm{colim}_{k\to\infty} g_k$ yields the retraction, which has the property that $r(\langle e \rangle_M [\Delta^n]) \subseteq \langle e \rangle_M [\Delta^n] \cap N_{-1}$, so it is E_M-controlled when E_M is the control of M.

If $G \neq \{1\}$ we can choose the above lifts equivariantly, e.g. by constructing first a lift for one cell in a G-orbit and then extending equivariantly. This shows the general case. □

As a special case it follows that cofibrations have the homotopy extension property, i.e., homotopies on A can be extended to M. The usual arguments show that being homotopic is an equivalence relation, and that the homotopy equivalences satisfy the 2-out-of-3 property, i.e., the saturation axiom.

We need a little bit more of homotopy theory.

Definition 6.2.2 (Deformation retraction). Let $i\colon A \rightarrowtail M$ be a cellular inclusion in $\mathcal{C}^G$, i.e., we can consider A as a submodule of M. A is a *deformation retract* of M if there is a map $r\colon M \to A$ such that $r \circ i$ is id_A and $i \circ r$ is homotopic to id_M relative A.

The map i is called the *inclusion* and r is called the *retraction* or *deformation retraction.*

For $f\colon A \to B$ the target B is a retract of the mapping cylinder $T(f)$.

Lemma 6.2.3. *$p\colon T(f) \to B$ is even a deformation retraction.*

Proof. We only have to prove that $\iota_1 \circ p\colon T(f) \to B \to T(f)$ is homotopic relative B to $\mathrm{id}_{T(f)}$. Recall that $T(f)$ is defined as the pushout of $B \leftarrow A[1] \to A[\Delta^1]$. We see that $\iota_1 \circ p$ is induced by $p_1\colon A[\Delta^1] \to A[1] \to A[\Delta^1]$. It suffices to give a homotopy from the identity to $A[\Delta^1] \to A[1] \to A[\Delta^1]$ which is relative to $A[1]$.

But there is a well-known map $\widehat{H}\colon \Delta^1 \times \Delta^1 \to \Delta^1$ of simplicial sets inducing such a map. Thus the homotopy H which is induced by $\widehat{H}$ is a homotopy relative to $A[1]$ which induces the desired homotopy. □

The lemma obviously implies that p is a homotopy equivalence, hence the cylinder functor satisfies the *cylinder axiom* of [Wal85, 1.6]. For the next part we need a diagram language.

6.2.4 Describing maps by diagrams In the following we will often have to describe maps of the form $A[\Delta^1 \times \Delta^1] \to B$ or similar. We give concise ways to describe them. The simplicial set $\Delta^1 \times \Delta^1$ comes from a simplicial complex, so it suffices to give compatible maps on the 0-, 1- and 2-simplices. We use the pictures

$$\begin{smallmatrix} \bullet & \to & \bullet \\ \uparrow & \nearrow & \uparrow \\ \bullet & \to & \bullet \end{smallmatrix}, \quad \begin{smallmatrix} \bullet & & \bullet \\ \uparrow & & \\ \bullet & \to & \bullet \end{smallmatrix}, \quad \begin{smallmatrix} \bullet & & \bullet \\ & & \uparrow \\ \bullet & & \bullet \end{smallmatrix}, \quad \begin{smallmatrix} \bullet & \to & \bullet \\ \uparrow & \nearrow & \\ \bullet & \to & \bullet \end{smallmatrix} \qquad \text{etc.}$$

to denote simplicial subsets of $\Delta^1 \times \Delta^1$ which are generated by the shown 1- and 2-simplices. A 2-simplex is in the subset if its boundary is. The dots are only drawn to specify the corresponding subset and are only in the subset if they are a boundary. As an example, we write a map

$$A[\Delta^1 \times \{0,1\}] \cup A[0 \times \Delta^1] \to B \quad \text{as} \quad A[\begin{smallmatrix} \bullet & \to & \bullet \\ \uparrow & & \\ \bullet & \to & \bullet \end{smallmatrix}] \to B.$$

We can use the same kind of diagrams for other simplicial sets like $\Delta^1 \cup_{\Delta^0, 1 \leftrightarrow 0} \Delta^1$ and products of them.

If we want to concisely describe such maps we often draw diagrams like the ones above and write the maps into it. Here are some examples. The left picture below shows a homotopy H from α to β, the middle one shows a horn $A[\Lambda_0^2] \to B$, and the right one the map $A[\Delta^2] \to B$ which arises by filling the horn in the middle. Tr denotes a constant ("Trivial") homotopy.

$\alpha \xrightarrow{H} \beta$ $\qquad$ (horn: $\alpha \xrightarrow{H} \beta$, $\alpha \xrightarrow{\mathrm{Tr}} \alpha$) $\qquad$ (filled: $\alpha \xrightarrow{H} \beta$, $\alpha \xrightarrow{\mathrm{Tr}} \alpha$, $\beta \xrightarrow{\overline{H}} \alpha$)

The diagram shows how one can prove the symmetry of the relation "homotopic" by horn-filling. We sometimes call $\overline{H}$ the "inverse homotopy" to H. It is usually not unique.

Sometimes we leave out the decorations for vertices, as they are uniquely determined by the decorations on the arrows, and draw dots instead. We usually leave out the decoration for the 2-simplices as well, as the actual maps are usually less important for us. All this works for more complicated simplicial sets as long as we can draw diagrams for them.

6.2.5 Rectifying homotopy commutative diagrams If we have the homotopy commutative diagram on the left below, we can turn it into a strictly commutative diagram on the right below.

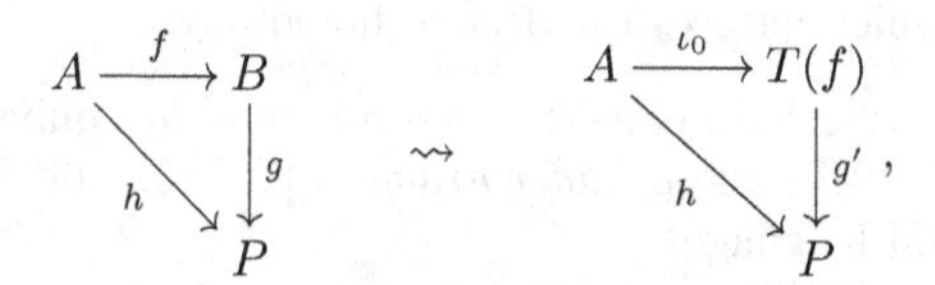

Define $g' \colon A[\Delta^1] \cup B \to P$ as induced by the homtopy and by g. The back inclusion $\iota_1 \colon B \to T(f)$ is a homotopy equivalence and $g = g' \circ \iota_1$.

If $f \colon A \to B$ is a homotopy equivalence, then the two-out-of-three property implies that then $i \colon A \to T(f)$ is a homotopy equivalence.

Proposition 6.2.6. *If $f \colon A \to B$ is a homotopy equivalence, then A is even a deformation retract of $T(f)$ via ι_0.*

We prove the proposition in the rest of this section. It is an adaption of the corresponding proof for topological spaces which the author learned from F. Waldhausen [Wal, pp. 140ff.]. Let g be the homotopy inverse of f, then we have a homotopy commutative diagram $\mathrm{id}_A = g \circ f$. By the argument above we get a map $T(f) \to A$. This is the retraction r.

Lemma 6.2.7. *The map $T(f) \xrightarrow{r} A \xrightarrow{\iota_0} T(f)$ is homotopic to the identity.*

Proof. By Lemma 6.2.3 the composition $T(f) \xrightarrow{p} B \xrightarrow{\iota_1} T(f)$ is a homotopic to the identity. So we pre- and postcompose $T(f) \to A \to T(f)$ with $T(f) \to B \to T(f)$ and get a map which is homotopic to it. This can be written as

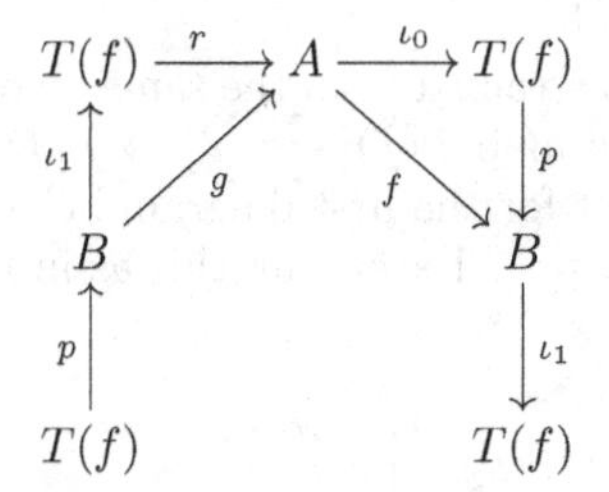

with compositions identified as f and g. But $f \circ g$ is homotopic to id_B by assumption. So we are left with

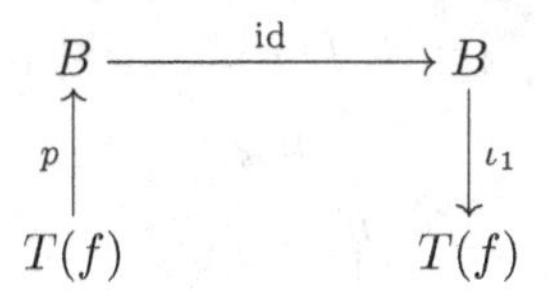

which is homotopic to $\mathrm{id}_{T(f)}$ again by Lemma 6.2.3. Being homotopic is an equivalence relation so $\iota_0 \circ r$ is homotopic to $\mathrm{id}_{T(f)}$. □

The homotopy does not need to be relative to A, but we can improve it as follows. Let $s := \iota_0 \circ r \colon T(f) \to A \to T(f)$ and let H be the homotopy from $\mathrm{id}_{T(f)}$ to s we get by Lemma 6.2.7. We have $s \circ \iota_0 = \iota_0$ as well as $\mathrm{id}_{T(f)} \circ \iota_0 = \iota_0$ so on the endpoints H is relative to the cellular inclusion $\iota_0 \colon A \to T(f)$. We want to make the whole homotopy relative to A, i.e., $A[\Delta^1] \xrightarrow{\iota_0[\Delta 1]} T(f)[\Delta^1] \xrightarrow{H} T(f)$ should be equal to $A[\Delta^1] \xrightarrow{p} A \xrightarrow{\iota_0} T(f)$.

Lemma 6.2.8. *Let s be the map $T(f) \xrightarrow{r} A \xrightarrow{\iota_0} T(f)$. There is a homotopy relative A from the identity on $T(f)$ to s.*

Proof. (We use the diagram notation of 6.2.4.) A is a retract of $T(f)$ and $\iota_0 \colon A \to T(f)$ has the homotopy extension property. We will use this homotopy extension property to construct a certain map $T(f)[\Delta^1 \times \Delta^1] \to T(f)$ which restricted to $1 \times \Delta^1$ will be the desired homotopy from $\mathrm{id}_{T(f)}$ to s relative to A.

Note that s is an idempotent, i.e., $s^2 = s$. We use the notation from above. The proof will proceed as follows. We will prescribe the map $T(f)[\Delta^1 \times \Delta^1] \to T(f)$ on the subspace $A[\Delta^1 \times \Delta^1] \rightarrowtail T(f)[\Delta^1 \times \Delta^1]$ and on the top, bottom and left part of $\Delta^1 \times \Delta^1 = \cdots$, i.e., on $T(f)[\ \cdots\ ]$. Then we check that the two maps are compatible. This will give a map

$$T(f)[\ \cdots\ ] \cup A[\ \cdots\ ] \to T(f)$$

which can be extended by the homotopy extension property to the desired map $T(f)[\ \cdots\ ] \to T(f)$.

Both maps will be constructed from the same map, which we describe first. Horn-filling gives for any map $T(f)[\ \cdots\ ] \to T(f)$ a map $T(f)[\ \cdots\ ] \to T(f)$, in particular we get for the first diagram below the second one, where $\overline{H}$ is the "inverse homotopy". Extending this as in the third diagram below gives a map $G\colon T(f)[\Delta^1 \times \Delta^1] \to T(f)$.

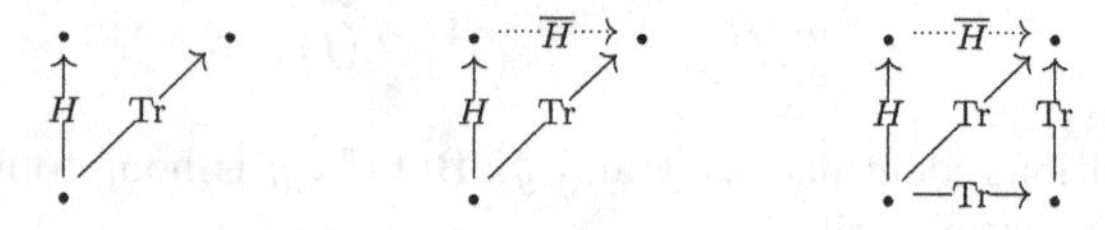

Define the map $A[\Delta^1 \times \Delta^1] \to T(f)$ as the restriction of G to $A[\Delta^1 \times \Delta^1]$. Define the map $T(f)[\ \cdots\ ] \to T(f)$ as

$$\begin{array}{ccc} \bullet & \xrightarrow{\overline{H}\circ s} & \bullet \\ {\scriptstyle H}\uparrow & & \\ \bullet & \xrightarrow{\mathrm{Tr}} & \bullet \end{array}$$

so on the $\cdots$ -part it is the restriction of G, but on the upper part $\cdots$ we replace the homotopy $\overline{H}$ by $\overline{H} \circ s$. This replacement is crucial for the proof.

We check that these maps are compatible. First $\overline{H}$ is a homotopy from s to id, hence $\overline{H} \circ s$ is a homotopy from s^2 to s; but $s^2 = s$ so it agrees with H on the upper left vertex. Second, restricted to A the map s is the inclusion $\iota_0\colon A \to T(f)$, hence $\overline{H} \circ s \circ \iota_0 = \overline{H} \circ \iota_0$. So this glues to a map

$$T(f)[\ \cdots\ ] \cup A[\ \cdots\ ] \to T(f).$$

This can be interpreted as a map $T(f)[0 \times \Delta^1] \to T(f)$ together with a homotopy on the submodule $T(f)[0 \times \{0,1\}] \cup A[0 \times \Delta^1]$. So using the homotopy extension property we get map $T(f)[\Delta^1 \times \Delta^1] \to T(f)$. This map in turn defines a homotopy when restricting along $T(f)[1 \times \Delta^1] \to T(f)[\Delta^1 \times \Delta^1]$ (which is $T(f)[\ \cdots\ ] \to T(f)[\ \cdots\ ]$). This homotopy starts at the identity, ends at the map s and is the constant homotopy on A. Hence it is the desired homotopy. □

6.2.9 Alternative proof Another way to prove this statement is the following idea, which I owe to a discussion with Wolfgang Steimle. Assume we have cofibrations $A \rightarrowtail X$, $A \rightarrowtail Y$ and a map $f\colon X \to Y$ respecting the inclusions.

Lemma 6.2.10. *Assume that f has a homotopy inverse. Then f has a homotopy inverse under A. This is, the inverse and all homotopies respect the inclusion of A.*

Here is a sketch of a proof. We can assume that f is homotopic to id_X. We have to find a left-inverse to f, such that the composition is homotopic to id_X relative to A. We use our diagram language. Let H be the homotopy $f \xrightarrow{H} \mathrm{id}_X$. Restricted to A it is a homotopy $\mathrm{id}_A \xrightarrow{H} \mathrm{id}_A$. Using homotopy extension we get the dotted map in the diagram below and define g as the shown composition

$$\begin{array}{ccc} A[\Delta^1] \cup X[0] & \xrightarrow{H_A \cup \mathrm{id}} & X \\ \downarrow & \overset{K}{\nearrow} & \uparrow g \\ X[\Delta^1] & \longleftarrow & X[1] \end{array} .$$

Let S be the simplicial set $\bullet \leftarrow \bullet \rightarrow \bullet \times \bullet \rightarrow \bullet$. We want to get a map $X[S] \to X$ via horn-filling. We prescribe it on $X[S']$ and $A[S]$ as shown below, where S' is the subset of S shown on the left.

$$\begin{array}{ccc} g \circ f & \xrightarrow{\mathrm{Tr}} & \bullet \\ \uparrow K \circ f & & \\ f & & \\ \downarrow H & & \\ \mathrm{id}_X & \xrightarrow{\mathrm{Tr}} & \bullet \end{array} \qquad \begin{array}{ccc} \mathrm{id}_A & \xrightarrow{\mathrm{Tr}} & \bullet \\ \uparrow H_A & \nearrow H_A & \uparrow \mathrm{Tr} \\ f & \xrightarrow{H_A} & \bullet \\ \downarrow H_A & \searrow H_A & \downarrow \mathrm{Tr} \\ \mathrm{id}_A & \xrightarrow{\mathrm{Tr}} & \bullet \end{array}$$

Filling it to $X[S] \to X$ and restricting to $X[\bullet \leftarrow \bullet \rightarrow \bullet \times 1]$ we get the desired homotopy from id to $g \circ f$ which is constant on A. The same argument gives a left-inverse to g, hence f has a two-sided inverse. □

We keep the original proof because it is more explicit. Also it generalizes directly in subsequent work, when we treat germs.

6.3 Pushouts of weak equivalences

Lemma 6.3.1. *Let*

$$\begin{array}{ccc} A & \longrightarrow & B \\ \downarrow & & \downarrow \\ C & \longrightarrow & D \end{array}$$

be a pushout diagram in $\mathcal{C}^G$ where $A \to C$ is a cofibration and a homotopy equivalence. Then $B \to D$ is a homotopy equivalence.

This is a key result on the way to prove the Glueing Lemma for homotopy equivalences. We remark that almost exactly the same proof works if we assume that $A \to B$ is a cofibration instead of $A \to C$.

We can factor $f\colon A \to C$ into $A \rightarrowtail T(f) \to C$. Taking the pushouts along the cellular inclusion $A \rightarrowtail T(f)$ and along the cofibration $A \rightarrowtail C$ gives a commutative diagram

$$\begin{array}{ccc} A & \longrightarrow & B \\ \downarrow & & \downarrow \\ T(f) & \longrightarrow & Q \\ \downarrow & & \downarrow \\ C & \longrightarrow & D \end{array}$$

and the induced map $Q \to D$ completes the lower square to a pushout square.

The following Lemma shows that both maps $B \to Q$ and $Q \to D$ are homotopy equivalences, so their composition $B \to D$ is one.

Lemma 6.3.2. *In the situation above the following holds.*

1. *The map $B \to Q$ is a homotopy equivalence. (This uses that $A \to C$ is a homotopy equivalence.)*
2. *The map $Q \to D$ is a homotopy equivalence. (This uses that $A \to C$ is a cofibration.)*

Proof of part (1). By Proposition 6.2.6, A is a deformation retract of $T(f)$ via the cellular inclusion $A \rightarrowtail T(f)$. One checks directly that then B is a deformation retract of Q. Hence in particular $B \to Q$ is a homotopy equivalence. □

Proof of part (2). Written out $Q \to D$ is the map

$$B \cup_{A[0]} A[\Delta^1] \cup_{A[1]} C \to B \cup_A C$$

induced by $A[\Delta^1] \to A$. We have to construct a homotopy inverse for this map. We will construct a homotopy equivalence $A[\Delta^1] \cup_{A[1]} C \to C$ and its inverse which is relative to $\iota_0^A\colon A[0] \rightarrowtail A[\Delta^1] \cup_{A[1]} C$, resp. to $j_A\colon A \rightarrowtail C$, hence glues along $A[0] \to B$ to the desired homotopy equivalence

$$B \cup_{A[0]} A[\Delta^1] \cup_{A[1]} C \xrightarrow{\simeq} B \cup_A C\,,$$

as $-[\Delta^1]$ commutes with pushouts. We therefore have to construct for

$$e\colon A[\Delta^1] \cup_{A[1]} C \to C$$

(induced by $A[\Delta^1] \to A$) maps

$$g\colon C \to A[\Delta^1] \cup_{A[1]} C$$

and homotopies

$$H\colon C[\Delta^1] \to C$$
$$G\colon \ (A[\Delta^1] \cup_{A[1]} C)[\Delta^1] \to A[\Delta^1] \cup_{A[1]} C$$

with the properties

$$\begin{aligned} H_0 &= e \circ g & H_1 &= \mathrm{id} \\ G_0 &= \mathrm{id} & G_1 &= g \circ e \\ G \circ \iota_0^A[\Delta^1] &= \iota_0^A & H \circ j_A[\Delta^1] &= j_A \\ g \circ j_A &= \iota_0^A & e \circ \iota_0^A &= j_A\,. \end{aligned}$$

Using the homotopy extension property of the cofibration $j_A\colon A \rightarrowtail C$ there is a retraction $R\colon C[\Delta^1] \to A[\Delta^1] \cup_{A[1]} C$. Define g as the composition

$$C \xrightarrow{\iota_0^C} C[\Delta^1] \xrightarrow{R} A[\Delta^1] \cup_{A[1]} C.$$

We get $g \circ j_A = \iota_0^A$.

Define H as the composition $e \circ R\colon C[\Delta^1] \to A[\Delta^1] \cup_{A[1]} C \to C$. One checks that H is a homotopy from $e \circ g$ to id_C relative to A.

For the other composition consider the commutative diagram

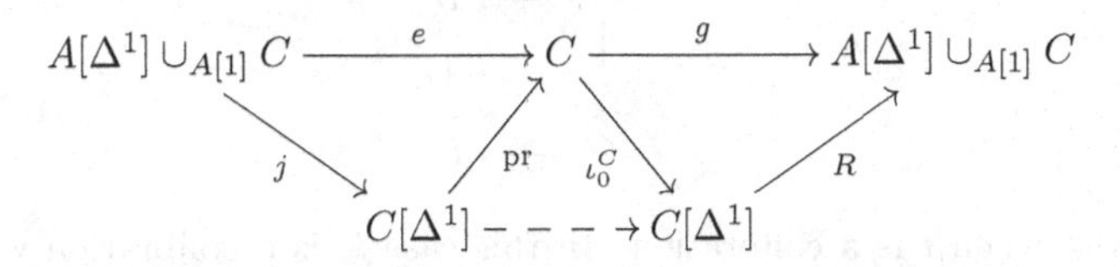

where dashed map is the projection to $C[0]$. It is homotopic relative $C[0]$ to the identity. This gives a homotopy G from the identity to the composition $g \circ e$, using that R is a retraction for j. One checks that G is relative to $A[0]$.

This shows that e is a homotopy equivalence and therefore makes $Q \to D$ into one. □

6.4 The Glueing Lemma

The Glueing Lemma is the following statement.

Lemma 6.4.1. *If we have the diagram in* $\mathcal{C}^G$

$$\begin{array}{ccccc} B & \longleftarrow\!\!\prec & A & \longrightarrow & C \\ \downarrow & & \downarrow & & \downarrow \\ B' & \longleftarrow\!\!\prec & A' & \longrightarrow & C' \end{array}$$

with $A \rightarrowtail B$ and $A' \rightarrowtail B'$ cofibrations and all three vertical arrows are homotopy equivalences, then the induced map

$$B \cup_A C \to B' \cup_{A'} C'$$

on the pushouts is also a homotopy equivalence.

Proof. It it shown in Lemma II.8.8 in [GJ99, p. 127] that a *category of cofibrant objects* satisfies the Glueing Lemma. We recall that notion from [GJ99, p. 122]. It was first introduced by Kenneth Brown in [Bro73], where he treats the dual version.

A *category of cofibrant objects* is a category $\mathcal{D}$ which satisfies the following axioms.

0. The category contains all finite coproducts.
1. The 2-out-of-3 property holds for weak equivalences.
2. The composition of cofibrations is a cofibration, isomorphisms are cofibrations.
3. Pushout diagrams of the form

$$\begin{array}{ccc} A & \longrightarrow & B \\ \downarrow i & & \downarrow i_* \\ C & \longrightarrow & D \end{array}$$

   exist when i is a cofibration. In this case i_* is a cofibration which is additionally a weak equivalence if i is one.

4. For each object there is a cylinder object.
5. For each X the unique map $* \to X$ from the initial object is a cofibration.

The notion of a cylinder object in [GJ99, p. 123] is slightly different from our notion, but if the Cylinder Axiom 6.2.3 holds our Cylinder Functor applied to the identity yields a cylinder object in the sense of [GJ99, p. 123].

We have shown that these axioms hold for $\mathcal{C}^G$. Hence the glueing lemma holds. □

Summarized, we have established that $\mathcal{C}^G(X, \mathcal{E}, \mathcal{F}; R)$ has the structure of a category with cofibrations and weak equivalences, where the cofibrations are isomorphic to cellular inclusions and the weak equivalences are the homotopy equivalences. That is, we finished the proof of Theorem 3.1.4.

6.5 The Extension Axiom

Next we want to prove the extension axiom for our category $\mathcal{C}^G(X; R)$. We will need to use explicitly that we can add maps. Unlike the results in the previous sections, the extension axiom does not hold in Waldhausen's category of spaces over a point, see [Wal85, 1.2].

Let $\mathcal{C}$ be a category with cofibrations. A *cofiber sequence* in $\mathcal{C}$ is a sequence $A \rightarrowtail B \twoheadrightarrow C$ in $\mathcal{C}$ where $A \rightarrowtail B$ is a cofibration and $B \twoheadrightarrow C$ is isomorphic to the map $B \twoheadrightarrow B/A := B \cup_A *$.

A subcategory $w\mathcal{C}$ of weak equivalences of $\mathcal{C}$ satisfies the *Extension Axiom* if for each map of cofiber sequences

$$\begin{array}{ccccc} A & \rightarrowtail & B & \twoheadrightarrow & C \\ \downarrow{\scriptstyle f_A} & & \downarrow{\scriptstyle f_B} & & \downarrow{\scriptstyle f_C} \\ A' & \rightarrowtail & B' & \twoheadrightarrow & C' \end{array}$$

where f_A and f_C are weak equivalences the map f_B is a weak equivalence. Sometimes B (resp. f_B) is called an *extension* of A by C (resp. of f_A by f_C).

We first need a relative homotopy lifting property. We directly prove a more general horn-filling property.

Lemma 6.5.1 (Horn-filling relative to a map). *Let $A \rightarrowtail M$ be a cellular inclusion in $\mathcal{C}^G$. Let $U \rightarrowtail P$ also be a cellular inclusion in $\mathcal{C}^G$ and let $P \twoheadrightarrow Q := P/U$ the quotient map. Then $A \to M$ has the* relative horn-filling property *with respect to $P \to Q$. This means, given a horn $\Lambda^n_i \subseteq \Delta^n$ and a solid commutative diagram of controlled maps*

$$\begin{array}{ccc} M[\Lambda^n_i] \cup A[\Delta^n] & \longrightarrow & P \\ \downarrow & \nearrow & \downarrow \\ M[\Delta^n] & \longrightarrow & Q \end{array} \qquad (7)$$

then the dashed lift exists.

Specializing to $n = 1$, $i = 1$ gives the homotopy lifting property with respect to $P \to Q$. The proof proceeds similarly to the proof of Lemma 6.2.1. It is not stated there in the full generality, as we will need the generalized version only in this section. So we choose to keep the already quite complicated proof of Lemma 6.2.1 a little bit simpler.

We will need the following extra ingredient: Any surjective map $B \twoheadrightarrow C$ of simplicial abelian groups is a Kan fibration. This follows e.g. from [GJ99, Corollary V.2.7, p. 263]. Consequently, for a cellular inclusion of simplicial R-modules $A \rightarrowtail B$, the map $B \twoheadrightarrow B/A$ is a Kan fibration of simplicial sets.

Proof of Lemma 6.5.1. The proof is very similar to the proof of Lemma 6.2.1, but more involved. The main point is that we need to find a lift relative to the map $P \to Q$. To still keep the control, we have to strengthen the induction hypothesis.

We first treat the case $G = \{1\}$. Let $B_k := A \cup M_k$, where M_k is the submodule of M generated by all cells of dimension $\leq k$. We do induction over k. We abbreviate $N_k := \left(M[\Lambda_i^n] \cup_{B_k[\Lambda_i^n]} B_k[\Delta^n]\right)$, $N_\infty := M[\Delta^n] = \bigcup_k N_k$. We have to find a lift in the diagram (which also fixes our notation for the maps)

$$\begin{array}{ccc} N_{-1} & \xrightarrow{f} & P \\ {\scriptstyle i}\downarrow & \nearrow & \downarrow{\scriptstyle p} \\ N_\infty & \xrightarrow{h} & Q \end{array} \quad . \tag{8}$$

We need to be able to restrict p "locally", such that it is still a fibration. It suffices that we construct "locally" maps which are surjections of abelian simplicial groups after forgetting control and R-module structure. We make the following choices. For each $e_Q \in \diamond_R Q$ choose an $\vartheta(e_Q) \in \diamond_R P$ with $p(\vartheta(e_Q)) = e_Q$. Such a map $\vartheta\colon \diamond_R Q \to \diamond_R P$ exists as $\diamond_R P \cong \diamond_R Q \cup \diamond_R U$.

We assume the following induction hypothesis:

1. There is a map $g_k\colon N_k \to P$ which extends f over h, i.e., is a partial lift in the diagram (8).

2. For each $e_0 \in \diamond_R M$ the map g_k restricts to

$$\begin{aligned} &\langle e_0\rangle_M [\Delta^n] \cap N_k \xrightarrow{g_k} \\ &\langle f(\langle e_0\rangle_M [\Delta^n] \cap N_{-1})\rangle_P \cup \bigcup \left\{ \langle \vartheta(e_Q)\rangle_P \;\middle|\; e_Q \in \diamond_R \langle h(\langle e_0\rangle_M [\Delta^n])\rangle_Q \right\} \end{aligned} \tag{9}$$

The second condition implies that g_k is $E_f \circ E_M \cup E_M \circ E_h \circ E_P$-controlled. Roughly speaking it ensures that the lift does not hit a module which is uncontrollably large. Here is a reason for why it has to be at least that size. First we must allow a cell e_0 to at least hit the image of f of the part of the cell intersecting N_{-1}. Second, the cell hits certain elements in Q, so we must have possible lifts for all of them.

We do induction over k. We can attach cells of the same dimension independently, so we only treat the case of attaching one cell e of dimension k.

As before the left square of the following diagram is a pushout.

$$\begin{array}{ccccc} R[\Delta^k \times \Lambda^n_i \cup \partial\Delta^k \times \Delta^n] & \xrightarrow{\partial e_*} & B_k[\Lambda^n_i] \cup_{B_{k-1}[\Lambda^n_i]} B_{k-1}[\Delta^n] & \xrightarrow{g_{k-1}} & P \\ \downarrow & & \downarrow & \nearrow & \downarrow p \\ R[\Delta^k \times \Delta^n] & \xrightarrow{e_*} & B_k[\Delta^n] & \xrightarrow{h} & Q \end{array} \qquad (10)$$

We can replace the middle column by $N_{k-1} \to N_k$ and the diagram remains commutative and the left square a pushout. So we only have to find a lift in the outer diagram of (10). We abbreviate

$$P^f(e) := \langle f\left(\langle e\rangle_M [\Delta^n] \cap N_{-1}\right)\rangle_P$$
$$P^h(e) := \bigcup\left\{ \langle\vartheta(e_Q)\rangle_P \;\middle|\; e_Q \in \diamond_R \left\langle h(\langle e\rangle_M [\Delta^n])\right\rangle_Q \right\}$$

Both are cellular submodules of P. We get a factorization of the outer diagram of (10)

$$\begin{array}{ccccc} R[\Delta^k \times \Lambda^n_i \cup \partial\Delta^k \times \Delta^n] & \xrightarrow{\partial e_*} & P^f(e) \cup P^h(e) & \longrightarrow & P \\ \downarrow & & \downarrow & & \downarrow p \\ R[\Delta^k \times \Delta^n] & \xrightarrow{e_*} & \left\langle h(\langle e\rangle_M [\Delta^n])\right\rangle_Q & \longrightarrow & Q \end{array}$$

by the induction hypothesis and it suffices to find a lift in the left diagram. By the fundamental lemma, and because the middle column in the diagram has bounded support on some $\{x\}^E$, it suffices to find a (dashed) lift in the diagram of simplicial sets

$$\begin{array}{ccc} \Delta^k \times \Lambda^n_i \cup \partial\Delta^k \times \Delta^n & \longrightarrow & P^f(e) \cup P^h(e) \\ \downarrow & \nearrow & \downarrow \\ \Delta^k \times \Delta^n & \longrightarrow & \left\langle h(\langle e\rangle_M [\Delta^n])\right\rangle_Q \end{array}$$

Such a lift exists if the right map is a Kan Fibration. But as it is a homomorphism of simplicial abelian groups it suffices to show that it is surjective. But $P^h(e) \to \left\langle h(\langle e\rangle_M [\Delta^n])\right\rangle_Q$ is already surjective by construction, as $P^h(e)$ has exactly one cell e_P for each cell e_Q of $\left\langle h(\langle e\rangle_M [\Delta^n])\right\rangle_Q$ and by definition of ϑ the cell e_P is mapped to e_Q. This gives the lift g_k, and by construction it satisfies the first condition of the induction hypothesis for k.

The second condition is satisfied for e by construction and for e_0 with $e \notin \langle e_0\rangle_M$ by the induction hypothesis for $k-1$. Otherwise $\langle e\rangle_M \subseteq \langle e_0\rangle_M$

which implies $P^f(e) \subseteq P^f(e_0)$ and $P^h(e) \subseteq P^h(e_0)$. Then g_k restricted to $\langle e_0 \rangle_M [\Delta^n]$ factors as

$$\begin{aligned} \langle e_0 \rangle_M [\Delta^n] \cap N_k &= \langle e_0 \rangle_M [\Delta^n] \cap (N_{k-1} \cup \langle e \rangle_M [\Delta^n]) \\ &\xrightarrow{g_k} P^f(e_0) \cup P^h(e_0) \ \cup \ P^f(e) \cup P^h(e) = P^f(e_0) \cup P^h(e_0) \end{aligned}$$

Therefore the second condition is also satisfied.

If $G \neq \{1\}$ we can choose the above lifts equivariantly, e.g. by constructing first a lift for one cell in a G-orbit and then extending equivariantly. This shows the general case. □

Lemma 6.5.2 (Extension axiom). *Let*

$$\begin{array}{ccccc} A & \rightarrowtail & B & \twoheadrightarrow & C \\ \downarrow{\sim} & & \downarrow & & \downarrow{\sim} \\ A' & \rightarrowtail & B' & \twoheadrightarrow & C' \end{array}$$

be a map of cofiber sequences in $\mathcal{C}^G$. Assume that $A \to A'$ and $C \to C'$ are homotopy equivalences. Then $B \to B'$ is a homotopy equivalence.

Proof. We can factor the vertical maps functorially by using the Cylinder Functor. As a Cylinder Functor is exact it respects the cofiber sequences. We get a diagram

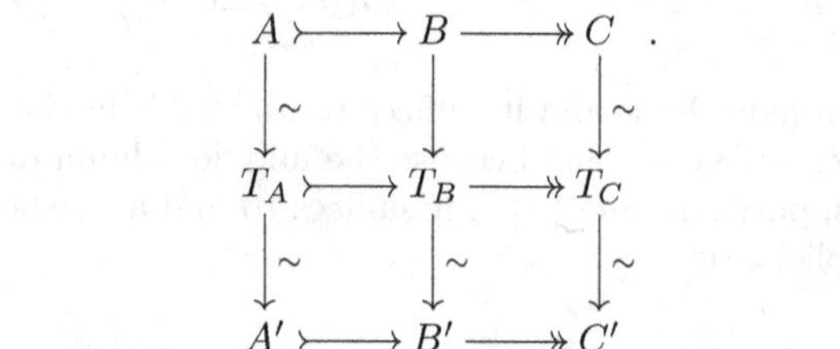

By Proposition 6.2.6 A and C are deformation retracts of T_A and T_C, respectively, with the inclusions being the left and the right vertical upper maps. What remains to be shown is that the vertical upper middle map is a homotopy equivalence. This is proved in Lemma 6.5.3 below, where it is shown that B is a deformation retract of T_B. □

Lemma 6.5.3. *Assume we have a cofiber sequence $A \rightarrowtail B \twoheadrightarrow \overline{B}$ in $\mathcal{C}^G$ where $\overline{B} = B/A$ for brevity. Suppose we have a diagram*

$$\begin{array}{ccccc} A & \rightarrowtail & B & \twoheadrightarrow & \overline{B} \\ \downarrow & & \downarrow & & \downarrow \\ T_A & \rightarrowtail & T_B & \twoheadrightarrow & T_{\overline{B}} \end{array}$$

in $\mathcal{C}^G$ where the horizontal lines are cofiber sequences and the vertical arrows are cellular inclusions. Suppose that A and $\overline{B}$ are deformation retracts of T_A and $T_{\overline{B}}$ with inclusions the left and right vertical maps. Then B is a deformation retract of T_B with inclusion the middle vertical map.

We prove a slightly stronger statement than Lemma 6.5.3:

Lemma 6.5.4. *Assume that we are in the situation of Lemma 6.5.3. Let D_0 be a cellular submodule of D. Then each controlled map $(D, D_0) \to (T_B, B)$ of pairs in $\mathcal{C}^G$ is controlled homotopic relative D_0 to a map into B.*

Proof of Lemma 6.5.3 using 6.5.4. By Lemma 6.5.4 the map id: $(T_B, B) \to (T_B, B)$ is controlled homotopic relative B to a map $T_B \to B$. This is the desired deformation retraction. □

Remark 6.5.5 (Toy situation). Assume we have a commutative diagram of abelian groups

$$\begin{array}{ccccc} A & \longrightarrow & B & \longrightarrow & \overline{B} \\ \downarrow & & \downarrow{\scriptstyle f_B} & & \downarrow \\ A' & \longrightarrow & B' & \longrightarrow & \overline{B}' \end{array}$$

where the horizontal lines are short exact sequences. Assume that the outer maps are surjective. We want to show that the middle map is surjective. The proof proceeds exactly like the proof of Lemma 6.5.4, but is easier. We give it to help with the general proof.

Let α be an element in B'. We will denote the constructed elements by consecutive Greek letters and denote projections to the quotient by a bar. So $\overline{\alpha}$ is an element in $\overline{B}'$. As $\overline{B} \to \overline{B}'$ is surjective there is an element β in $\overline{B}$ which maps to $\overline{\alpha}$. As $B \to \overline{B}$ is surjective there is an element γ in B which maps to $\beta \in \overline{B}$. The elements $f_B(\gamma)$ and α do not need to be equal in B', but they become equal when projected to $\overline{B}'$, so $\alpha - f_B(\gamma)$ factors through $A' \rightarrowtail B'$. As $A \to A'$ is surjective there is an element δ in A which maps to $\alpha - f_B(\gamma)$ in B'. Hence, considered in B, $f_B(\delta + \gamma)$ equals α.

Lemma 6.5.1 applies to the maps $B \to \overline{B}$ and $T_B \to T_{\overline{B}}$, so we have the relative homotopy lifting property with respect to these maps.

Proof of Lemma 6.5.4. Let $\alpha\colon (D, D_0) \to (T_B, B)$ be a controlled map. This gives a map $\overline{\alpha}$ into $(\overline{B}, T_{\overline{B}})$. As $\overline{B}$ is a deformation retract of $T_{\overline{B}}$ we get a homotopy $\overline{H}\colon D[\Delta^1] \to T_{\overline{B}}$ from $\overline{\alpha}$ to a map into $\overline{B}$ which is constant on D_0. It comes from the deformation of $(\overline{B}, T_{\overline{B}})$ precomposed with α. Lemma 6.5.1

applies to the map $T_B \to T_{\overline{B}}$. So we get a lift H of $\overline{H}$, relative to α and D_0.

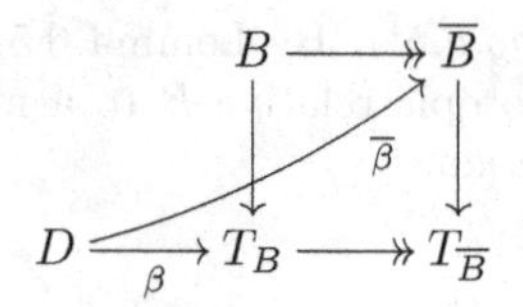

This is a homotopy from α to a better map, call it $\beta\colon D \to T_B$. However, β might not yet factor through B in which case the lemma would follow. But composition with $T_B \to T_{\overline{B}}$ gives a map $\overline{\beta}$ to $T_{\overline{B}}$ which factors through $\overline{B}$.

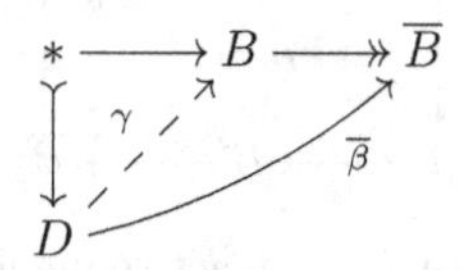

Using Lemma 6.5.1 again this time for $B \twoheadrightarrow \overline{B}$ and the constant homotopy of $\overline{\beta}$ in $\overline{B}$ we get some lift of $\overline{\beta}$ to B, call it γ.

$$\begin{array}{ccccc} * & \longrightarrow & B & \twoheadrightarrow & \overline{B} \\ \downarrow & \overset{\gamma}{\nearrow} & & \overset{\overline{\beta}}{\nearrow} & \\ D & & & & \end{array}$$

It follows that the difference $\beta - \gamma\colon D \to T_B$ is zero when composed with $T_B \to T_{\overline{B}}$. Hence it factors through T_A. As the restrictions of β and γ to D_0 both lie in B the restriction of $\beta - \gamma$ to D_0 factors through A. So $\beta - \gamma$ gives a map $(D, D_0) \to (T_A, A)$. We can show the situation by the following commuting diagrams.

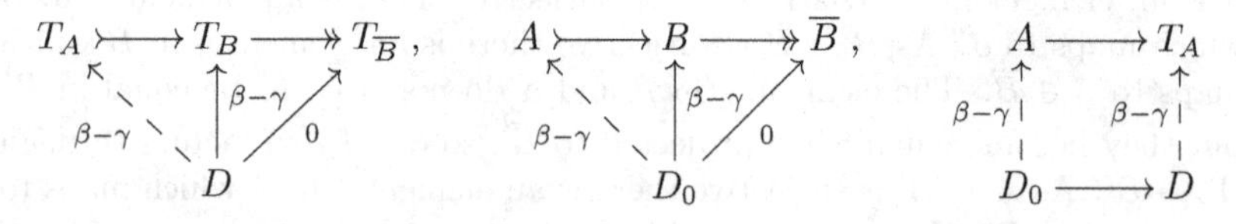

Hence, as $A \to T_A$ is a deformation retraction, there is a homotopy G relative to D_0 of $\beta - \gamma$ to a map into A. It comes from the deformation of (A, T_A) precomposed with $\beta - \gamma$. Call the resulting map $\delta\colon D \to A$. Via the inclusion $(T_A, A) \to (T_B, B)$ the map G can be viewed as a homotopy to T_B with:

$$\begin{aligned} G\colon &\quad D[\Delta^1] \to T_B \\ G_{|0} &\quad = \beta - \gamma \\ G_{|1} &\quad = \delta \\ G_{|D_0[\Delta^1]} &\quad = \beta - \gamma_{|D_0} \end{aligned}$$

Therefore $G+\gamma\colon D[\Delta^1] \to T_B$ is a homotopy from β to $\delta+\gamma$, where δ and γ factor through B so the sum also factors through B. Furthermore the homotopy is constant on D_0. Concatenating the two homotopies H and G thus gives a homotopy relative D_0 from α to a map into B. This is what we wanted to show.

Note that all maps above are in fact in $\mathcal{C}^G$, because maps in $\mathcal{C}^G$ form an abelian group and being homotopic relative a subspace is an equivalence relation in $\mathcal{C}^G$. □

7 Proofs III: Finiteness conditions

7.1 Finiteness conditions

Let $(X, \mathcal{E}, \mathcal{F})$ be a G-equivariant control space. We have shown that the category $\mathcal{C}^G(X, \mathcal{E}, \mathcal{F}; R)$ has the structure of a category with cofibrations and weak equivalences. Also, the weak equivalences satisfy the saturation and extension axiom. We have a cylinder functor which satisfies the cylinder axiom.

A full subcategory $\mathcal{C}^G_?$ of $\mathcal{C}^G(X; R)$ inherits all that structure if the following two conditions as satisfied:

(C1) For $C \leftarrow A \rightarrowtail B$ in $\mathcal{C}^G_?$ the pushout is in $\mathcal{C}^G_?$.

(C2) For A in $\mathcal{C}^G_?$, $A[\Delta^1]$ is in $\mathcal{C}^G_?$.

We show that the finite, homotopy finite and homotopy finitely dominated objects satisfy these conditions.

7.2 Finite modules

There is the obvious notion of a set over X and controlled maps of sets over X. Let $(M, \diamond_R M, \kappa)$ be a controlled module over X. Our prime example of a set over X is $(\diamond_R M, \kappa)$. If (M, κ^1_R), (M, κ^2_R) are controlled modules over X such that $(\diamond_R M, \kappa^1_R)$ and $(\diamond_R M, \kappa^2_R)$ are controlled isomorphic as sets over X then (M, κ^1_R) and (M, κ^2_R) are controlled isomorphic.

A controlled module $(M, \diamond_R M, \kappa)$ over X is locally finite if the set $(\diamond_R M, \kappa)$ is locally finite over X, i.e., each $x \in X$ has a neighborhood U with $\kappa^{-1}(U) \subseteq \diamond_R M$ being finite.

Remark 7.2.1. Note that modules isomorphic to finite modules do not need to be finite again, if the control space is not "good". Take as example $\mathbb{R} \smallsetminus \{0\}$ with metric control.

A control space $(X, \mathcal{E}, \mathcal{F})$ is called proper, if for each compact subset K and $E \in \mathcal{E}$, $F \in \mathcal{F}$ we have that $(F \cap K)^E \cap F$ is contained in a compact set. If the control space is proper then modules isomorphic to finite modules are again finite, see also Remark 7.2.4.

We denote the full subcategory of finite objects by $\mathcal{C}^G_f$. We want to show that it is indeed a category with cofibrations and weak equivalences. Is is clear that $A[\Delta^1]$ is again finite if A is finite. The main part is to show that the pushout of $C \leftarrow A \rightarrowtail B$ exists when A, B, C are finite modules and $A \rightarrowtail B$ is a cofibration *in* $\mathcal{C}^G$. If we know that it is isomorphic to a cellular inclusion of finite modules we are done.

But that is not obvious, we mentioned above that not every module which is isomorphic to a finite module needs to be finite again. Further, the problem is not only that we could change the map κ to X, but we could have a different cellular structure.

Lemma 7.2.2. *Let $f' \colon A \to B$ be a cofibration in $\mathcal{C}^G_f$. Then it is isomorphic to a cellular inclusion in $\mathcal{C}^G_f$.*

The proof is fairly complicated when done in detail. The first step is the following:

By definition f' is only isomorphic to a cellular inclusion in $\mathcal{C}^G$, which does not need to be in $\mathcal{C}^G_f$ by the Remark above.

By Lemma 6.1.1 the pushout of f' along id_A can be chosen as

$$\begin{array}{ccc} A & \xrightarrow{f'} & B \\ \downarrow{\scriptstyle \mathrm{id}_A} & & \downarrow \\ A & \xrightarrow{f} & D \end{array}$$

such that f is a cellular inclusion. Then D is isomorphic to the finite module B, but need not be finite itself. We show in the next lemma that D can indeed be made into a finite module. That lemma finishes the proof.

Lemma 7.2.3. *Let $f \colon A \to D$ be a cellular inclusion in $\mathcal{C}^G$ such that A is finite and (D, κ^D) is isomorphic to a finite module (B, κ^B). Then there is a control map $\overline{\kappa}^D \colon \diamond_R D \to X$ such that $(D, \overline{\kappa})$ is a finite module which is isomorphic to (D, κ^D).*

It follows that $A \to (D, \overline{\kappa})$ is isomorphic to a cellular inclusion in $\mathcal{C}^G_f$.

Proof. We only sketch the proof. The difficult part is, of course, that D and B might have different cellular structures.

We have to find for $(\diamond_R D, \kappa^D)$ a set over X which is controlled isomorphic to it and locally finite. We will that in two steps, first improve $\kappa_0 := \kappa^D$ to κ_1, and then to κ_2. All maps $\kappa_1, \kappa_2 \colon \diamond_R D \to X$ are controlled isomorphic to κ_0 and "improve" κ_0, in particular κ_2 is a locally finite set over X. Than we can take $\overline{\kappa} := \kappa_2$. We prove the non-equivariant case first, i.e., assume $G = \{e\}$.

First we define κ_1 such that its image is contained in the image of κ^B. As (B, κ^B) and (D, κ_0) are controlled isomorphic there is an $E \in \mathcal{E}$ such that for each $e \in \diamond_R D$ there is an $x(e) \in \operatorname{Im} \kappa^B \subseteq X$ such that $(\kappa_0(e), x(e)) \in E$. Set $\kappa_1(e) := x(e)$.

Set

$$T := \{x \in X \mid \kappa_1^{-1}(x) \text{ is infinite}\}.$$

As X is Hausdorff we have that (D, κ_1) is finite if and only if T is empty. So T are the "trouble points". We change κ_1 on $\kappa_1^{-1}(T)$. We can proceed degreewise. The rough idea is, that if $\kappa_1^{-1}(T)$ is infinite in that degree, it needs to come from an infinite submodule of B.

Let $\theta\colon B \to D$ be the controlled isomorphism. So we change κ_1 on, say, $d \in \diamond_R D$ to map to $\kappa_B(b)$ for some $b \in \diamond_R B$ with $d \in \langle\theta(b)\rangle_D$. Careful checking shows that the changed map is locally finite, at least in that degree. Also, the new map is controlled isomorphic to the old one.

For G being non-trivial, we can make all choices G-equivariant and are done. □

Remark 7.2.4. The proof of the Lemma implies the following for the control space X if T was not empty: There are points $x \in X$ and $E \in \mathcal{E}$ such that $\{x\}^E$ is not contained in a compact subset. Namely i_n^x must hit infinitely many cells of B over points in $\{x\}^E$, but B is locally finite. In particular X is not a proper control space in the sense of Section 2.2.

If X and Y are G-equivariant control spaces, then any G-equivariant map $f\colon X \to Y$ induces a functor $\mathcal{C}^G(X) \to \mathcal{C}^G(Y)$. For the finite objects we have the following obvious criterion.

Lemma 7.2.5. *Let $\varphi\colon (X, \mathcal{E}_X, \mathcal{F}_X) \to (Y, \mathcal{E}_Y, \mathcal{E}_Y)$ be a map of control spaces which maps locally finite sets over X to locally finite sets over Y. Then φ induces a functor $\mathcal{C}_f^G(X, R, \mathcal{E}_X, \mathcal{F}_X) \to \mathcal{C}_f^G(Y, R, \mathcal{E}_Y, \mathcal{F}_Y)$.* □

Remark 7.2.6. Note that inclusions of subspaces do not map locally finite sets to locally finite sets in general. A counterexample is the inclusion $\mathbb{R} \smallsetminus \{0\} \to \mathbb{R}$. However closed inclusions do map locally finite sets to locally finite set and hence do induce a functor of categories of controlled modules.

7.3 Homotopy finite objects

Showing that homotopy finite objects form a Waldhausen category follows formally from the facts that the finite ones are an exact subcategory. It is follows directly from the cylinder axiom that the cylinder functor of homotopy finite objects is again homotopy finite. Assume that $C \leftarrow A \rightarrowtail B$ is a diagram of homotopy finite objects and $A \rightarrowtail B$ a cofibration. To show that the pushout is again homotopy finite one uses factorizations given by the cylinder functor and the glueing lemma several times to obtain a homotopy equivalences to the pushout of a diagram of finite objects. Here is a detailed proof:

Proof. So assume that there are finite objects A', B', C' weakly equivalent to A, B, C. Note that we have inverses for weak equivalences, which we will use freely. Below we denote mapping cylinders by M_A, M_B, etc. and cofibrations by $\rightarrowtail$.

We get a chain of maps of diagrams. In the following the arrows marked with $\xrightarrow{\bullet}$ are defined by composition. The first step is

$$\begin{array}{ccccc} C & \longleftarrow & A & \rightarrowtail & B \\ \| & & \uparrow\sim & & \uparrow\sim \\ C & \xleftarrow{\bullet} & A' & \rightarrowtail & M_B \end{array},$$

where A' is finite and M_B is the mapping cylinder of $A' \to A \rightarrowtail B$, which still is homotopy finite. Next we get a map

$$\begin{array}{ccccc} C & \longleftarrow & A' & \rightarrowtail & M_B \\ \sim\downarrow & & \| & & \| \\ C' & \xleftarrow{\bullet} & A' & \rightarrowtail & M_B \end{array}$$

by C being homotopy finite. Then take

$$\begin{array}{ccccc} C' & \longleftarrow & A' & \rightarrowtail & M_B \\ \sim\uparrow & & \| & & \| \\ M_{C'} & \leftarrowtail & A' & \rightarrowtail & M_B \end{array}$$

with $M_{C'}$ being the cylinder of $A' \to C'$ which is finite as A' and C' are finite. Finally we get a map

$$\begin{array}{ccccc} M_{C'} & \leftarrowtail & A' & \rightarrowtail & M_B \\ \| & & \| & & \downarrow\sim \\ M_{C'} & \leftarrowtail & A' & \xrightarrow{\bullet} & B' \end{array}$$

as M_B is weakly equivalent to B'. Using the Glueing Lemma four times gives that $C \cup_A B$ is weakly equivalent to the finite object $M_{C'} \cup_{A'} B'$. □

7.4 Homotopy finitely dominated objects

We give a three different characterizations of homotopy finitely dominated objects.

Definition 7.4.1. Let M, M' be objects in $\mathcal{C}^G$.

1. M is called a *retract* of M' if there are maps $i\colon M \to M'$, $r\colon M' \to M$ such that $r \circ i = \mathrm{id}_M$.

2. M is called a *homotopy retract* of M', or *dominated by* M', if there are maps $i\colon M \to M'$, $r\colon M' \to M$ and a homotopy $H\colon M[\Delta^1] \to M$ from $r \circ i$ to id_M.

Lemma 7.4.2. *Let $A \in \mathcal{C}^G$. Then the following are equivalent.*

1. *A is a homotopy retract of a finite module A'.*

2. *A is a retract of a homotopy finite module A''.*

3. *A is a homotopy retract of a homotopy finite module A'''.*

Proof. Clearly (1) $\Rightarrow$ (3) and (2) $\Rightarrow$ (3) hold. We show (3) $\Rightarrow$ (1) first.

As A''' is homotopy finite, there is a finite module B and maps $f\colon A''' \to B$, $g\colon B \to A'''$ such that $g \circ f \simeq \mathrm{id}_{A'''}$, so A is a homotopy retract of B via $A \xrightarrow{i} A''' \xrightarrow{f} B$ and $B \xrightarrow{g} A''' \xrightarrow{r} B$.

Now we show (1) $\Rightarrow$ (2). We have maps $i\colon A \to A'$, $r\colon A' \to A$ with $r \circ i \simeq \mathrm{id}_A$. We can make the homotopy commutative diagram

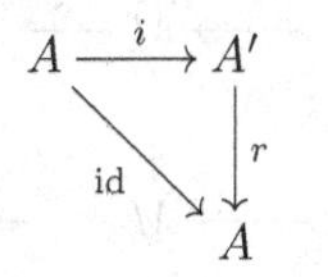

into a strict commutative one, namely

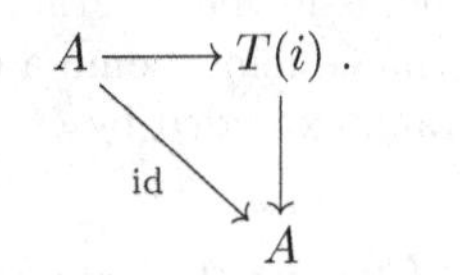

Hence A is a retract of $T(i)$ and as $T(i) \xrightarrow{\sim} A'$ is a homotopy equivalence, $T(i)$ is homotopy finite. $\square$

Lemma 7.4.3. *$\mathcal{C}^G_{hfd}$ is a category with cofibrations and weak equivalences. It has a Cylinder Functor satisfying the Cylinder Axiom and the class of weak equivalences satisfies the Extension and the Saturation Axiom.*

Proof. Again we only show (C1) and (C2) from before. Assume that A, B, C are retracts of homotopy finite objects A', B', C'. Note that we can make the co-retraction into a cofibration by replacing A' with the mapping cylinder of $A \to A'$, so we will assume that the co-retractions i_A, i_B, i_C are actually cofibrations.

We want to show that $C \cup_A B$ is a retract of a homotopy finite object. We reduce this to the case where A is homotopy finite. Consider the commutative

diagram

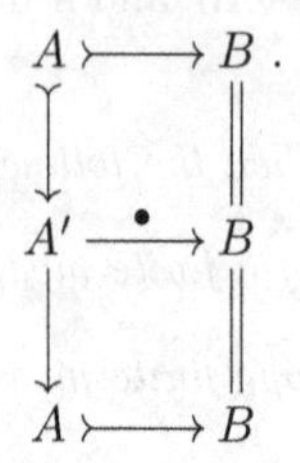

(As before $\xrightarrow{\bullet}$ denotes a map defined by composition.) We can factor the horizontal maps into cofibrations simultaneously using the Cylinder Functor. We obtain

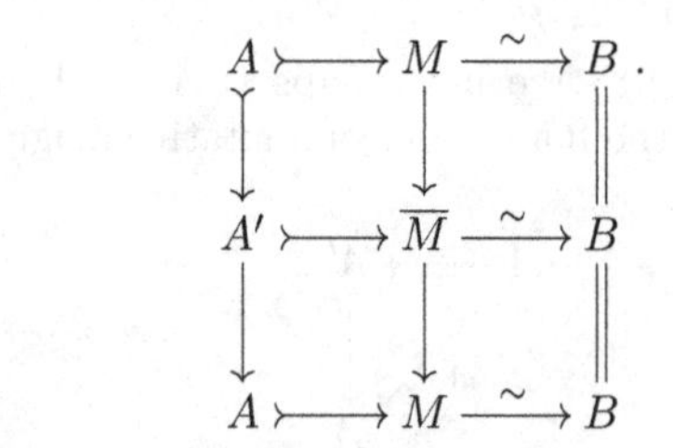

There and in all following diagrams the composition of the vertical arrows is always the identity, which holds in the diagram above by the functoriality of the Cylinder Functor. By the Glueing Lemma $C \cup_A M$ is weakly equivalent to $C \cup_A B$. Then the diagram, extended by C,

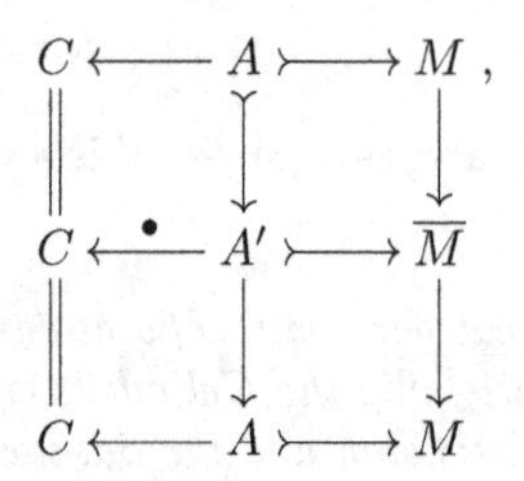

shows that $C \cup_A M$ is a retract of $C \cup_{A'} \overline{M}$. We are done if we show that $C \cup_{A'} \overline{M}$ is finitely dominated. As M is homotopy equivalent to the homotopy finitely dominated object B, Lemma 7.4.2 shows that $\overline{M}$ is again a retract of a homotopy finite module M'.

Now we can use that we have co-retractions $C \rightarrowtail C'$, $\overline{M} \rightarrowtail M'$ with C', M' homotopy finite objects, which are also cofibrations. This gives a commuting

retraction diagram

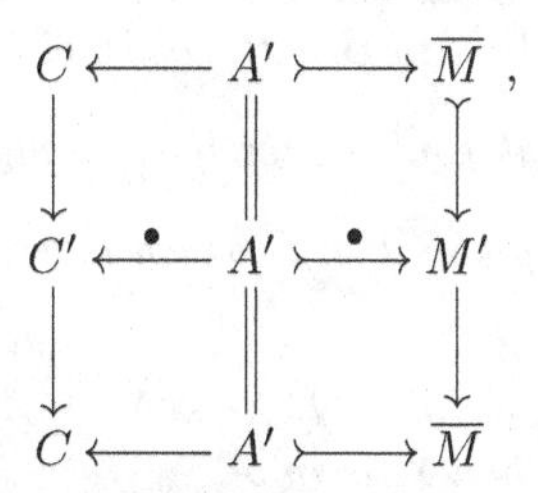

where we want to emphasize, that the map $A' \rightarrowtail M'$, defined by composition, is a cofibration. Thus $C \cup_{A'} \overline{M}$ is a retract of $C' \cup_{A'} M'$, which is homotopy finite, as being a pushout of homotopy finite objects along a cofibration.

For (C2), $A[\Delta^1]$ is dominated by $A'[\Delta^1]$. □

8 Proofs IV: Connective algebraic K-theory of categories of controlled simplicial modules

8.1 Algebraic K-theory

In the last section we showed that the categories $\mathcal{C}^G_f$, $\mathcal{C}^G_{hf}$ and $\mathcal{C}^G_{hfd}$ are all categories with cofibrations and weak equivalences, so we can use Waldhausen's $\mathcal{S}_\bullet$-construction from [Wal85] to produce an algebraic K-Theory spectrum $K(\mathcal{C}^G_?)$ and therefore also the corresponding infinite loop space. Define $K_n(\mathcal{C}^G_?)$ for $n \geq 0$ as the nth homotopy group $\pi_n K(\mathcal{C}^G_?)$. This algebraic K-Theory spectrum is always connective so we do not assign any name to its negative homotopy groups.

Remark 8.1.1. There is a slight set-theoretical problem, as $\mathcal{C}^G_a$ is not a small category according to our definition but it needs be one to apply the K-theory construction. However, we take the usual approach (see e.g. [Wal85, Remark before 2.1.1]) and fix a suitable large set-theoretical small category of simplicial R-modules to begin with. Then all the categories we consider are again small. (We could get such a category by fixing a large cardinal and require all elements to lie in it.) We will assume such a choice from now on.

We want to show that $K_n(\mathcal{C}^G_f)$ $K_n(\mathcal{C}^G_{hf})$ and $K_n(\mathcal{C}^G_{hfd})$ agree for $n \geq 1$. For this we need a cofinality theorem first.

8.2 A cofinality theorem

For comparison of homotopy finite and homotopy finitely dominated objects we need a cofinality result.

Theorem 8.2.1 (Waldhausen-Thomason cofinality). *Let $\mathcal{A}$ and $\mathcal{B}$ be Waldhausen categories. Suppose that $\mathcal{A}$ is a full subcategory of $\mathcal{B}$ which satisfies the following conditions*

1. *$\mathcal{A}$ is a Waldhausen subcategory: $f\colon X \to B$ in $\mathcal{A}$ is a cofibration if and only if it is a cofibration in $\mathcal{B}$ with cokernel in $\mathcal{A}$.*

2. *A map in $\mathcal{A}$ is a weak equivalence if and only if it is one in $\mathcal{B}$.*

3. *$\mathcal{A}$ is saturated: Every object in $\mathcal{B}$ which is weakly equivalent to an object in $\mathcal{A}$ is itself in $\mathcal{A}$.*

4. *$\mathcal{A}$ is closed under extensions: If $X \rightarrowtail Y \twoheadrightarrow Z$ is a cofiber sequence in $\mathcal{B}$ and X, Z are in $\mathcal{A}$, then Y is in $\mathcal{A}$.*

5. *$\mathcal{B}$ has mapping cylinders satisfying the cylinder axiom and $\mathcal{A}$ is closed under them.*

6. *$\mathcal{A}$ is cofinal in $\mathcal{B}$: For every object X in $\mathcal{B}$ there is an object $\overline{X}$ in $\mathcal{B}$ such that $X \vee \overline{X}$ is in $\mathcal{A}$.*

Then

$$K(\mathcal{A}) \to K(\mathcal{B}) \to \text{“}K_0(\mathcal{B})/K_0(\mathcal{A})\text{”}$$

is a homotopy fiber sequence of connective spectra. Here “$K_0(\mathcal{B})/K_0(\mathcal{A})$” denotes the Eilenberg-MacLane spectrum with the group $K_0(\mathcal{B})/K_0(\mathcal{A})$ in degree 0.

Remark 8.2.2 (Similar results). The result is inspired from Thomason-Trobaugh [TT90, Exercise 1.10.2] and Vogell [Vog90, Theorem 1.6]. However, we have slightly different assumptions. In particular in [TT90] the saturatedness assumption is missing. We will provide a counterexample in Section 8.2.5. Parts of the proof were also inspired by Weibel's K-Book [Wei13, Corollary V.2.3.1 and prerequisites], which unfortunately suffers from the same missing assumption, see again Section 8.2.5. Staffeldt, [Sta89, Thm. 2.1], has a similar result as ours in the context of exact categories. It is strictly less general because it only treats isomorphisms as weak equivalences.

Remark 8.2.3. Instead of proving the theorem directly, we prove a lemma about K_0 and then rely on the cofinality theorems of Thomason-Trobaugh [TT90, Thm. 10.1] and Waldhausen's strict cofinality theorem [Wal85, 1.5.9]. (The latter has an implicit assumption that the subcategory is full, see 8.2.6.)

The following lemma will provide a crucial step.

Lemma 8.2.4. *Assume we are in the same situation as in Theorem 8.2.1. Assume further that $\mathcal{A} \to \mathcal{B}$ induces a surjection $K_0(\mathcal{A}) \to K_0(\mathcal{B})$. Then $\mathcal{A}$ is* strictly cofinal *in $\mathcal{B}$ in the sense of [Wal85, 1.5.9], i.e., for each $B \in \mathcal{B}$ there is a $A \in \mathcal{A}$ such that $B \vee A \in \mathcal{A}$. (In this situation we do not need the cylinder functor assumption, but all the other ones are used.)*

In the situation of the lemma Waldhausen's strict cofinality theorem applies and shows that $K(\mathcal{A}) \simeq K(\mathcal{B})$.

Proof. Recall (e.g. from [TT90, 1.5.6]) that $K_0(\mathcal{A})$ is the abelian group generated by isomorphism classes of objects $[A]$, $A \in \mathcal{A}$ with relations

1. $[A] = [B]$ if there is a weak equivalence $A \xrightarrow{\sim} B$
2. $[A] + [C] = [B]$ if there is a cofiber sequence $A \rightarrowtail B \twoheadrightarrow C$.

Let $K_0'(\mathcal{A})$ be the group where we ignore the weak equivalences and consider only split cofiber sequences. That is, it is the group generated by isomorphism classes of objects $[A]$, $A \in \mathcal{A}$ with relation

1. $[A] + [C] = [A \vee C]$ for $A, C \in \mathcal{A}$.

We do the same for $\mathcal{B}$. From the inclusion $i \colon \mathcal{A} \to \mathcal{B}$ we get a (solid) commutative diagram

$$\begin{array}{ccccc} K_0'(\mathcal{A}) & \longrightarrow & K_0'(\mathcal{B}) & \dashrightarrow & G' \\ \downarrow & & \downarrow & & \downarrow \\ K_0(\mathcal{A}) & \longrightarrow & K_0(\mathcal{B}) & \dashrightarrow & G \end{array}$$

which we can extend to the cokernels G and G' as shown. We claim $G' \to G$ is an isomorphism.

First, G is the abelian group with generators $[B]$ for $B \in \mathcal{B}$ and relations

1. $[A] + [C] = [B]$ if there is a cofiber sequence $A \rightarrowtail B \twoheadrightarrow C$.
2. $[A] = 0$ if $A \in \mathcal{A}$.
3. $[A] = [B]$ if there is a weak equivalence $A \xrightarrow{\sim} B$

We claim that (3) is redundant: By cofinality there is an $\overline{A} \in \mathcal{B}$ such that $A \vee \overline{A} \in \mathcal{A}$. By the Glueing Lemma $A \vee \overline{A} \xrightarrow{\sim} B \vee \overline{A}$ is still a weak equivalence. By saturation therefore $B \vee \overline{A}$ is also in $\mathcal{A}$. Therefore, by (2), $[A \vee \overline{A}] = 0$ and $[B \vee \overline{A}] = 0$ in G. By (1) then $[A] = -[\overline{A}] = [B]$ in G, which implies (3).

Further, G' is the abelian group with generators $[B]$ for $B \in \mathcal{B}$ and relations

1. $[A] + [B] = [A \vee B]$.
2. $[A] = 0$ if $A \in \mathcal{A}$.

We show that the stronger relation (1) for G comes from G' and therefore the groups are isomorphic. Let $A \rightarrowtail B \twoheadrightarrow C$ be a cofiber sequnce in $\mathcal{B}$. Then, by cofinality, there are $\overline{A}, \overline{C}$ in $\mathcal{B}$ such that $\overline{A} \vee A$ and $\overline{C} \vee C$ are in $\mathcal{A}$. Then

$$A \vee \overline{A} \to B \vee \overline{A} \vee \overline{C} \to C \vee \overline{C} \tag{11}$$

is a cofiber sequence in $\mathcal{B}$ by the pushout axiom, and the first and last object are in $\mathcal{A}$. By being closed under extensions, therefore the middle term $B \vee \overline{A} \vee \overline{C}$ is also in $\mathcal{A}$. It follows that in G' we have

$$[B] + [\overline{A}] + [\overline{C}] = 0$$
$$[A] + [\overline{A}] = 0$$
$$[C] + [\overline{C}] = 0$$

and therefore $[B] = [A] + [C]$ in G'. If follows $G \cong G'$.

We now prove that $\mathcal{A} \subseteq \mathcal{B}$ is strictly cofinal. If $K_0(\mathcal{A}) \to K_0(\mathcal{B})$ is a surjection, G and therefore G' is trivial. Hence $K_0'(\mathcal{A}) \to K_0'(\mathcal{B})$ is surjective. Let $B \in \mathcal{B}$, then there is an $A \in \mathcal{A}$ such that $[A] = [B] \in K_0'(\mathcal{B})$. Now $K_0'(\mathcal{B})$ is just a group completion with respect to $\vee$ after taking isomorphism classes. Therefore $[A] = [B]$ if and only if there is a $C \in \mathcal{B}$ with $A \vee C \cong B \vee C$. (This is the usual algebraic argument.) As $\mathcal{A}$ is cofinal, there is a $\overline{C} \in \mathcal{B}$ such that $C \vee \overline{C} \in \mathcal{A}$. Therefore

$$A \vee (C \vee \overline{C}) \cong B \vee C \vee \overline{C}.$$

with, of course, $B \vee C \vee \overline{C} \in \mathcal{A}$. This shows strict cofinality. □

Proof of Thm. 8.2.1. We want to apply [TT90, 1.10.1]. For the convenience of the reader we quote the result:

> 1.10.1. **Cofinality Theorem.** Let $v\mathcal{B}$ be a Waldhausen category with a cylinder functor satisfying the cylinder axiom. Let G be an abelian group, and $\pi\colon K_0(v\mathcal{B}) \to G$ an epimorphism. Let $\mathcal{B}^w$ be the full subcategory of those B in $\mathcal{B}$ for which the class $[B]$ in $K_0(v\mathcal{B})$ has $\pi[B] = 0$ in G. Make $\mathcal{B}^w$ a Waldhausen category with $v(\mathcal{B}^w) = \mathcal{B}^w \cap v(B)$, $\mathrm{co}(\mathcal{B}^w) = \mathcal{B}^w \cap \mathrm{co}(B)$. Let "$G$" denote G considered as Eilenberg-MacLane spectrum whose only non-zero homotopy group is G in dimension 0.
>
> Then there is a homotopy fibre sequence
>
> $$K(v\mathcal{B}^w) \to K(v\mathcal{B}) \to \text{“}G\text{”}$$

Define G as $K_0(\mathcal{B})/K_0(\mathcal{A})$. So we get the above fiber sequence and in particular $K_0(v\mathcal{B}^w) = \ker K_0(v\mathcal{B}) \to G$. Clearly the inclusion $\mathcal{A} \to \mathcal{B}$ factors as $\mathcal{A} \to \mathcal{B}^w$, as $\pi[A] = 0$ for $A \in \mathcal{A}$ by definition. It suffices to show that $K(\mathcal{A}) \to K(\mathcal{B}^w)$ is a weak equivalence. Like in $\mathcal{B}$, $\mathcal{A}$ is cofinal in $\mathcal{B}^w$, as one can choose the same complement. Then $\mathcal{A} \subseteq \mathcal{B}^w$ satisfies the assumptions of Theorem 8.2.1. But $K_0(\mathcal{A}) \to K_0(\mathcal{B}^w)$ is surjective, so by Lemma 8.2.4 $\mathcal{A}$ is strongly cofinal in $\mathcal{B}^w$. By Waldhausen's strict cofinality Theorem [Wal85, 1.5.9] there is a homotopy equivalence $K(\mathcal{A}) \to K(\mathcal{B}^w)$. This shows that we get the desired homotopy fiber sequence

$$K(\mathcal{A}) \to K(\mathcal{B}) \to \text{“}K_0(\mathcal{B})/K_0(\mathcal{A})\text{”}.$$ □

8.2.5 Saturated is necessary: A counterexample The assumption (3) in Theorem 8.2.1 "saturatedness" in Theorem 5.1 is necessary. Here is a counterexample, which I owe to a discussion with Chuck Weibel. Let $\mathcal{C}$ be the following Waldhausen category:

1. Objects are finite pointed sets X with decomposition $X = A \vee B \vee C$. (Think of the elements as being colored, while the basepoint is black.)

2. Morphisms are maps $A \vee B \vee C \to A' \vee B' \vee C'$ such that they restrict to maps $A \to A'$, $B \to A' \vee B'$, $C \to A' \vee C'$. That is, you can change any color to A or map to the basepoint, or do not change the color.

3. Let Cofibrations be split injections. That is maps $i \colon X \to X'$ such that there is a map $p \colon X' \to X$ with $p \circ i = \mathrm{id}$. It follows that $X' \cong X \vee Y$ for some Y in $\mathcal{C}$. This is, they come from the direct sum "$\vee$" in pointed sets.

4. Let a weak equivalence be a bijection of pointed sets.

This category has $K_0(\mathcal{C}) \cong \mathbb{Z}$, as each object is weakly equivalent to an object $A \vee * \vee *$.

Consider the full subcategory $\mathcal{B}$ of objects $A \vee B \vee C$ with $|A| = |C|$. This is a cofinal subcategory in $\mathcal{C}$. It is not saturated: While $A \vee * \vee C$ is equivalent to $(A \vee C) \vee * \vee *$ in $\mathcal{C}$, the latter is not in $\mathcal{B}$. It satisfies all the other assumptions of Theorem 5.1 except for the cylinder functor (v), so Lemma 8.2.4 would apply and show that $K_0(\mathcal{B}) \subseteq K_0(\mathcal{C})$. However, one sees that $K_0(\mathcal{B}) = \mathbb{Z} \oplus \mathbb{Z}$, with generators represented by $A \vee * \vee C$ and $* \vee B \vee *$, as all weak equivalences in $\mathcal{B}$ are isomorphisms. Hence $K_0(\mathcal{B}) \to K_0(\mathcal{C})$ cannot be an injection.

This counterexample shows that the saturatedness assumption is necessary and missing in Exercise 1.10.2 in [TT90], as well as in Corollary V.2.3.1 in [Wei13]. The latter is deduced in [Wei13] from Theorem II.9.4 through a chain along exercise II.9.14 ("Grayson's trick"), Theorem IV.8.9 and Remark IV.8.9.1, (which states $K_0(\mathcal{B}) = K_0(\mathcal{C})$ is equivalent to $\mathcal{B}$ being strictly cofinal in $\mathcal{C}$). The gap is in the proof of Theorem II.9.4, which claims that the proof of Lemma II.7.2 applies verbatim. The above counterexample in particular applies to Theorem II.9.4.

8.2.6 Fullness is necessary For completeness, let us remark that the fullness assumption in 8.2.1 is also necessary: In Waldhausen's cofinality theorem [Wal85, 1.5.9] he does not mention that one needs to assume that the subcategory $\mathcal{A} \subseteq \mathcal{B}$ is full. There is a counterexample due to Inna Zakharevich [Zak10] which shows that one has to assume it, which we reproduce below.

> Consider the following example. Let C be the category of pairs of pointed finite sets, whose morphisms $(A, B) \to (A', B')$ are

pointed maps $A \vee B \to A' \vee B'$, and let B be the category of pairs of pointed finite sets whose morphisms $(A, B) \to (A', B')$ are pairs of pointed maps $A \to B$ and $A' \to B$. We make C a Waldhausen category by defining the weak equivalences to be the isomorphisms, and the cofibrations to be the injective maps. B is clearly cofinal in C, but $K_0(B) = \mathbb{Z} \times \mathbb{Z}$, while $K_0(C) = \mathbb{Z}$.[...]

8.3 Change of finiteness conditions

We turn to the proofs for Subsection 4.1. We need Waldhausen's Approximation Theorem, which we recall for convenience.

Definition 8.3.1 (Approximation Property [Wal85, 1.6],[TT90, 1.9.1]). Let $F\colon \mathcal{A} \to \mathcal{B}$ be an exact functor of categories with cofibrations and weak equivalences. F has the *Approximation Property* if the following two axioms hold.

(App 1) A map f in $\mathcal{A}$ is a weak equivalence if (and only if) its image $F(f)$ in $\mathcal{B}$ is a weak equivalence.

(App 2) Given any object A in $\mathcal{A}$ and a map $x\colon F(A) \to B$ in $\mathcal{B}$ there exists a map $a\colon A \to A'$ in $\mathcal{A}$ and a weak equivalence $x'\colon F(A') \to B$ in $\mathcal{B}$ such that the triangle

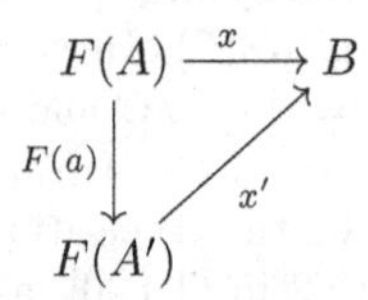

commutes.

Theorem 8.3.2 (Approximation Theorem [Wal85, 1.6.7],[TT90, 1.9.1]). *Let $\mathcal{A}$, $\mathcal{B}$ be categories with cofibrations and weak equivalences which satisfy the Saturation Axiom. Assume $\mathcal{A}$ has a Cylinder Functor satisfying the Cylinder Axiom. Let $F\colon \mathcal{A} \to \mathcal{B}$ be an exact functor with the Approximation Property. Then F induces an equivalence*

$$K(F)\colon K(w\mathcal{A}) \to K(w\mathcal{B})$$

on connective algebraic K-theory spectra.

We used Thomason-Trobaugh's remark in [TT90, 1.9.1] that we can use a weaker version of the approximation property. In [Wal85] there is the further requirement in (App 2) that a is a cofibration, which we can always be arrange due to the existence of a Cylinder Functor.

Proof of Proposition 4.1.3. To prove (1) we use Waldhausen's Approximation Theorem 8.3.2 and apply it to the inclusion functor. We check the conditions. First, all our categories satisfy the Saturation Axiom and have a Cylinder functor satisfying the Cylinder Axiom.

A map is a homotopy equivalence in $\mathcal{C}_f^G$ if and only if it is one in $\mathcal{C}_{hf}^G$, so (App 1) is satisfied.

So given $A \in \mathcal{C}_f^G$, $B \in \mathcal{C}_{hf}^G$ and a map $f: A \to B$. For B there is by definition a finite object $B_f \in \mathcal{C}_f^G$ which is homotopy equivalent to B, i.e., there are maps $g: B_f \to B$ and $\overline{g}: B \to B_f$ with both compositions being homotopic to the identity. Define $j: A \to B_f$ as $j := \overline{g} \circ f$. Then $g \circ j$ is homotopic to f. Using the cylinder functor (and Remark 6.2.5) we can rectify the homotopy commutative diagram on the left below to the strict commutative diagram on the right:

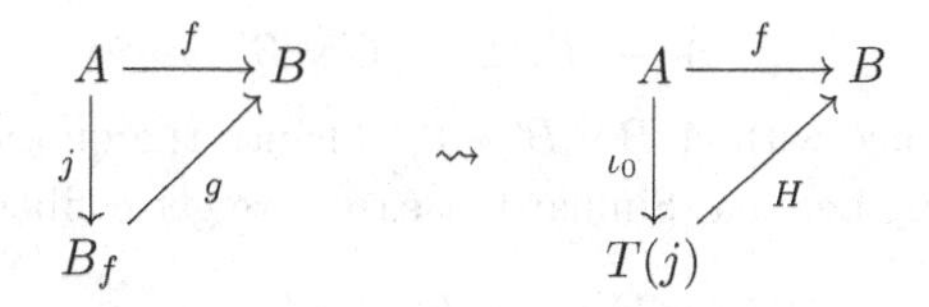

As $B_f \to T(j)$ is a homotopy equivalence, by the Saturation Axiom H is a homotopy equivalence. This shows (App 2) and therefore (1).

For (2) we use the Waldhausen-Thomason cofinality Theorem 8.2.1. We have to check the conditions (1) to (6). Most of them are clear or shown in the previous sections.

In particular a map $A \to A'$ in $\mathcal{C}_{hf}^G$ which is a cofibration in $\mathcal{C}_{hfd}^G$ is a cofibration in $\mathcal{C}_{hf}^G$, and therefore its quotient is again in $\mathcal{C}_{hf}^G$. This is (1). By definition $\mathcal{C}_{hf}^G$ is full in $\mathcal{C}_{hfd}^G$. A map in the former is a weak equivalence if and only if it is in the latter. Similar the cylinder functor is just inherited. This shows (2) and (5). An object homotopy equivalent to an object in $\mathcal{C}_{hf}^G$ is homotopy finite, hence itself in $\mathcal{C}_{hf}^G$, this shows (3).

We are left with first showing the cofinality (6) and then that $\mathcal{C}_{hf}^G$ is closed under extensions in $\mathcal{C}_{hfd}^G$ (4).

For $B \in \mathcal{C}_{hfd}^G$ there is an $A \in \mathcal{C}_{hf}^G$ such that B is a retract of A, i.e., there are maps $r: A \to B$, $i: B \to A$ such that $r \circ i = \mathrm{id}_B$. By replacing A with $T(i)$ we can assume that i is a cofibration, hence there is a cofiber sequence

$$B \overset{i}{\rightarrowtail} A \overset{p}{\twoheadrightarrow} C := A/B.$$

The retraction $r: A \to B$ and the map $* \to C$ give a map $A \to B \vee C$, and $* \to B$ and $A \to C$ give another one. The sum of these maps makes the

diagram

$$\begin{array}{ccccc} B & \rightarrowtail & A & \twoheadrightarrow & C \\ \downarrow = & & \downarrow & & \downarrow = \\ B & \rightarrowtail & B \vee C & \twoheadrightarrow & C \end{array}$$

commutative and both rows are cofiber sequences. By the Extension Axiom 6.5.2 the map $A \to B \vee C$ is a homotopy equivalence, hence $B \vee C \in \mathcal{C}^G_{hf}$. This show cofinality.

Next we need to show that $\mathcal{C}^G_{hf}$ is closed under extensions in $\mathcal{C}^G_{hfd}$. So let

$$A \rightarrowtail B \twoheadrightarrow C$$

be a cofiber sequence in $\mathcal{C}^G_{hfd}$ with $A, C \in \mathcal{C}^G_{hf}$ and B ("the extension of A by C") in $\mathcal{C}^G_{hfd}$. As $\mathcal{C}^G_{hf}$ is cofinal there is a $B' \in \mathcal{C}^G_{hfd}$ such that $B \vee B' \in \mathcal{C}^G_{hf}$. Then

$$A \rightarrowtail B \vee B' \twoheadrightarrow C \vee B'$$

is a cofiber sequence with $A, B \vee B' \in \mathcal{C}^G_{hf}$, hence the quotient $C \vee B'$ is in $\mathcal{C}^G_{hf}$ by the Glueing Lemma. Similar but easier we get cofiber sequences

$$C \rightarrowtail C \vee B' \twoheadrightarrow B'$$

showing $B' \in \mathcal{C}^G_{hf}$ and

$$B' \rightarrowtail B \vee B' \twoheadrightarrow B$$

showing $B \in \mathcal{C}^G_{hf}$, what we wanted to show.

The Cofinality Theorem 8.2.1 therefore gives us a homotopy fiber sequence of connective spectra

$$K(\mathcal{C}^G_{hf}) \to K(\mathcal{C}^G_{hfd}) \to K_0(\mathcal{C}^G_{hfd})/K_0(\mathcal{C}^G_{hf})$$

As $\pi_n(K_0(\mathcal{C}^G_{hfd})/K_0(\mathcal{C}^G_{hf})) = 0$ for $n \neq 0$ part (2) of the proposition follows. □

Remark 8.3.3. In view of the proposition one can consider $\mathcal{C}^G_{hfd}$ as kind of "idempotent completion" of $\mathcal{C}^G_{hf}$. (Recall that for algebraic K-theory of rings the idempotent completion [Fre03, 2.B, p.61] of the category of finitely generated free R-modules gives the category of finitely generated projective R-modules, which has the correct K_0, cf. also e.g. [CP97].)

Corollary A.2.2 from the appendix shows that idempotents and certain homotopy idempotents split in $\mathcal{C}^G_{hfd}$. The author does not know if every homotopy idempotent splits in $\mathcal{C}^G_{hfd}$. Hence it is not clear that $K_0(\mathcal{C}^G_{hfd})$ is the "correct" group from this point of view. However, the difference and interplay between $K_0(\mathcal{C}^G_{hf})$ and $K_0(\mathcal{C}^G_{hfd})$ is crucial in later work to construct a non-connective delooping of $K(\mathcal{C}^G_?)$ for all $? = f, hf, hfd$. The deloopings will then be equivalent non-connective K-theory spectra for all three finiteness conditions.

8.4 Easy examples

If X is a point, the category $\mathcal{C}(X, R)$ is just the category of simplicial R-modules. We use the definition of algebraic K-Theory of simplicial rings we from [Wal85, 2.3]. Then $\mathcal{C}^G_{hfd}(G/1, R)$ is equivalent (as category with cofibrations and weak equivalences) to the category of homotopy finitely dominated $R[G]$-modules. Corollary 4.1.5 follows.

8.5 Change of rings

Let $f\colon R \to S$ be a map of simplicial rings.

If M is a cellular R-module then $S \otimes_R M$ is a cellular S-module and we get a natural bijection $\diamond_R M \cong \diamond_S(S \otimes_R M)$ which makes $S \otimes_R M$ into a controlled S-module. This construction respects all finiteness conditions and cofibrations, so we get an exact functor $S \otimes_R -\colon \mathcal{C}^G_?(X, R) \to \mathcal{C}^G_?(X, S)$.

Theorem 8.5.1 (Change of rings). *Let $f\colon R \to S$ be map of simplicial rings which is a weak equivalence. Then f induces a map $\mathcal{C}^G_f(X, R) \to \mathcal{C}^G_f(X, S)$ which is an equivalence on algebraic K-Theory.*

For technical reasons we assume all modules to be finite-dimensional in this section. Therefore we only have this theorem for the finiteness condition f. Note that the theorem for hf and hfd, except the K_0-part of hfd, is already implied by the Theorem 8.5.1 using Proposition 4.1.3.

The proof takes the rest of this section, we need some preparations first. A map of simplicial rings which is a weak equivalence of the underlying simplicial sets is called a *weak equivalence of simplicial rings* for short.

Lemma 8.5.2. *Let $R \to S$ be a weak equivalence of simplicial rings and P a cellular (uncontrolled) R-module. Let $\eta\colon P \to \operatorname{res}_R S \otimes_R P$ be the unit of the adjunction between the induction $S \otimes_R -$ and the restriction res_R. Then η is a weak equivalence of simplicial R-modules and in particular a homotopy equivalence of simplicial sets.*

Proof. This follows from the Glueing Lemma and induction over the dimension of P. We get a pushout-diagram

$$\begin{array}{ccccc}
\coprod R[\Delta^n] & \longleftarrow & \coprod R[\partial\Delta^n] & \longrightarrow & P_{n-1} \\
\downarrow \simeq & & \downarrow \simeq & & \downarrow \simeq \\
\coprod S[\Delta^n] & \longleftarrow & \coprod S[\partial\Delta^n] & \longrightarrow & S \otimes_R P_{n-1}
\end{array}$$

where the vertical maps are weak equivalences of simplical R-modules, hence by the Glueing Lemma for simplicial R-modules (cf. [GJ99, II.8.12;III.2.14]) the pushout $P_n \to S \otimes_R P_n$ is a weak equivalence. We have $P = \bigcup_n P_n$ and the n-skeleton of P and P_n agree. Also the n-skeleton of $S \otimes_R P$ and $S \otimes_R P_n$

agree and $S \otimes_R P = \bigcup_n S \otimes_R P_n$. Therefore $P \to S \otimes_R P$ is a weak equivalence. As simplicial abelian groups are fibrant as simplicial sets the weak equivalence is a homotopy equivalence of simplicial sets. □

Lemma 8.5.3. *Let $M, P \in \mathcal{C}_a^G(X, R)$. Let $i\colon A \subseteq M$ be a cellular submodule. Assume M is finite-dimensional. Let $g\colon A \rightarrowtail P$ and $\widehat{f}\colon S \otimes_R M \to S \otimes_R P$ be maps such that the diagram*

$$\begin{array}{ccc} S \otimes_R A & \xrightarrow{S \otimes g} & S \otimes_R P \\ {\scriptstyle S\otimes i}\big\downarrow & \nearrow_{\widehat{f}} & \\ S \otimes_R M & & \end{array} \tag{12}$$

commutes. Then there is a map $f\colon M \to P$ such that $\widehat{f}$ is homotopic to $S \otimes_R f$ relative to the cellular submodule $S \otimes_R A$ of $S \otimes_R M$.

Proof. Assume M, $\widehat{f}$, g, P are E-controlled. We do induction over the dimension of cells of M which are not in A. As usual it suffices to consider only one cell. Let $e\colon R[\Delta^n] \to M$ be attached to A via $\partial\colon R[\partial\Delta^n] \to A$.

Looking at the smallest submodules containing e, ∂e and $\widehat{f}(S \otimes_R e)$ we get the following commutative diagram. (We denote by $S \otimes_R e$ the cell in $S \otimes_R M$ corresponding to e via the isomorphism $\diamond_R M \cong \diamond_S(S \otimes_R M)$.)

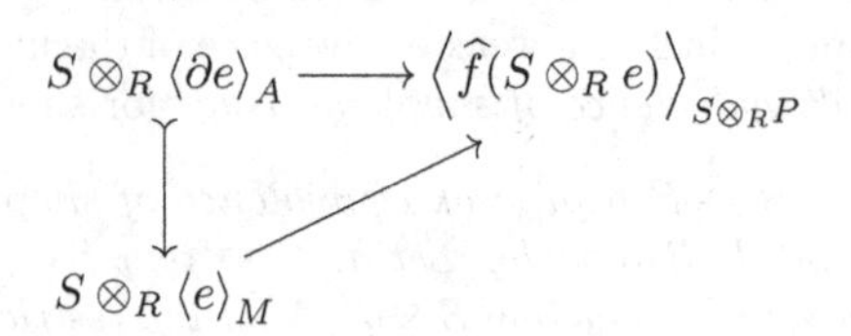

Because everything is E-controlled the support of every module is contained in $\{\kappa(e)\}^E$. Note that $\left\langle \widehat{f}(S \otimes_R e) \right\rangle_{S \otimes_R P}$ is isomorphic to $S \otimes_R P'$ for $P' \subseteq P$ a cellular R-submodule. Using the adjunction $S \otimes_R -$ and restriction res_R we obtain the solid commutative diagram below.

$$\begin{array}{ccccc} R[\partial\Delta^n] & \dashrightarrow & \langle \partial e \rangle_A & \longrightarrow & P' \\ \big\downarrow & & \big\downarrow & & \big\downarrow {\scriptstyle \sim\ \eta} \\ R[\Delta^n] & \dashrightarrow & \langle e \rangle_M & \longrightarrow & \mathrm{res}_R\, S \otimes_R P' \end{array}$$

Here η is the unit of the adjunction. We want to find a lift up to homotopy relative to $\langle \partial e \rangle_A$ in the solid diagram. We can extend the diagram to the left by the dashed square, which is a pushout square. Hence, using the adjunction

$R[-]$ and forgetful functor, it suffices to construct a lift up to homotopy relative to $\partial\Delta^n$ in the diagram of simplicial sets

$$\begin{array}{ccc} \partial\Delta^n & \xrightarrow{g} & P' \\ \downarrow i & & \sim \downarrow \eta \\ \Delta^n & \xrightarrow{\widehat{f}} & \operatorname{res}_R S \otimes_R P' \end{array}.$$

There is a lift because by Lemma 8.5.2 η is a homotopy equivalence of simplicial sets. With more effort we can arrange that we lift to a map $f\colon \Delta^n \to P'$ with $f \circ i = g$ and $\eta \circ f$ is homotopic to $\widehat{f}$ with the homotopy begin constant when restricted via i.

So we get a map

$$\langle e \rangle_M \xrightarrow{f} P'$$

such that $S \otimes_R f$ is homotopic to $\widehat{f}\colon S \otimes_R \langle e \rangle_M \to S \otimes_R P'$ relative to $S \otimes_R \langle \partial e \rangle_A$. As $S \otimes_R P'$ has support on $\{\kappa_R(e)\}^E\}$ the map and the homotopy are E-controlled.

So, assuming the first cells of M which are not in A are of dimension n, we can use this procedure and the homotopy extension property to produce a map $S \otimes_R M \to S \otimes_R P$ which satisfies the assumption of the lemma for an $A' = A \cup \mathrm{sk}_n M$. Induction and the finite-dimensionality of M finishes the proof. □

Proof of Theorem 8.5.1. We want to apply Waldhausen's Approximation Theorem 8.3.2 to the functor $F := S \otimes_R -\colon \mathcal{C}_f^G(X,R) \to \mathcal{C}_f^G(X,S)$. We prove (App 1) first. Let $\alpha\colon M \to M'$ be a map in $\mathcal{C}_f^G(X,R)$ such that $S \otimes_R \alpha$ is a homotopy equivalence in $\mathcal{C}_f^G(X,S)$. By Lemma 8.5.3 there is a map $\beta'\colon M' \to M$ such that the homotopy inverse $\beta\colon S \otimes_R M' \to S \otimes_R M$ of $S \otimes_R \alpha$ in $\mathcal{C}_?^G(X,S)$ is homotopic to $S \otimes_R \beta'$. Hence there is a homotopy $H\colon S \otimes_R M[\Delta^1] \to S \otimes_R M$ from $S \otimes_R \mathrm{id}_R$ to $S \otimes_R (\beta' \circ \alpha)$ in $\mathcal{C}_f^G(X,S)$ which is homotopic relative to $M[\partial\Delta^1]$ to a homotopy $S \otimes_R H'$ where H' is a homotopy from id_R to $\beta' \circ \alpha$, using Lemma 8.5.3 again. Vice versa for $\alpha \circ \beta'$, so α is also a homotopy equivalence in $\mathcal{C}_f^G(X,R)$.

For (App 2) take $M \in \mathcal{C}_f^G(X,R)$ and $N \in \mathcal{C}_f^G(X,S)$ and a map $f\colon S \otimes_R M \to N$. Assume that it is a cellular inclusion by taking the mapping cylinder. We show that N is homotopy equivalent relative $S \otimes_R M$ to a module $S \otimes_R \overline{M}^2$, with $\overline{M}^2 \in \mathcal{C}_f^G(X,R)$.

Assume that the n-skeleton of $S \otimes_R M$ and N agree. Let N^{n+1} be the $(n+1)$-skeleton of N relative to $S \otimes_R M$, i.e., $N^{n+1} = \mathrm{sk}_{n+1} N \cup M$. Then N^{n+1} is the pushout

$$S[\textstyle\coprod \Delta^{n+1}] \longleftarrow S[\textstyle\coprod \partial\Delta^{n+1}] \xrightarrow{\varphi^{n+1}} S \otimes_R M\ .$$

where φ^{n+1} is the attaching map for the cells. By Lemma 8.5.3 there is a map $\psi^{n+1} \colon R[\coprod \partial\Delta^{n+1}] \to M$ such that $S \otimes_R \psi^{n+1}$ is homotopic to a φ^{n+1}. Call the homotopy H^{n+1}. Applying the Glueing Lemma to the diagram (where all vertical maps are homotopy equivalences)

$$\begin{array}{ccccc} S[\coprod \Delta^{n+1}] & \longleftarrow & S[\coprod \partial\Delta^{n+1}] & \xrightarrow{\varphi^{n+1}} & S \otimes_R M \\ \downarrow & & \downarrow & & \downarrow \\ S[\coprod \Delta^{n+1}][\Delta^1] & \longleftarrow & S[\coprod \partial\Delta^{n+1}][\Delta^1] & \xrightarrow{H^{n+1}} & S \otimes_R M[\Delta^1] \\ \uparrow & & \uparrow & & \uparrow \\ S \otimes_R R[\coprod \Delta^{n+1}] & \longleftarrow & S \otimes_R R[\coprod \partial\Delta^{n+1}] & \xrightarrow{S\otimes\psi^{n+1}} & S \otimes_R M \end{array}$$

shows that the pushout of the first row is homotopy equivalent to the pushout of the last row. (This is a simplicial version of the topological fact that homotopic attaching maps yield homotopy equivalent CW-complexes.) Choose such a homotopy equivalence ξ. In the last row $S \otimes_R -$ commutes with the pushout, define $\overline{M}$ as the pushout of

$$R[\coprod \Delta^{n+1}] \longleftarrow R[\coprod \partial\Delta^{n+1}] \xrightarrow{\psi^{n+1}} M\ .$$

Then take the pushout along $\xi \colon N^{n+1} \to S \otimes_R M$ and the inclusion $N^{n+1} \rightarrowtail N$ to obtain $\overline{N}$:

$$\begin{array}{ccc} N^{n+1} & \rightarrowtail & N \\ \xi \downarrow \simeq & & \downarrow \simeq \\ S \otimes_R \overline{M} & \underset{\overline{f}}{\rightarrowtail} & \overline{N} \end{array}$$

Now the $(n+1)$-skeleton of $S \otimes_R \overline{M}$ isomorphic to $\overline{N}$ via $\overline{f}$. By induction and because N is finite-dimensional we get a diagram

$$\begin{array}{ccc} S \otimes_R M & \rightarrowtail & N \\ \downarrow & & \downarrow \simeq \\ S \otimes_R \overline{M}^1 & \xrightarrow{\cong} & \overline{N}^1 \end{array}$$

which we can make into the desired diagramm

$$\begin{array}{ccc} S \otimes_R M & \longrightarrow & N\ . \\ S\otimes - \downarrow & \nearrow \sim & \\ S \otimes_R \overline{M}^2 & & \end{array}$$

using a homotopy inverse for the right map and defining $\overline{M}^2$ as the mapping cylinder of $M \to \overline{M}^1$ to make the diagram strictly commutative. This proves (App 2). The theorem follows by the Approximation Theorem 8.3.2. □

9 Applications

We outline some application and the relevance of our category $\mathcal{C}^G(X;R)$. We will not provide proofs because they require considerably more technology.

9.1 Controlled algebra for discrete rings

Each (discrete) ring R_d can be made into a simplicial ring taking the constant functor $[n] \mapsto R_d$ where all structure maps are the identity. Then a simplicial module over R_d is essentially, by the Dold-Kan-Theorem (see [GJ99]), a non-negatively graded chain complex. (More precisely there is an adjunction between simplicial modules and non-negative chain complexes and this adjunction is a equivalence on homotopy categories.) Even more, a cellular simplicial R_d-module with cells only in dimension 0 is just a free R_d-module.

Therefore we can look at the subcategory of 0-dimensional controlled R_d-modules, we suppress the control space in the following. This has no longer the nice homotopical properties, but it is an additive category. In fact, this is essentially the category of controlled R_d-modules of e.g. [BFJR04] or [PW85]. (There is a small technical difference to [BFJR04] in the definition of morphisms, but that does not affect the algebraic K-theory of that category.)

Therefore the category $\mathcal{C}^G(X;R)$ we present here can be viewed as a homotopical generalization of the category of controlled R_d-modules. Unfortunately it comes with a price: The arguments to treat $\mathcal{C}^G(X;R)$ get more involved. There are a lot cases where we need to invoke Waldhausen's approximation theorem where in the case of discrete modules and rings it would sufficient to prove two categories at hand are equivalent. However, we believe that most arguments have an analogue for $\mathcal{C}^G(X;R)$.

9.2 The Farrell-Jones Conjecture

9.2.1 Statement and Significance Let R be a ring or a simplicial ring and G a (discrete) group. The Farrell-Jones Conjecture provides a "calculation" of $K_n(R[G])$, $n \in \mathbb{Z}$, the algebraic K-theory of the group ring $R[G]$, in terms of the algebraic K-theory of R and the "geometry" of the group G. More precisely, it claims that the so-called assembly map

$$H_n^G(E_{\mathcal{VC}}G; \mathbf{K}_R) \to K_n(R[G])$$

is an isomorphism for every $n \in \mathbb{Z}$. Here the right-hand side is the (non-connective) algebraic K-theory of the group ring $R[G]$, while the left-hand side is the G-equivariant homology theory with coefficients in the G-equivariant non-connective K-theory spectrum, evaluated at the classifying space of G for virtual cyclic subgroups. We refrain from discussing more details, as the Farrell-Jones Conjecture is not our main focus in this article and refer to [Bar13] or the slightly outdated survey [LR05].

The Farrell-Jones Conjecture implies a plethora of other usually long-standing conjectures. This includes the vanishing of the Whitehead Group for torsionfree groups and the Borel conjecture about the rigidity of aspherical manifolds. We refer to [LR05, BLR08b] for details. Therefore it is interesting to know the Farrell-Jones Conjecture for as many rings R, called the "coefficients", and groups G as possible.

9.2.2 Status and Proofs There is recent and ongoing progress on class of groups for which the Farrell-Jones Conjecture is known. Recent approaches in fact prove a more general version, the "Farrell-Jones Conjecture with wreath products", see Section 6 of [BLRR14]. Also, that version even allows any additive category $\mathcal{A}$ as coefficients. If $\mathcal{A}$ is the category of finitely generated free R-modules, we obtain back the version we stated above.

Recent work by Bartels, Farrell, Lück, Reich, Rüping, Wegner, Wu and others shows the "Farrell-Jones Conjecture with wreath products and coefficient in an additive category" for large classes of groups, most recently for $GL_n(\mathbb{Z})$ an some related groups in [BLRR14], solvable Baumslag-Solitar groups in [FW13] and, more generally, solvable groups in [Weg15].

All recent proofs have in common, that they start by translating the Farrell-Jones Conjecture to a problem in the algebraic K-theory of controlled algebra, a strategy first formulated in this way in [BLR08a].

9.2.3 A reformulation in terms of controlled algebra Let Z be a G-CW-complex. We want to make $X \times G \times [1, \infty)$ into a G-control space. Recall the continuous control conditions $\mathcal{E}_{cc}$ on $X \times [1, \infty)$ from Example 2.2.7. We can "pull back" this morphism control conditions along the projection $p \colon X \times G \times [1, \infty) \to X \times [1, \infty)$ by setting

$$p^{-1}\mathcal{E}_{cc} := \{(p \times p)^{-1}(E) \mid E \in \mathcal{E}_c c\}.$$

Then $p^{-1}\mathcal{E}_{cc}$ is a morphism control structure on $X \times G \times [1, \infty)$. We get object support conditions by setting

$$\mathcal{F}_{Gc}(X \times G) := \{G.K \times [1, \infty) \mid K \subseteq X \times G \text{ is compact}\}$$

where $G.K$ is the G-orbit of K. If X is $E_{\mathcal{VC}}G$, the classifying space for G (cf. [tD87, I.6]) and the family $\mathcal{VC}$ of virtually cyclic subgroups, we obtain a category

$$\mathcal{O}^G = \mathcal{C}^G_f(E_{\mathcal{VC}}G \times G \times [1, \infty), p^{-1}\mathcal{E}_{cc}, \mathcal{F}_{Gc}; R)$$

of controlled simplicial R-modules. As explained above, for a discrete ring R_d there is a similar category of discrete controlled modules which we call $\mathcal{O}_d^G$ for brevity. As it is an additive category, its algebraic K-theory is defined.

Theorem ([BLR08a, 3.8]). *$K_i(\mathcal{O}_d^G) = 0$ for all $i \in \mathbb{N}$ if and only if the Farrell-Jones Conjecture holds for G.*

Hence one can use controlled algebra and manipulation of the control space to prove the Farrell-Jones Conjecture. This is in fact the strategy carried out by recent proofs.

For simplicial rings an analogue of the theorem holds by unpublished work of the author in [Ull11]. Thus this article should be viewed as first step to carry out the successful program of proving the Farrell-Jones Conjecture for discrete rings in the settings of simplicial rings.

9.3 Non-connective algebraic K-theory

There is a second direct application we can give, but again not prove here. Take the control space $(\mathbb{R}^n, \mathcal{E}_d)$ arising from the euclidean metric. Then there is a map

$$K(R) \to \Omega^n K(\mathcal{C}_f(\mathbb{R}^n; R))$$

of connective K-theory spectra which is an isomorphism on π_i for $i \geq 1$ and an injection on π_0. This deloops $K(R)$, that means the $K(\mathcal{C}_f(\mathbb{R}^n; R))$ for varying n can be made into a spectrum which may have interesting negative homotopy groups and where the positive homotopy group are the ones of $K(R)$. This is the first construction of a non-connective K-theory spectrum for simplicial rings. It generalizes the delooping construction of $K(R_d)$ of [PW85]. However, is it known that $\pi_i K(R) = K_i(\pi_0 R)$ for $i = 0, 1$. Because a Bass-Heller-Swan theorem is expected to hold for algebraic K-theory of simplicial rings this means that the negative algebraic K-groups of a simplicial ring are just the ones of the discrete ring $\pi_0 R$. But of course having spectrum is more information than just knowing its homotopy groups.

The proofs of both applications need considerably more technology in $\mathcal{C}^G(X; R)$, namely a notion of germs, which were developed in [Ull11]. We will come back to these in later work.

9.4 Ring spectra

There are generalizations of rings for which the Farrell-Jones Conjecture should give interesting results with implications to manifold theory. The details are hard to explain in brief, but homotopy theorist know since a long time that the so-called "ring spectra" provide a natural generalization of rings, and simplicial rings are an intermediate step between rings and ring spectra. Algebraic K-theory can be defined for ring spectra, this done in [EKMM97]. The statement of the Farrell-Jones Conjecture makes sense for connective ring spectra as coefficients. In fact, when Farrell and Jones in [FJ93, FJ87] originally stated and proved a version of their conjecture (for a certain class of groups), they also treated the case of pseudoisotopies, which is more or less the case where the sphere spectrum, the "initial ring spectrum", are the coefficients.

We hope that the theory presented here can be adapted for ring spectra. We do not want to go into details, but let us remark that in the 1990's a bunch of different models for ring spectra and categories of modules over ring spectra where discovered. The main models are symmetric spectra [HSS00], orthogonal spectra [MMSS01] and S-modules [EKMM97]. The category of S-modules is special among these as it has the nice property that it has a model structure such that every object is fibrant and [EKMM97, III.2] provides a nice theory of cellular objects. Unfortunately the category is rather hard to define. It looks like a suitable candidate to carry our the program presented here, but certainly a lot of work still needs to be done for that.

A A simplicial mapping telescope

A map $\eta\colon K \to K$ in $\mathcal{C}_a^G$ is called a *homotopy idempotent* if η^2 is homotopic to η. Here we provide the necessary tools we need about homotopy idempotents in this and later work. This gives some insight into the category $\mathcal{C}_{hfd}^G(X; R)$, for any control space X and simplicial ring R.

We defined the category with cofibrations and weak equivalences $\mathcal{C}_a^G = \mathcal{C}_a^G(X, R, \mathcal{E}, \mathcal{F})$ for a control space $(X, \mathcal{E}, \mathcal{F})$ and a simplicial ring R in Section 2.5.3.

A.1 Coherent homotopy idempotents

Some parts of Theorem A.2.1 below need an extra assumption on the idempotent, which we will define now.

Definition A.1.1. A homotopy idempotent $\eta\colon K \to K$ with homotopy H from η^2 to η is called *coherent* if there is a map $G\colon K[\Delta^1 \times \Delta^1] \to K$ whose restrictions to the boundary look as in the following diagram

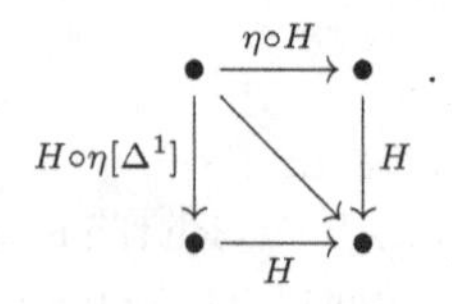

Note that we used the diagram language of 6.2.4 to describe the map. We will use it in following without further comment. If $\eta^2 = \eta$ then η is coherent.

Lemma A.1.2. *If η arises from a homotopy domination, it is coherent.*

Proof. The assumption means that there are maps $i\colon K \to L$, $p\colon L \to K$ such that $\eta = p \circ i$ and $i \circ p \simeq \mathrm{id}$ via a homotopy H'. Then η is an idempotent with homotopy $p \circ H' \circ i$ from η^2 to η.

The coherence homotopy G can be given by the composition

$$K[\Delta^1 \times \Delta^1] \cong K[\Delta^1][\Delta^1] \xrightarrow{i[\Delta^1][\Delta^1]} L[\Delta^1][\Delta^1] \xrightarrow{H'[\Delta^1]} L[\Delta^1] \xrightarrow{H'} K \xrightarrow{p} K. \qquad \square$$

We will prove that every coherent homotopy idempotent in $\mathcal{C}_a^G$ splits up to homotopy.

Remark A.1.3. The author does not know if every homotopy idempotent in $\mathcal{C}_a^G$ is coherent. For the topogical case it is known that there are unpointed homotopy idempotents of infinite-dimensional CW-complexes which do not split, however every pointed homotopy idempotent as well as every homotopy idempotent of finite-dimensional CW-complexes splits, see [HH82].

A.2 Existence and properties of a mapping telescope

The results we show in this appendix are summarized in the following Theorem. Its proof will take the rest of this appendix.

Theorem A.2.1. *Let $\eta\colon K \to K$ be a homotopy idempotent in $\mathcal{C}_a^G(X)$. There is a construction* $\mathrm{Tel}(-)$ *which assigns to any homotopy idempotent η an object* $\mathrm{Tel}(\eta)$ *in $\mathcal{C}_a^G(X)$. It has the following properties.*

1. *There is a cellular inclusion $\iota\colon K \rightarrowtail \mathrm{Tel}(\eta)$*

2. *Let*

$$\begin{array}{ccc} A & \xrightarrow{\mu} & A \\ \downarrow f & & \downarrow f \\ K & \xrightarrow{\eta} & K \end{array}$$

be a strict commutative diagram of homotopy idempotents. Then f induces a map $f_*\colon \mathrm{Tel}(\mu) \to \mathrm{Tel}(\eta)$. This is functorial in f. In particular if f is an isomorphism then $\mathrm{Tel}(f)$ is an isomorphism.

3. *If $\eta, \mu\colon K \to K$ are homotopic homotopy idempotents then there is a homotopy equivalence*
$$\mathrm{Tel}(\eta) \xrightarrow{\simeq} \mathrm{Tel}(\mu).$$

4. *Consider the telescope $\mathrm{Tel}(\mathrm{id}_K)$ of the homotopy idempotent $\mathrm{id}_K\colon K \to K$. There is a map*
$$\mathrm{Tel}(\mathrm{id}_K) \to K$$
which is a homotopy equivalence.

5. *All maps in* (2) *to* (4) *are relative to* $\iota\colon K \to \mathrm{Tel}(\eta)$, *i.e., they commute with this cellular inclusion.*

6. *From* (2) *we get for* $\mu = \eta = f$ *an induced map* $\eta_*\colon \mathrm{Tel}(\eta) \to \mathrm{Tel}(\eta)$. *This map is a homotopy equivalence. If* η *is coherent,* η_* *is homotopic to* id.

7. *If* η *is coherent then there is a map* $c\colon \mathrm{Tel}(\eta) \to K$ *such that* $\iota \circ c$ *is homotopic to* $\eta_*\colon \mathrm{Tel}(\eta) \to \mathrm{Tel}(\eta)$ *and hence, by (6), to the identity on* $\mathrm{Tel}(\eta)$. *Therefore* $\mathrm{Tel}(\eta)$ *is a homotopy retract of* K. *Further* $c \circ \iota$ *is homotopic to* η *itself.*

Corollary A.2.2 (Coherent homotopy idempotents split). *Let* $\eta\colon K \to K$ *be a coherent homotopy idempotent in* $\mathcal{C}^G_{\mathrm{a}}$. *Then there is a* $B \in \mathcal{C}^G_{\mathrm{a}}$ *such that* K *is homotopy equivalent to* $\mathrm{Tel}(\eta) \vee B$. *Moreover under this equivalence* η *corresponds to the projection* $\mathrm{pr}\colon \mathrm{Tel}(\eta) \vee B \to \mathrm{Tel}(\eta) \to \mathrm{Tel}(\eta) \vee B$, *i.e., there is a homotopy commutative diagram*

$$\begin{array}{ccc} K & \xrightarrow{f} & \mathrm{Tel}(\eta) \vee B \\ \big\downarrow{\scriptstyle \eta} & & \big\downarrow{\scriptstyle \mathrm{pr}} \\ K & \xrightarrow{f} & \mathrm{Tel}(\eta) \vee B \end{array}$$

where f *is the homotopy equivalence* $K \xrightarrow{\simeq} \mathrm{Tel}(\eta) \vee B$.

Proof. We know by A.2.1 (7) that $\mathrm{Tel}(\eta)$ is a homotopy retract of K. We can make the homotopy commutative diagram

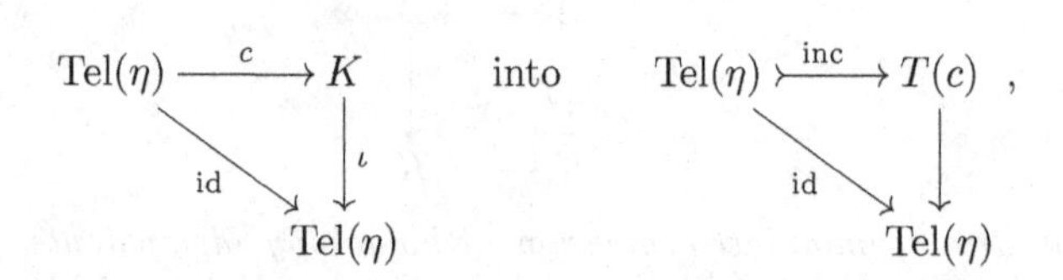

which is a strict commutative diagram, where $T(c)$ is the mapping cylinder of c and inc is a cellular inclusion. Take the cofiber of inc and call it B. The sum of the retraction $T(c) \to \mathrm{Tel}(\eta)$ and the quotient map $T(c) \to B$ gives a map $s\colon T(c) \to \mathrm{Tel}(\eta) \vee B$ (using that $\mathcal{C}^G$ is an additive category). The map makes the diagram of cofiber sequences

$$\begin{array}{ccccc} \mathrm{Tel}(\eta) & \rightarrowtail & T(c) & \twoheadrightarrow & B \\ \big\downarrow & & \big\downarrow{\scriptstyle s} & & \big\downarrow \\ \mathrm{Tel}(\eta) & \rightarrowtail & \mathrm{Tel}(\eta) \vee B & \twoheadrightarrow & B \end{array}$$

commutative and the Extension Axiom 6.5.2 shows that s is a homotopy equivalence. This gives the homotopy equivalence $f\colon K \to T(c) \to \mathrm{Tel}(\eta) \vee B$.

By A.2.1 (7) the map $\eta\colon K \to K$ factorizes up to homotopy as $c \circ i\colon K \to \mathrm{Tel}(\eta) \to K$. Hence the upper triangle in

$$\begin{array}{ccccc} K & \xrightarrow{\eta} & K & & \\ {\scriptstyle \iota}\downarrow & \overset{c}{\nearrow} & \downarrow & & \\ \mathrm{Tel}(\eta) & \rightarrowtail & T(c) & \twoheadrightarrow & B \end{array}$$

is homotopy commutative, whereas the lower one commutes strictly. It follows that $K \xrightarrow{\eta} K \to T(c) \to B$ is homotopic to the zero map.

Further $K \to T(c) \to \mathrm{Tel}(\eta)$ equals ι, hence by adding homotopies the map

$$f \circ \eta\colon K \to K \xrightarrow{\simeq} T(c) \xrightarrow{\simeq} \mathrm{Tel}(\eta) \vee B$$

is homotopic to $K \xrightarrow{\iota} \mathrm{Tel}(\eta) \rightarrowtail \mathrm{Tel}(\eta) \vee B$. As

$$\begin{array}{ccc} K & \xrightarrow{\iota} & \mathrm{Tel}(\eta) \vee B \\ {\scriptstyle \eta}\downarrow & & \downarrow{\scriptstyle \eta_* \vee 0_B} \\ K & \xrightarrow{\iota} & \mathrm{Tel}(\eta) \vee B \end{array}$$

is strictly commutative by A.2.1 (2) and (5), where 0_B denotes the zero map on B, and as $\eta_* \simeq \mathrm{id}$ by A.2.1 (6) it follows that

$$\begin{array}{ccc} K & \xrightarrow[f]{\simeq} & \mathrm{Tel}(\eta) \vee B \\ {\scriptstyle \eta}\downarrow & & \downarrow{\scriptstyle \mathrm{id} \vee 0_B} \\ K & \xrightarrow[f]{\simeq} & \mathrm{Tel}(\eta) \vee B \end{array}$$

is homotopy commutative and the claim follows. $\square$

The proof of Theorem A.2.1 is a bit involved. Because we need to handle concatenation of homotopies well, we need an analogue of "Moore homotopies", i.e., homotopies where we allow the "intervals" to have different lengths. We will introduce it in the next section.

A.3 Simplicial Intervals

We start by defining what we will mean by an *interval* in the category of simplicial sets. This is no common notion there. The name is chosen to stress the analogies to the topological setting. Basically we want a nice formal description of simplicial set of the form $\to\leftarrow$ etc.

Definition A.3.1 (Simplicial intervals).

1. Let $i \in \mathbb{N}$. A one-point simplicial set $I(i)$, $I(i)_k = \{i\}$, together with a bijection $l\colon I(i)_0 \to \{i\}$ from its zero simplices is a called a *point at* i or *interval of length* 0 *from* i *to* i.

2. An *interval of length* 1 *from* i *to* $(i+1)$, denoted $I(i,i+1)$, is a simplicial set isomorphic to Δ^1 together with a bijection of its zero simplices to the set $\{i,i+1\}$, $l\colon I(i,i+1)_0 \to \{i,i+1\}$. The map l is called the *labeling.*

3. Let $i,j \in \mathbb{N}$, $i+2 \leq j$. An *interval of length* $(i-j)$ *from* i *to* j is a simplicial set $I(i,j)$ together with a bijection $l\colon I(i,j)_0 \to \{i,i+1,\ldots,j\}$ such that there is a pushout diagram

$$\begin{array}{ccc} I(j-1) & \rightarrowtail & I(j-1,j) \\ \downarrow & & \downarrow \\ I(i,j-1) & \longrightarrow & I(i,j) \end{array} \tag{13}$$

where the maps are compatible with the labelings and $I(j-1) \rightarrowtail I(j-1,i)$, $I(j-1) \rightarrowtail I(i,j-1)$ are the obvious inclusions.

4. The *standard interval from* i *to* $(i+1)$ is the simplicial set Δ^1 together with the labeling $l(0)=i$, $l(1)=i+1$. The *standard interval from* i *to* j for $i+2 \leq j$ is the simplicial set arising from the standard interval from i to $j-1$ by the pushout (13) with $I(j-1,j)$ being the standard interval of length 1.

5. An interval $I(i,j)$ from i to j is called *ordered* if it is isomorphic to the standard interval from i to j and the isomorphism respects the labeling.

Remark A.3.2. We sometimes draw pictures for intervals. The standard interval is $0 \to 1$. The four intervals for $I(0,2)$ are the following ones:

$$\begin{array}{cc} 0 \to 1 \to 2 & 0 \to 1 \leftarrow 2 \\ 0 \leftarrow 1 \to 2 & 0 \leftarrow 1 \leftarrow 2. \end{array}$$

That the notion $I(i,j)$ is ambiguous is intentional, as we want to allow all those cases.

A.3.3 Concise notation We often just write $I(i,j)$ for an interval from i to j leaving all the other data understood. For $A \in \mathcal{C}_a^G$ we also often abbreviate $A[I(i,j)]$ as $A[i,j]$ and $A[I(i)]$ as $A[i]$, slightly misusing notation.

A.3.4 The infinite interval Define an simplicial set $I(i,\infty)$ to be an *interval from* i *to* ∞ if it is the filtered colimit (or union) of intervals $I(i,j)$ for $j \to \infty$. It is called *ordered* if each of the $I(i,j)$ is.

A.4 Long Homotopies

Our notion of interval gives rise to a notion of homotopy.

Definition A.4.1 (Long Homotopy). Let $I(0,j)$ be an interval from 0 to j. Let $f_0, f_j \colon A \to B$ be two maps in $\mathcal{C}_a^G$. A *(long) homotopy* from f_0 to f_j is a map $H \colon A[I(0,j)] \to B$ such that the restriction to $A[0]$ is f_0 and the restriction to $A[j]$ is f_j. We say that H has *length* j.

Example A.4.2. If $f \colon A \to B$ is a map in $\mathcal{C}_a^G$ and $I(0,i)$ any interval we always have the *constant* or *trivial homotopy* $\mathrm{Tr} \colon A[0,i] \to B$ induced by the map $A[0,i] \to A \to B$. We also define it for $i = 0$ and therefore call the map $\mathrm{Tr} \colon A[0,0] = A[0] = A \xrightarrow{f} B$ the trivial homotopy of length 0.

A.4.3 Ordinary and long homotopies Every homotopy in the usual sense in a long homotopy of length 1. Every long homotopy gives a homotopy in the usual sense by the Kan property. This is not functorial, which is the reason why we need to consider long homotopies. We will omit the "long" in the following.

A.4.4 Concatenation of intervals If $I(0,i)$ and $I(0,j)$ are intervals we define the concatenation $I(0,i) \,\square\, I(0,j)$ to be the pushout

$$\begin{array}{ccc} I(i) & \longrightarrow & I(i,i+j) \\ \downarrow & & \downarrow \\ I(0,i) & \longrightarrow & I(0,i) \,\square\, I(0,j) \end{array}$$

where $I(i,i+j)$ is defined as a "relabeling" of $I(0,j)$, replace the labeling l of $I(0,j)$ by $l(k) = i + k$.

A.4.5 Concatenation of homotopies Homotopies which agree on the start resp. endpoint can be concatenated. For $H_1 \colon A[0,i] \to B$, $H_2 \colon A[0,j] \to B$ with $H_{1|A[i]} = H_{2|A[0]}$ define the *concatenation*

$$H_1 \,\square\, H_2 \colon A[0, i+j] \to B$$

as the map induced by the identification on the pushout $I(0,i) \,\square\, I(0,j)$. The concatenation of homotopies is strictly associative.

A.4.6 Inverse homotopies If $I(0,j)$ is an interval, define the *reversed interval* $\overline{I(0,j)}$ as the same simplicial set with the labeling l replaced by $\bar{l}(k) := j - l(k)$. If $H \colon A[I(0,j)] \to B$ is a homotopy the *inverse homotopy* $\overline{H}$ is the obvious map $\overline{H} \colon A[\overline{I(0,j)}] \to B$.

If $j = 1$ and $I(0,j)$ is an ordered interval we draw the homotopy as $\xrightarrow{\;H\;}$ and the inverse homotopy as $\xleftarrow{\;H\;}$.

Lemma A.4.7 (Concating a homotopy and its inverse). *Let $H\colon A[0,1] \to B$ be a homotopy. The concatenation $H \square \overline{H}$ is homotopic, relative boundary, to the constant (or trivial) homotopy* $\mathrm{Tr}\colon A[0,2] \to A \to B$.

Proof. Assume that $I(0,1)$ is the standard interval, the other case proceeds similar. The homotopy $A[0,2][\Delta^1] \to B$ is given by the left diagram below. It is constructed by glueing the 2-simplices together which are shown on the right. These arise from the 2nd degeneracy map $\Delta^2 \to \Delta^1$.

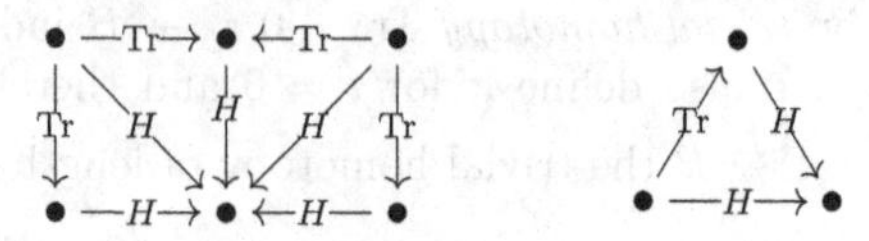

□

A.4.8 Variations of the previous lemma In the proof of the lemma we gave a homotopy from the constant homotopy to the given one. We can give one in the other direction by a similar proof where we use a map $A[\Delta^2] \to B$ arising from the Kan extension property.

Of course also $\overline{H} \square H$ is homotopic to the trivial homotopy. The lemma also holds if we allow an interval of length n instead of length 1. To prove this one does induction over n and starts building the homotopy from the middle, patching in in each induction step some of four trivial homotopies of the remaining homotopies H' of length 1, like

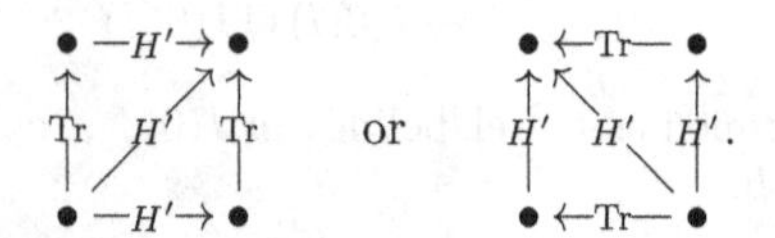

These techniques will work in the more complicated situations later, hence we will tacitly only draw the diagrams for length 1 homotopies in the following.

A.4.9 Intervals of different length are homotopy equivalent For $A \in \mathcal{C}_a^G$ the modules $A[0,i]$, $(i \in \mathbb{N})$, are homotopy equivalent to A and the homotopies are relative to the endpoints. This means, that the restriction of both homotopies to the endpoints gives the constant homotopy there. This allows us to glue the homotopy equivalences together later. The result will follow from the following lemma.

Lemma A.4.10. *Let Λ_i^2 be the ith horn and d_i the ith face of Δ^2. Then $A[d_i]$ and $A[\Lambda_i^2]$ are homotopy equivalent relative the 0-simplices of d_i. The homotopy equivalence can be chosen to be one of the maps $A[\Lambda_i^2] \to A[d_i]$ which induced by collapsing one 1-simplex.*

Proof. Consider the composition

$$A[\Lambda_i^2] \to A[\Delta^2] \to A[d_i]$$

where the first map is the inclusion and the last map (and hence the composition) can is induced by any map collapsing a 1-simplex not equal to d_1. It is not hard to see that the first map has a deformation retraction by horn-filling. The second map is induced by a deformation retraction of simplicial sets. Therefore the composition is a homotopy equivalence relative to the 0-simplices of d_i. □

Corollary A.4.11. *Let* $I(0,1)$ *be the standard interval and* $I(0,i)$ *any interval. Then we have a homotopy equivalence relative endpoints*

$$A[0,1] \simeq A[0,i].$$

Proof. Lemma A.4.10 implies $A[0,2] \simeq A[0,1]$ relative endpoints if there is a projection $I(0,2) \to I(0,1)$. It also implies $A[\rightarrow] \simeq A[\leftarrow]$ relative endpoints by the chain $A[\rightarrow] \simeq A[\rightarrow\leftarrow] \simeq A[\leftarrow]$ of homotopy equivalences relative endpoints. The corollary follows by induction. □

A.4.12 Homotopies of infinite length In the following we assume the infinite interval $I(0,\infty)$ to be ordered for simplicity (cf. A.3.4). We abbreviate $A[I(0,\infty)]$, $A \in \mathcal{C}_a^G$, as $A[0,\infty)$. We suggestively call a map $A[0,\infty) \to B$ a homotopy *of infinite length.* From such a homotopy we want to get a homotopy of length 1. In general, this is of course impossible. But if the homotopy is "convergent" in the sense below this can be done.

Lemma A.4.13 (Convergent Homotopy Lemma). *Let* $H\colon A[0,\infty) \to B$ *be a* convergent homotopy, *this means we assume:*

1. *There is a filtration* $A_0 \subseteq A_1 \subseteq \cdots \subseteq A_n \subseteq \cdots \subseteq A$ *by cellular submodules such that* $\bigcup_i A_i = A$.
2. *For each* A_i *there is an* n_i *such that* $H_{|A_i[n_i,\infty)}$ *is the constant homotopy* Tr *(cf. A.4.2).*

Then there exists a homotopy $G\colon A[\Delta^1] \to B$ *with* $G_{|A[0]} = H_{|A[0]}$ *and* $G_{|A_i[1]} = H_{|A_i[n_i]}$.

Remark A.4.14. Recall that $A_i[n_i]$ and $A_i[n_i,\infty)$ denote obvious cellular submodules of $A[0,\infty)$. We can and will assume in the proof that $n_{i+1} \geq n_i$.

This lemma is well-known in the topological case. G may be called the "limit" of the homotopy H.

Proof of Lemma A.4.13. Recall that we assumed $I(0,\infty)$ to be an ordered interval. We first enlarge $I(0,\infty)$ to a new simplicial set $\widehat{I(0,\infty)}$ by filling some horns.

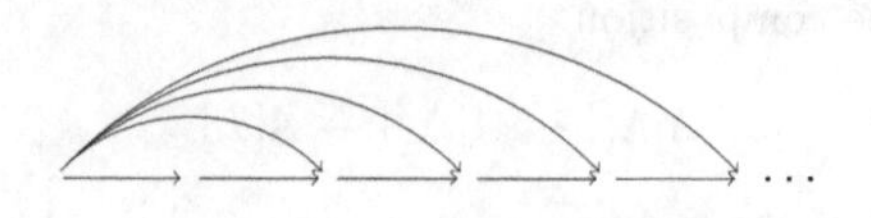

Figure 1: A sketch of $\widehat{I(0,\infty)}$.

The subsimplicial set $I(0,2)$ is isomorphic to the horn Λ^2_1. We take the pushout of $\Delta^2 \leftarrowtail \Lambda^2_1 \rightarrowtail I(0,\infty)$ and call it $\widehat{I(0,2)}$. It has an extra 1-simplex with boundaries $I(0)$ and $I(2)$ in $I(0,\infty)$, which we call $(0 \to 2)$.

In $\widehat{I(0,2)}$ the 1-simplices $(0,\to 2)$ and $I(2,3)$ constitute a horn Λ^2_1. Like before we define $\widehat{I(0,3)}$ to be the following pushout

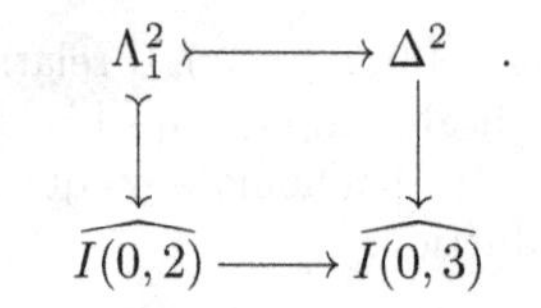

Again it has an extra 1-simplex with boundaries $I(0)$ and $I(3)$ in $I(0,\infty)$, which we call $(0 \to 3)$. Now we proceed by induction and define $\widehat{I(0,\infty)}$ as the filtered colimit $\widehat{I(0,\infty)} := \bigcup_n \widehat{I(0,n)}$. Figure 1 sketches a picture of $\widehat{I(0,\infty)}$ with $I(0,\infty)$ being the bottom line.

We have $\widehat{A[0,\infty)} = \bigcup_n \widehat{A[0,n]}$, with $\widehat{A[0,n]}$ and $\widehat{A[0,\infty)}$ being abbreviations for $A[\widehat{I(0,n)}]$ and $A[\widehat{I(0,\infty)}]$, respectively. Note that $\widehat{A[0,n]}$ arises from $\widehat{A[0,n-1]}$ by horn-filling. We want to construct a certain map $\widehat{H}\colon \widehat{A[0,\infty)} \to B$ which extends H.

We do induction over i. Assume that we have constructed a homotopy $G_i\colon A_i\widehat{[0,\infty)} \to B$ which extends $H\colon A_i[0,\infty) \to B$ and has the property that $G_{i|A_i[(0\to n)]} = G_{i|A_i[(0\to n_i)]}$ for all $n \geq n_i$. By iterating the relative horn-filling property 6.2.1 we can extend G_i to a map $A_{i+1}\widehat{[0,n_{i+1}]} \cup A_i\widehat{[0,\infty)} \to B$.

For $n \geq n_{i+1}$ we do not want to apply the relative horn-filling property as we need special fillings. By assumption $H_{|A_{i+1}[n,n+1]}$ is the constant homotopy Tr. Hence the (relative) horn spanned by $A_{i+1}[(0 \to n)]$ and $A_{i+1}[n,n+1]$, given by $A_{i+1}[\Lambda^2_1] \to A_{i+1}\widehat{[0,n]} \cup A_i\widehat{[0,\infty)}$, can be filled in the following way

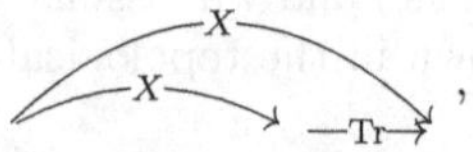

$$\text{(triangle with edges } X,\ X,\ \mathrm{Tr}) \, , \tag{14}$$

where X is the homotopy coming from the previous horn-fillings. This defines a map $G_{i+1}\colon A_{i+1}\widehat{[0,\infty)}$ with $G_{i+1|A[(0\to n)]} = G_{i+1|A[(0\to n_{i+1})]}$ for all $n \geq n_{i+1}$. This shows the induction step.

Taking the colimit over G_i we get a map $\widehat{H}\colon \widehat{A[0,\infty)} \to B$. We now define $G_{|A_j}\colon A_j[\Delta^1] \to B$ as the restriction of $\widehat{H}$ (or equivalently G_j) to $A_j[(0 \to n_j)]$, i.e., to the 1-simplex from 0 to n_j. This is compatible with the inclusion $A_j \to A_{j+1}$ and thus the colimit over j gives the desired homotopy $G\colon A[\Delta^1] \to B$. □

Corollary A.4.15. *The map $i\colon A \to A[0,\infty)$ has a deformation retraction, so in particular i is a homotopy equivalence.*

Proof. The map $[0,\infty) \to 0$ induces a retraction $r\colon A[0,\infty) \to A$ for i. We have to prove that the composition $i \circ r\colon A[0,\infty) \to A \to A[0,\infty)$ is homotopic to the identity. We use the Convergent Homotopy Lemma A.4.13. Define the convergent homotopy $H\colon A[0,\infty)[0,\infty) \to A[0,\infty)$ as the map induced by the map

$$(i,j) \mapsto \min(i,j)$$

where we use that map $[0,\infty) \times [0,\infty) \to [0,\infty)$ is determined on the 0-simplices. We regard j as the homotopy direction. This map has the following properties:

1. For $j = 0$ it is the projection to 0, hence the map $i \circ r$.
2. For any $j \geq i$ the map $A[0,i][j] \to A[0,i]$ is the identity.
3. $\bigcup_i A[0,i]$ is a filtration of $A[0,\infty)$ by cellular modules.

Now the Convergent Homotopy Lemma A.4.13 applies and hence we get a homotopy $G\colon A[0,\infty)[\Delta^1] \to A[0,\infty)$ from $i \circ r$ to the identity. □

A.5 On the mapping telescope

In the following $I = I(0,i)$ is always an interval and $f\colon A \to A$ a map in $\mathcal{C}_a^G$.

A.5.1 Long mapping cylinder Define the mapping cylinder $M^I(f)$ for I of f like for $I = \Delta^1$ in 6.1.3. We have the analogous notions of front and back inclusion as well as projection. We modified the notation because the mapping cylinder will play a slightly different role than in 6.1.3 and because we need to keep track of the interval.

Definition A.5.2 (Long Mapping Telescope). If $I = I(0,1)$ define $\mathrm{Tel}^I(f)$, the *mapping telescope of $f\colon A \to A$ for the interval I* as the pushout

$$\begin{array}{ccc} \coprod_{i=1}^{\infty} A[i] \amalg \coprod_{i=0}^{\infty} A[i+1] & \xrightarrow{\iota_0 \amalg \iota_1} & \coprod_{i=0}^{\infty} M^{I(i,i+1)}(f) . \\ \big\downarrow c & & \big\downarrow \\ \coprod_{i=1}^{\infty} A[i] & \longrightarrow & \mathrm{Tel}^I(f) \end{array}$$

In the diagram ι_1 is the back inclusion of $A[i+1]$ into $M^{I(i,i+1)}(f)$, ι_0 the front inclusion, both from the ith summand to the ith summand. c is the map which maps the summand $A[i]$ to $A[i]$.

For a general interval I the definition works analogously, if we delete "$[i]$" etc. in the diagram above. (Which makes it less sugestive.)

A.5.3 Front inclusion The front inclusion into the first mapping cylinder $\iota_0 A\colon M^I(f)$ (which is not used in the diagram above) gives a map

$$\iota\colon A \to \mathrm{Tel}^I(f)$$

which is a called the *front inclusion* of the mapping cylinder.

Remark A.5.4. The telescope consists of infinitely many mapping cylinders plugged together on the right. Each mapping cylinder has the same interval structure. We used that the countable coproducts exists in $\mathcal{C}_a^G$.

Using Corollaries A.4.11 and A.4.15 we obtain the following lemma.

Lemma A.5.5. *Recall Δ^1 is the standard interval.*

1. *The mapping cylinders for I and Δ^1 are homotopy equivalent relative to the front and the back inclusion: $M^I(f) \simeq M^{\Delta^1}(f) = T(f)$.*

2. *The mapping telescopes for I and Δ^1 are homotopy equivalent: $\mathrm{Tel}^I(f) \simeq \mathrm{Tel}^{\Delta^1}(f)$.*

3. *The telescope $\mathrm{Tel}^I(\mathrm{id}_A)$ of the identity is homotopy equivalent to A, the homotopy equivalence is given by the projection to A.*

Each map $I \to \Delta^1$ respecting the endpoints can be chosen to induce the first two homotopy equivalences. □

Remark A.5.6. While we can give the homotopy equivalences quite explicit, the inverse is not canonical and not easy to write down as we used the Kan Extension property to construct it.

A.5.7 First part of Theorem A.2.1 We can apply this section to an idempotent $\eta\colon K \to K$. Define

$$\mathrm{Tel}(\eta) := \mathrm{Tel}^{\Delta^1}(\eta)$$

Then Theorem A.2.1 (1) and (2) are satisfied, by the definition of ι in A.5.3 and the functoriality of mapping cylinders and pushouts along cellular inclusions.

A.5.8 Homotopy commutative squares For the rest of Theorem A.2.1 we need to know more about homotopy commutative diagrams. They will induce a map of (long) mapping telescopes, but it is only "functorial", in some sense we make precise, if we allow to change the intervals.

Definition A.5.9 (Homotopy commutative square). A square in $\mathcal{C}_a^G$

$$\begin{array}{ccc} A & \xrightarrow{f} & A \\ \downarrow{\scriptstyle a} & & \downarrow{\scriptstyle a} \\ B & \xrightarrow{g} & B \end{array} \tag{15}$$

is *homotopy commutative* if there is an interval $I = I(0,i)$ and a *specified homotopy* $H^a\colon A[0,i] \to B$ which goes from $g \circ a$ to $a \circ f$. This should mean $H^a{}_{|A[0]} = g \circ a$ and $H^a{}_{|A[i]} = a \circ f$.

Remark A.5.10. The homotopy of a homotopy commutative square always goes from the lower left corner to the upper right, it is helpful to visualize this as

$$\begin{array}{ccc} A & \xrightarrow{f} & A \\ \downarrow{\scriptstyle a} & \nearrow^{H} & \downarrow{\scriptstyle a} \\ B & \xrightarrow{g} & B \end{array}$$

when thinking about the homotopies. We chose the direction of the homotopy such that it will fit together with our definition of mapping cylinder.

The next observation is central for the rest of the proof.

Lemma/Definition A.5.11 (Stacking squares). Homotopy commutative squares can be composed (stacked). Given two homotopy commutative squares

$$\begin{array}{ccc} A & \xrightarrow{f} & A \\ \downarrow{\scriptstyle a} & & \downarrow{\scriptstyle a} \\ B & \xrightarrow{g} & B \end{array}\ , \qquad \begin{array}{ccc} B & \xrightarrow{g} & B \\ \downarrow{\scriptstyle b} & & \downarrow{\scriptstyle b} \\ C & \xrightarrow{h} & C \end{array} \tag{16}$$

with homotopies H^a, H^b using intervals I^a, I^b, then composed (stacked) square

$$\begin{array}{ccc} A & \xrightarrow{f} & A \\ \downarrow{\scriptstyle b\circ a} & & \downarrow{\scriptstyle b\circ a} \\ C & \xrightarrow{h} & C \end{array}$$

is homotopy commutative with homotopy

$$(H^b \circ a[I^b]) \square (b \circ H^a)\colon A[I^b \square I^a] \to C. \quad \square$$

A.5.12 Stacking is associative Composition (stacking) of homotopy commutative squares is strictly associative, because concatenation of homotopies is. The length of the homotopies add. If the square is strictly commutative we can and hence will assume that the "homotopy" has length 0.

Note that we only "compose in one direction" of the two possible directions in which the square could be "stacked". The reason is simply, that this is the only case we are interested in.

Lemma/Definition A.5.13 (Homotopy commutative squares and mapping cylinders)**.** Let I be an interval. A homotopy commutative square

$$\begin{array}{ccc} A & \xrightarrow{f} & A \\ \downarrow{\scriptstyle a} & & \downarrow{\scriptstyle a} \\ B & \xrightarrow{g} & B \end{array} \tag{17}$$

with homotopy $H\colon A[I] \to B$ induces a map called $(H,a)_*\colon M^I(f) \to B$ such that the diagram

$$\begin{array}{ccccc} A & \xrightarrow{\iota_0} & M^I(f) & \xleftarrow{\iota_1} & A \\ \downarrow{\scriptstyle a} & & \downarrow & & \downarrow{\scriptstyle a} \\ B & \xrightarrow{g} & B & \xleftarrow{\mathrm{id}_B} & B \end{array} \tag{18}$$

commutes (strictly). Here ι_0 is the front and ι_1 the back inclusion. Each such diagram determines uniquely the homotopy of (17).

Proof. The pushout of the strictly commutative diagram

$$\begin{array}{ccccc} A[I] & \xleftarrow{\iota_1} & A[i] & \xrightarrow{f} & A \\ \downarrow{\scriptstyle H} & & \downarrow{\scriptstyle a\circ f} & & \downarrow{\scriptstyle a} \\ B & \xleftarrow{\mathrm{id}_B} & B & \xrightarrow{\mathrm{id}_B} & B \end{array}$$

gives the map $(H,a)_*\colon M^I(f) \to B$ and then diagram (18) commutes. Conversely taking the map $A[I] \to M^I(f) \to B$ gets back the homotopy H and the commutativity of (18) shows that H makes the square (17) homotopy commutative. □

A.5.14 Induced map on longer mapping cylinders Let J be another interval. Taking the mapping cylinder with J of the rows of the left-hand square of (18) gives a map $a[J] \,\square (H,a)_*\colon M^{J \,\square\, I}(f) \to M^J(g)$ such that the

diagram

$$\begin{array}{ccccc} A & \xrightarrow{\iota_0} & M^{J \Box I}(f) & \xleftarrow{\iota_1} & A \\ \Big\downarrow{\scriptstyle a} & & \Big\downarrow & & \Big\downarrow{\scriptstyle a} \\ B & \xrightarrow{\iota_0} & M^{J}(g) & \xleftarrow{\iota_1} & B \end{array} \tag{19}$$

commutes. We call the induced maps the *cylinder maps* of the homotopy commutative diagram, resp. the *cylinder maps with respect to* J. Basically, we make the cylinders longer by glueing in $A[J]$.

A.5.15 Composition of induced map We want to compose the map $(H, a)_*$, which should correspond to stacking homotopy commutative squares. Stacking squares makes the homotopies longer, so we will not have a composition on the nose, but only after changing the first map. The composition is visualized in Figure 2 below.

Definition A.5.16 (Composition). Given maps $f\colon A \to A$, $g\colon B \to B$ and $h\colon C \to C$ as well as $a\colon A \to B$ and $b\colon B \to C$. Assume we have cylinder maps $(H^a, a)_*\colon M^{I^a}(f) \to B$ and $(H^b, b)_*\colon M^{I^b}(g) \to C$ like in Definition A.5.13 satisfying diagrams like (18). Define the *"composition"* $(H^a, a)_* \boxdot (H^b, b)_*$ as

$$M^{I^b \Box I^a}(f) \xrightarrow{a[I^b] \Box (H^a, a)_*} M^{I^b}(g) \xrightarrow{(H^b, b)_*} C$$

More generally let J be another interval. Assume we have cylinder maps with respect to J

$$a[J] \Box (H^a, a)_*\colon M^{J \Box I^a}(f) \to M^J(g)$$

and

$$b[J] \Box (H^b, b)_*\colon M^{J \Box I^b}(g) \to M^J(h)$$

Define the *"composition"* as

$$\Big(a[J] \Box (H^a, a)_*\Big) \boxdot \Big(b[J] \Box (H^b, b)_*\Big)\colon$$
$$M^{J \Box I^b \Box I^a}(f) \xrightarrow{a[J] \Box a[I^b] \Box (H^a, a)_*} M^{J \Box I^b}(g) \xrightarrow{b[J] \Box (H^b, b)_*} M^J(h)$$

Lemma A.5.17. *Given two homotopy commutative squares*

$$\begin{array}{ccc} A & \xrightarrow{f} & A \\ \Big\downarrow{\scriptstyle a} & & \Big\downarrow{\scriptstyle a} \\ B & \xrightarrow{g} & B \end{array} , \qquad \begin{array}{ccc} B & \xrightarrow{g} & B \\ \Big\downarrow{\scriptstyle b} & & \Big\downarrow{\scriptstyle b} \\ C & \xrightarrow{h} & C \end{array}$$

with homotopies $H^a\colon A[I^a] \to B$, $H^b\colon B[I^b] \to C$. Then the cylinder map of the stacked homotopy commutative square (cf. A.5.11)

$$\begin{array}{ccc} A & \xrightarrow{f} & A \\ \downarrow{\scriptstyle b\circ a} & & \downarrow{\scriptstyle b\circ a} \\ C & \xrightarrow{h} & C \end{array}$$

is equal to the "composition" of the cylinder maps of the individual squares, i.e.,

$$\big((H^b \circ a[I^b]) \,\Box (b \circ H^a), b \circ a\big)_* = (H^b, b)_* \circ \Big(a[I^b] \,\Box (H^a, a)_*\Big)$$

The same is true for cylinder maps with respect to J.

Proof. We have to check the equality of two maps $M^{J \Box I^b \Box I^a}(f) \to M^J(h)$. Figure 2 shows the situation. With its help for the bookkeeping the equality can be checked directly. □

A.5.18 Induced maps on telescopes Everything from A.5.13 on transfers immediately to mapping telescopes, by glueing the parts together. In particular if we have homotopy commutative squares like in (A.5.17) we obtain a maps

$$(H^a, a)_*\colon \mathrm{Tel}^{J \Box I^a}(f) \to \mathrm{Tel}^J(g),$$
$$(H^b, b)_*\colon \mathrm{Tel}^{J \Box I^b}(g) \to \mathrm{Tel}^J(h).$$

and their "composition"

$$(H^b, b)_* \boxdot (H^a, a)_*\colon \mathrm{Tel}^{J \Box I^b \Box I^a}(f) \to \mathrm{Tel}^J(h)$$

which is the same as the induced map of the stacking of the homotopy commutative squares. The map $(H, a)_*$ commutes with the front inclusion ι from A.5.3.

We can specialize to H being the trivial homotopy of length 0 and $J := \Delta^1$. Then we get the strict functoriality of $\mathrm{Tel}^{\Delta^1}(-)$ from Theorem A.2.1 (2).

One the other hand we can consider the square given by $f = g$ and $a = \mathrm{id}_A$ being homotopy commutative with trivial homotopy $\mathrm{Tr}\colon A[I] \to A$, for I any interval. The resulting map $(\mathrm{Tr}, \mathrm{id})_*\colon \mathrm{Tel}^{\Delta^1 \Box I}(f) \to \mathrm{Tel}^{\Delta^1}(f)$ is induced by the projection $\Delta^1 \Box I \to \Delta^1$ mapping I to $I(1) \subseteq \Delta^1$.

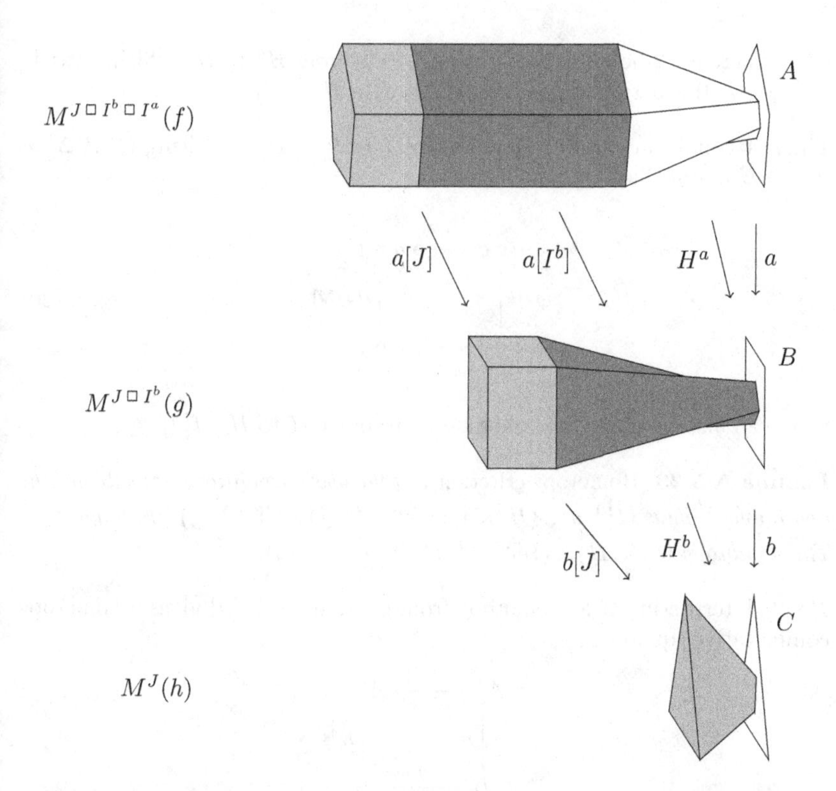

Figure 2: Composition of maps of long mapping cylinders. Shows the mapping cylinder construction is “functorial” if performed with long cylinders.

A.5.19 A homotopy criterion We need a criterion when two homotopy commutative squares

$$\begin{array}{ccc} A & \xrightarrow{f} & A \\ \downarrow{\scriptstyle a} & & \downarrow{\scriptstyle a} \\ B & \xrightarrow{g} & B \end{array} \qquad \text{and} \qquad \begin{array}{ccc} A & \xrightarrow{f} & A \\ \downarrow{\scriptstyle \widetilde{a}} & & \downarrow{\scriptstyle \widetilde{a}} \\ B & \xrightarrow{g} & B \end{array}$$

with homotopies H^a and $H^{\widetilde{a}}$ induce homotopic maps on mapping telescopes.

We make the following assumptions.

1. H^a and $H^{\widetilde{a}}$ have the same length and are indexed over the same interval $I = I(0, i)$. We can arrange this by extending with trivial homotopies.

2. There is a homotopy $H \colon A[J] \to B$ from a to $\widetilde{a}$.

3. There is "2-homotopy" $G\colon A[I][J] \to B$ from H^a to $H^{\widetilde{a}}$ which restricts on $I(0) \times J$ to $g \circ H$ and on $I(i) \times J$ to $H \circ f[J]$.

The last condition can be visualized for $I = J = \Delta^1$ by writing $G\colon A[\Delta^1 \times \Delta^1] \to B$ in our diagram language as

$$\begin{array}{ccc} g \circ a & \xrightarrow{H^a} & a \circ f \\ {\scriptstyle g\circ H}\downarrow & \searrow & \downarrow{\scriptstyle H\circ f[J]} \\ g \circ \widetilde{a} & \xrightarrow[H^{\widetilde{a}}]{} & \widetilde{a} \circ f \end{array} . \tag{20}$$

So G can also be viewed as a homotopy from $g \circ H$ to $H \circ f[J]$.

Lemma A.5.20 (Homotopy criterion)**.** *If the above conditions are satisfied the two induced maps* $(H^a, a)_*, (H^{\widetilde{a}}, \widetilde{a})_*\colon \mathrm{Tel}^{\Delta^1 \square I}(f) \to \mathrm{Tel}^{\Delta^1}(g)$ *are homotopic. The homotopy is* $(G, H)_*\colon \mathrm{Tel}^{\Delta^1 \square I}(f)[J] \to \mathrm{Tel}^{\Delta^1}(g)$.

Proof. Interpreting G as homotopy from $g \circ H$ to $H \circ f[J]$ gives a homotopy commutative square

$$\begin{array}{ccc} A[J] & \xrightarrow{f[J]} & A[J] \\ {\scriptstyle H}\downarrow & & \downarrow{\scriptstyle H} \\ B & \xrightarrow{g} & B \end{array}$$

with homotopy $G\colon (A[J])[I] \to B$. We get an induced map

$$(G, H)_*\colon \mathrm{Tel}^{\Delta^1 \square I}(f[J]) \to \mathrm{Tel}^{\Delta^1}(g).$$

As the telescope is a colimit it commutes with adjoining an interval, hence we can write the domain of the induced map as $\mathrm{Tel}^{\Delta^1 \square I}(f)[J]$. Therefore $(G, H)_*$ is a homotopy. We leave it to the reader to check that it is the desired one. □

Remark A.5.21. Lemma A.5.20 is a main tool in the following to analyze maps between telescopes. Thanks to the lemma we only need give diagrams like (20) to prove that certain maps on telescopes are homotopic. To simplify the diagram we will usually assume that all intervals have length 1 and are ordered. In A.4.8 we explained how to get to longer intervals from this.

Lemma A.5.22 (Homotopic maps between telescopes)**.** *Let* $f, g\colon A \to A$ *be homotopic maps. Then* $\mathrm{Tel}^{\Delta^1}(f)$ *and* $\mathrm{Tel}^{\Delta^1}(g)$ *are homotopy equivalent. The homotopy equivalences are relative to the front inclusions.*

Proof. Let $H\colon A[I] \to A$ be the homotopy from g to f. We get two homotopy commutative squares

$$\begin{array}{ccc} A & \xrightarrow{f} & A \\ \downarrow{\scriptstyle \mathrm{id}} & & \downarrow{\scriptstyle \mathrm{id}} \\ A & \xrightarrow{g} & A \end{array} \quad \text{and} \quad \begin{array}{ccc} A & \xrightarrow{g} & A \\ \downarrow{\scriptstyle \mathrm{id}} & & \downarrow{\scriptstyle \mathrm{id}} \\ A & \xrightarrow{f} & A \end{array}$$

with homotopies $H\colon A[I] \to A$ and $\overline{H}\colon A[\overline{I}] \to A$ (where the latter is the "inverse" homotopy, cf. Section A.4.5). This gives maps $(H, \mathrm{id})_*\colon \mathrm{Tel}^{\Delta^1 \square I}(f) \to \mathrm{Tel}^{\Delta^1}(g)$ and $(\overline{H}, \mathrm{id})_*\colon \mathrm{Tel}^{\Delta^1 \square \overline{I}}(g) \to \mathrm{Tel}^{\Delta^1}(f)$. The "composition"

$$(\overline{H}, \mathrm{id})_* \boxdot (H, \mathrm{id})_*\colon \mathrm{Tel}^{\Delta^1 \square \overline{I} \square I}(f) \to \mathrm{Tel}^{\Delta^1}(f)$$

(cf. A.5.18) is induced by the homotopy commutative square

$$\begin{array}{ccc} A & \xrightarrow{f} & A \\ \downarrow{\scriptstyle \mathrm{id}} & & \downarrow{\scriptstyle \mathrm{id}} \\ A & \xrightarrow{f} & A \end{array}$$

with homotopy $\overline{H} \square H\colon A[\overline{I} \square I] \to A$. By Lemma A.4.7 that homotopy is homotopic relative endpoints to the trivial homotopy, hence Lemma A.5.20 shows that $(\overline{H}, \mathrm{id})_* \boxdot (H, \mathrm{id})_*$ is homotopic to $(\mathrm{Tr}, \mathrm{id})_*$. The same holds for the other composition.

As $(\mathrm{Tr}, \mathrm{id})_*\colon \mathrm{Tel}^{\Delta^1 \square I}(f) \to \mathrm{Tel}^{\Delta^1}(f)$ is a homotopy equivalence induced by the projection $\Delta^1 \square I \to \Delta^1$ we get two homotopy commutative triangles

$$\begin{array}{ccc} \mathrm{Tel}^{\Delta^1 \square I}(f) & \xrightarrow{\simeq} & \mathrm{Tel}^{\Delta^1}(f) \\ \downarrow{\scriptstyle (H,\mathrm{id})_*} & \swarrow{\scriptstyle \varphi} & \\ \mathrm{Tel}^{\Delta^1}(g) & & \end{array} \quad \text{and} \quad \begin{array}{ccc} \mathrm{Tel}^{\Delta^1 \square \overline{I}}(g) & \xrightarrow{\simeq} & \mathrm{Tel}^{\Delta^1}(g) \\ \downarrow{\scriptstyle (\overline{H},\mathrm{id})_*} & \swarrow{\scriptstyle \psi} & \\ \mathrm{Tel}^{\Delta^1}(f) & & \end{array}$$

where φ is defined using a chosen homotopy inverse of the horizontal map and ψ similarly. We claim both compositions of these maps are homotopic to the identity. Together with these triangles we get a large diagram

$$\begin{array}{ccccc} \mathrm{Tel}^{\Delta^1 \square \overline{I} \square I}(f) & \xrightarrow{\simeq} & \mathrm{Tel}^{\Delta^1 \square I}(f) & \xrightarrow{\simeq} & \mathrm{Tel}^{\Delta^1}(f) \\ \downarrow{\scriptstyle (H,\mathrm{id})_*} & & \downarrow{\scriptstyle (H,\mathrm{id})_*} & \swarrow{\scriptstyle \varphi} & \\ \mathrm{Tel}^{\Delta^1 \square \overline{I}}(g) & \xrightarrow{\simeq} & \mathrm{Tel}^{\Delta^1}(g) & & \\ \downarrow{\scriptstyle (\overline{H},\mathrm{id})_*} & \swarrow{\scriptstyle \psi} & & & \\ \mathrm{Tel}^{\Delta^1}(f) & & & & \end{array}$$

where the square is strictly commutative. The left vertical composition is the "composition" $(\overline{H}, \mathrm{id})_* \boxdot (H, \mathrm{id})_*$ which is homotopic to $(\mathrm{Tr}, \mathrm{id})_*$, which is exactly the composition of the upper horizontal maps. It follows that $\psi \circ \varphi \simeq \mathrm{id}$ and similar for the other composition.

As the homotopy inverses in the definition of ψ and φ can be chosen to respect the front inclusion by Lemma A.5.5 and all other maps and homotopies are relative to it φ is a homotopy equivalence relative to the front inclusion. □

A.5.23 The shift map Recall that $\mathrm{Tel}^I(f)$ is a quotient of $\coprod_{n\in\mathbb{N}} M^I(f)$. The map taking the nth component to the $(n+1)$st component is compatible with the quotient, hence induces a map $\mathrm{Tel}^I(f) \to \mathrm{Tel}^I(f)$, which we will call the *shift map* and denote it by sh.

Lemma A.5.24. *Let I be an interval, $f\colon A \to A$ a self-map. The maps* sh *and* $(\mathrm{Tr}, f)_*$ *from* $\mathrm{Tel}^I(f)$ *to* $\mathrm{Tel}^I(f)$ *are homotopy inverse:*

$$(\mathrm{Tr}, f)_* \circ \mathrm{sh} = \mathrm{sh} \circ (\mathrm{Tr}, f)_* \simeq \mathrm{id}\colon \mathrm{Tel}^I(f) \to \mathrm{Tel}^I(f)$$

Sketch of proof. The first equality is clear. For the homotopy one restricts the map of telescopes to a map $M^I(f) \to M^I(f) \cup_A M^I(f)$, which maps into the second summand. Then one can construct a simplicial homotopy from this map to the map "inclusion of the first summand". Namely, the two inclusions $I \to I \square I$ give two maps $A[I] \to A[I \square I]$ which are homotopic by "sliding". The desired homotopy arises from this homotopy. The details are left to the reader. □

A.5.25 Telescopes of coherent homotopy idempotents Parts (1) to (5) of Theorem A.2.1 follow from the previous discussion. We will give a summary later on, but for the moment note that we have not even used that $f\colon A \to A$ is a homotopy idempotent. For the following we need to consider coherent homotopy idempotents.

Lemma A.5.26. *Let $\eta\colon K \to K$ be a* coherent *homotopy idempotent in $\mathcal{C}^G_a$. Then the induced map $\eta_* = (\mathrm{Tr}, \eta)_*\colon \mathrm{Tel}^{\Delta^1}(\eta) \to \mathrm{Tel}^{\Delta^1}(\eta)$ is not only a homotopy equivalence but even homotopic to the identity* id.

Proof. We show $\eta_* \circ \eta_* \simeq \eta_*$, then by Lemma A.5.24 we have $\eta_* \circ \mathrm{sh} \simeq \mathrm{id}$ and therefore $\eta_* \simeq \eta_* \circ \eta_* \circ \mathrm{sh} \simeq \eta_* \circ \mathrm{sh} \simeq \mathrm{id}$ and we are done.

So assume that $H\colon A[I] \to A$ is the homotopy from η^2 to η and for simplicity assume $I = \Delta^1$. As η is coherent we have a diagram

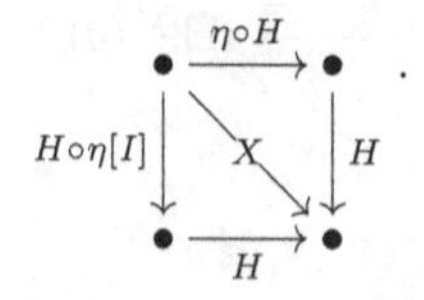

where X is is just a name to the denote the restriction to the diagonal. Using that diagram we can build the diagram

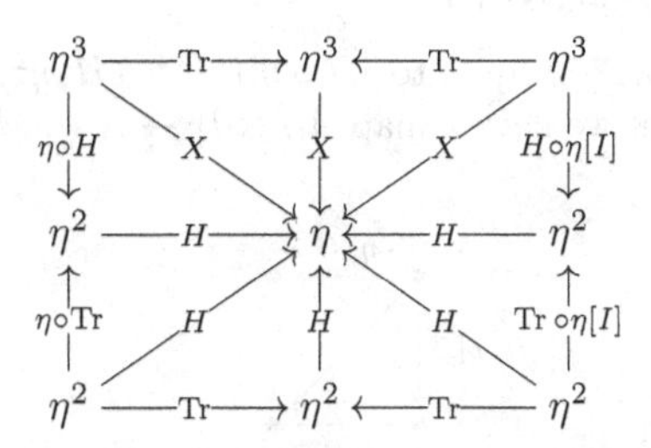

which gives a 2-homotopy G such that Lemma A.5.20 shows that $(G,H)_*$ is a homotopy from $(\mathrm{Tr},\eta^2)_*$ to $(\mathrm{Tr},\eta)_*$ as maps $\mathrm{Tel}^{\Delta^1\square I\square\overline{I}}(\eta)\to\mathrm{Tel}^{\Delta^1}(\eta)$. But $\Delta^1\square I\square\overline{I}\to\Delta^1$ induces a homotopy equivalence on telescopes such that the triangle for η_*

$$\begin{array}{ccc} \mathrm{Tel}^{\Delta^1\square I\square\overline{I}}(\eta) & \xrightarrow{\simeq} & \mathrm{Tel}^{\Delta^1}(\eta)\ , \\ {\scriptstyle\eta_*}\big\downarrow & \swarrow{\scriptstyle\eta_*} & \\ \mathrm{Tel}^{\Delta^1}(\eta) & & \end{array}$$

as well as the one for η^2_*, commutes strictly. Therefore also the maps on the cylinder of the same lengths are homotopic. □

Lemma A.5.27. *Let $\eta\colon K\to K$ be a coherent homotopy idempotent in $\mathcal{C}^G_a$. Then there is a map $c\colon \mathrm{Tel}^{\Delta^1}(\eta)\to K$ such that the composition $\iota\circ c$ with the inclusion $\iota\colon K\to\mathrm{Tel}^{\Delta^1}(\eta)$ is homotopic to the identity on $\mathrm{Tel}^{\Delta^1}(\eta)$, whereas the other composition $c\circ\iota$ is homotopic to $\eta\colon K\to K$.*

Proof. Let $H\colon A[I]\to A$ be the homotopy from η^2 to η. We get two homotopy commutative squares

$$\begin{array}{ccc} K & \xrightarrow{\mathrm{id}} & K \\ {\scriptstyle\eta}\big\downarrow & & \big\downarrow{\scriptstyle\eta} \\ K & \xrightarrow{\eta} & K \end{array} \qquad \text{and} \qquad \begin{array}{ccc} K & \xrightarrow{\eta} & K \\ {\scriptstyle\eta}\big\downarrow & & \big\downarrow{\scriptstyle\eta} \\ K & \xrightarrow{\mathrm{id}} & K \end{array}$$

with homotopies $H\colon A[I]\to A$ and $\overline{H}\colon A[\overline{I}]\to A$. Hence we get two induced maps

$$(H,\eta)_*\colon\ \mathrm{Tel}^{\Delta^1\square I}(\mathrm{id}_K)\to\mathrm{Tel}^{\Delta^1}(\eta)$$
$$(\overline{H},\eta)_*\colon\ \mathrm{Tel}^{\Delta^1\square\overline{I}}(\eta)\to\mathrm{Tel}^{\Delta^1}(\mathrm{id}_K).$$

Consider the "composition" (A.5.16)

$$(H,\eta)_* \boxdot (\overline{H},\eta)_*\colon\ \mathrm{Tel}^{\Delta^1 \square I \square \overline{I}}(\eta) \to \mathrm{Tel}^{\Delta^1}(\eta)$$

which by Lemma A.5.17 is equal to $(H \circ \eta[I] \square \eta \circ \overline{H}, \eta^2)_*$. As the homotopy idempotent is coherent we have a map $A[I \times I] \to A$ which is on the boundary of I^2:

$$\begin{array}{ccc} \bullet & \xrightarrow{\eta\circ H} & \bullet \\ {\scriptstyle H\circ\eta[I]}\downarrow & \searrow & \downarrow{\scriptstyle H} \\ \bullet & \xrightarrow[H]{} & \bullet \end{array} \quad . \tag{21}$$

Thus by pasting two copies of the above square together as shown below we get the 2-homotopy G

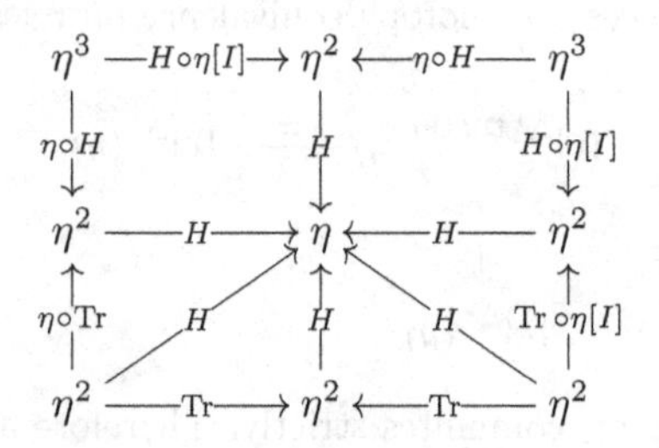

from $H \circ \eta[I] \square \eta \circ \overline{H}$ to Tr and by Lemma A.5.20 $(G,H)_*$ gives a homotopy from the composition $(H,\eta)_* \boxdot (\overline{H},\eta)_*$ to the map $(\mathrm{Tr},\eta)_*$. Similar the other "composition" $(\overline{H},\eta)_* \boxdot (H,\eta)_*$ is homotopic to $(\mathrm{Tr},\eta)_*\colon \mathrm{Tel}^{\Delta^1 \square \overline{I} \square I}(\mathrm{id}_K) \to \mathrm{Tel}^{\Delta^1}(\mathrm{id}_K)$ using the 2-homotopy

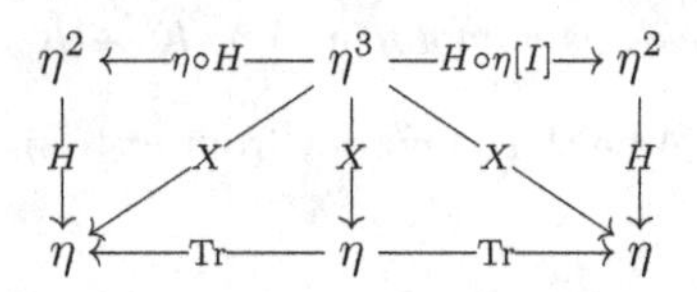

where X is the diagonal in diagram (21) and the upper left and right triangles are also from (21).

Now we make $(H,\eta)_*$ and $(\overline{H},\eta)_*$ into maps of telescopes of the same length as in the proof of Lemma A.5.22. Define $\widetilde{c}$ and $\widetilde{\iota}$ by choosing a homotopy inverse in the top row of the following diagrams

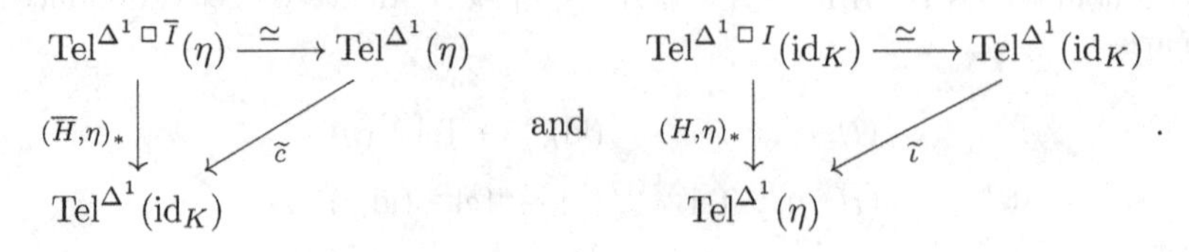

A similar argument as in the proof of Lemma A.5.22 using a big triangle shows that $\widetilde{c} \circ \widetilde{\iota}$ is homotopic to $\eta_*\colon \operatorname{Tel}^{\Delta^1}(\mathrm{id}_K) \to \operatorname{Tel}^{\Delta^1}(\mathrm{id}_K)$ and $\widetilde{\iota} \circ \widetilde{c}$ is homotopic to $\eta_*\colon \operatorname{Tel}^{\Delta^1}(\eta) \to \operatorname{Tel}^{\Delta^1}(\eta)$. Lemma A.5.26 shows that on $\operatorname{Tel}^{\Delta^1}(\eta)$ the map η_* is homotopic to the identity.

Now $\iota_{\mathrm{id}_K}\colon K \to \operatorname{Tel}^{\Delta^1}(\mathrm{id}_K)$ is a homotopy equivalence and even an inclusion for a deformation retraction pr by Lemma A.5.5. Now define $c := \mathrm{pr} \circ \widetilde{c}\colon \operatorname{Tel}^{\Delta^1}(\eta) \to K$, note $\widetilde{\iota} \circ \iota_{\mathrm{id}_K} = \iota_\eta$. Then $\iota \circ c = \widetilde{\iota} \circ \iota_{\mathrm{id}_K} \circ \mathrm{pr} \circ \widetilde{c}$ is homotopic to $\widetilde{\iota} \circ \widetilde{c}$ and hence to $\mathrm{id}_{\operatorname{Tel}^{\Delta^1}(\eta)}$ and $c \circ \iota = \mathrm{pr} \circ \widetilde{c} \circ \widetilde{\iota} \circ \iota_{\mathrm{id}_K}$ is homotopic to $\eta\colon K \to K$. This shows the lemma. □

A.5.28 Proof of Theorem A.2.1 We proved all parts of Theorem A.2.1. We defined $\operatorname{Tel}(\eta)$ in A.5.7, and settled (A.2.1(1)) and (A.2.1(2)). The inclusion $\iota\colon K \to \operatorname{Tel}(\mathrm{id}_K)$ is a homotopy equivalence (A.2.1(4)) by Lemma A.5.5 and homotopic maps gives homotopy equivalent telescopes (A.2.1(3)) by Lemma A.5.22. The compatibility with the inclusion (A.2.1(5)) was noted along the lemmas.

A homotopy idempotent induces a homotopy equivalence on its own telescope by Lemma A.5.24, and a coherent homotopy idempotent even induces a map homotopic to the identity by Lemma A.5.26 (A.2.1(6)). Finally the retraction up to homotopy to the inclusion (A.2.1(7)) is provided in Lemma A.5.27.

References

[Bar13] Arthur Bartels. On proofs of the Farrell-Jones Conjecture. *arXiv*, math.GT, March 2013.

[BFJR04] A. Bartels, T. Farrell, L. Jones, and H. Reich. On the isomorphism conjecture in algebraic K-theory. *Topology*, 43(1):157–213, 2004.

[BLR08a] Arthur Bartels, Wolfgang Lück, and Holger Reich. The K-theoretic Farrell-Jones conjecture for hyperbolic groups. *Invent. Math.*, 172(1):29–70, 2008.

[BLR08b] Arthur Bartels, Wolfgang Lück, and Holger Reich. On the Farrell-Jones conjecture and its applications. *J. Topol.*, 1(1):57–86, 2008.

[BLRR14] Arthur Bartels, Wolfgang Lück, Holger Reich, and Henrik Rüping. K- and L-theory of group rings over $GL_n(\mathbf{Z})$. *Publ. Math. Inst. Hautes Études Sci.*, 119:97–125, 2014.

[Bor94] F. Borceux. *Handbook of categorical algebra. 1*, volume 50 of *Encyclopedia of Mathematics and its Applications*. Cambridge University Press, Cambridge, 1994. Basic category theory.

[Bro73] Kenneth S. Brown. Abstract homotopy theory and generalized sheaf cohomology. *Trans. Amer. Math. Soc.*, 186:419–458, 1973.

[CP97] M. Cárdenas and E. K. Pedersen. On the Karoubi filtration of a category. *K-Theory*, 12(2):165–191, 1997.

[EKMM97] A. D. Elmendorf, I. Kriz, M. A. Mandell, and J. P. May. *Rings, modules, and algebras in stable homotopy theory*, volume 47 of *Mathematical Surveys and Monographs*. American Mathematical Society, Providence, RI, 1997. With an appendix by M. Cole.

[FJ87] F. T. Farrell and L. E. Jones. *K*-theory and dynamics. II. *Ann. of Math. (2)*, 126(3):451–493, 1987.

[FJ93] F. T. Farrell and L. E. Jones. Isomorphism conjectures in algebraic *K*-theory. *J. Amer. Math. Soc.*, 6(2):249–297, 1993.

[Fre03] Peter J. Freyd. Abelian categories. *Repr. Theory Appl. Categ.*, 3:1–190, 2003. Reprint of the 1964 edition.

[FW13] Tom Farrell and Xiaolei Wu. Farrell-Jones Conjecture for the solvable Baumslag-Solitar groups. *arXiv*, math.GT, April 2013.

[GJ99] P. G. Goerss and J. F. Jardine. *Simplicial homotopy theory*, volume 174 of *Progress in Mathematics*. Birkhäuser Verlag, Basel, 1999.

[Hat02] Allen Hatcher. *Algebraic topology*. Cambridge University Press, Cambridge, 2002.

[HH82] H. M. Hastings and A. Heller. Homotopy idempotents on finite-dimensional complexes split. *Proc. Amer. Math. Soc.*, 85(4):619–622, 1982.

[HSS00] Mark Hovey, Brooke Shipley, and Jeff Smith. Symmetric spectra. *J. Amer. Math. Soc.*, 13(1):149–208, 2000.

[LR05] W. Lück and H. Reich. The Baum-Connes and the Farrell-Jones conjectures in *K*- and *L*-theory. In *Handbook of K-theory. Vol. 1, 2*, pages 703–842. Springer, Berlin, 2005.

[MMSS01] M. A. Mandell, J. P. May, S. Schwede, and B. Shipley. Model categories of diagram spectra. *Proc. London Math. Soc. (3)*, 82(2):441–512, 2001.

[Ped00] Erik Kjær Pedersen. Controlled algebraic *K*-theory, a survey. In *Geometry and topology: Aarhus (1998)*, volume 258 of *Contemp. Math.*, pages 351–368. Amer. Math. Soc., Providence, RI, 2000.

[PW85] E. K. Pedersen and C. A. Weibel. A nonconnective delooping of algebraic K-theory. In *Algebraic and geometric topology (New Brunswick, N.J., 1983)*, volume 1126 of *Lecture Notes in Math.*, pages 166–181. Springer, Berlin, 1985.

[Sta89] R. E. Staffeldt. On fundamental theorems of algebraic K-theory. *K-Theory*, 2(4):511–532, 1989.

[tD87] Tammo tom Dieck. *Transformation groups*, volume 8 of *de Gruyter Studies in Mathematics*. Walter de Gruyter & Co., Berlin, 1987.

[TT90] R. W. Thomason and T. Trobaugh. Higher algebraic K-theory of schemes and of derived categories. In *The Grothendieck Festschrift, Vol. III*, volume 88 of *Progr. Math.*, pages 247–435. Birkhäuser Boston, Boston, MA, 1990.

[Ull11] Mark Ullmann. Controlled algebra for simplicial rings and the algebraic K-theory assembly map. Thesis, Düsseldorf, available from `http://docserv.uni-duesseldorf.de/servlets/DocumentServlet?id=17133`, 2011.

[Vog90] W. Vogell. Algebraic K-theory of spaces, with bounded control. *Acta Math.*, 165(3-4):161–187, 1990.

[Wal] F. Waldhausen. Lecture: Algebraische Topologie. Unpublished lecture notes (german), available from `www.math.uni-bielefeld.de/~fw`.

[Wal78] Friedhelm Waldhausen. Algebraic K-theory of topological spaces. I. In *Algebraic and geometric topology (Proc. Sympos. Pure Math., Stanford Univ., Stanford, Calif., 1976), Part 1*, Proc. Sympos. Pure Math., XXXII, pages 35–60. Amer. Math. Soc., Providence, R.I., 1978.

[Wal85] F. Waldhausen. Algebraic K-theory of spaces. In *Algebraic and geometric topology (New Brunswick, N.J., 1983)*, volume 1126 of *Lecture Notes in Math.*, pages 318–419. Springer, Berlin, 1985.

[Weg15] Christian Wegner. The Farrell-Jones conjecture for virtually solvable groups. *J. Topol.*, 8(4):975–1016, 2015.

[Wei02] M. Weiss. Excision and restriction in controlled K-theory. *Forum Math.*, 14(1):85–119, 2002.

[Wei13] Charles A. Weibel. *The K-book*, volume 145 of *Graduate Studies in Mathematics*. American Mathematical Society, Providence, RI, 2013. An introduction to algebraic K-theory.

[Zak10] Inna Zakharevich. Cofinal inclusions of waldhausen categories. `http://mathoverflow.net/questions/23515/cofinal-inclusions-of-waldhausen-categories`, May 2010. Question on mathoverflow.net.

Printed in the United States
By Bookmasters